Texts and Monographs in Physics

Springer
Berlin
Heidelberg
New York
Barcelona
Hong Kong
London
Milan
Paris
Singapore
Tokyo

Texts and Monographs in Physics

From Microphysics to Macrophysics I + II Methods and Applications of Statistical Physics By R. Balian

Variational Methods in Mathematical Physics A Unified Approach By P. Blanchard and E. Brüning

Multi-Hamiltonian Theory of Dynamical Systems By M. Błaszak

Quantum Mechanics: Foundations and Applications 3rd enlarged edition By A. Böhm

The Early Universe Facts and Fiction 3rd corrected and enlarged edition By G. Börner

Operator Algebras and Quantum Statistical Mechanics I + II 2nd edition By O. Bratteli and D. W. Robinson

Statistical Methods in Quantum Optics 1. Master Equations and Fokker–Planck Equations By H. J. Carmichael

Geometry of the Standard Model of Elementary Particles By. A. Derdzinski

Scattering Theory of Classical and Quantum *N*-Particle Systems By. J. Derezinski and C. Gérard

Effective Lagrangians for the Standard Model By A. Dobado, A. Gómez-Nicola, A. L. Maroto and J. R. Peláez

Quantum The Quantum Theory of Particles, Fields, and Cosmology By E. Elbaz

Quantum Relativity A Synthesis of the Ideas of Einstein and Heisenberg By D. R. Finkelstein

Quantum Mechanics I + II By A. Galindo and P. Pascual

The Elements of Mechanics By G. Gallavotti

Local Quantum Physics Fields, Particles, Algebras 2nd revised and enlarged edition By R. Haag

***CP* Violation Without Strangeness** Electric Dipole Moments of Particles, Atoms, and Molecules By I. B. Khriplovich and S. K. Lamoreaux

Quantum Groups and Their Representations By A. Klimyk and K. Schmüdgen

Inverse Schrödinger Scattering in Three Dimensions By R. G. Newton

Scattering Theory of Waves and Particles 2nd edition By R. G. Newton

Quantum Entropy and Its Use By M. Ohya and D. Petz

Generalized Coherent States and Their Applications By A. Perelomov

Essential Relativity Special, General, and Cosmological Revised 2nd edition By W. Rindler

Path Integral Approach to Quantum Physics An Introduction 2nd printing By G. Roepstorff

Renormalization An Introduction By M. Salmhofer

Finite Quantum Electrodynamics The Causal Approach 2nd edition By G. Scharf

From Electrostatics to Optics A Concise Electrodynamics Course By G. Scharf

The Mechanics and Thermodynamics of Continuous Media By M. Šilhavý

Large Scale Dynamics of Interacting Particles By H. Spohn

The Theory of Quark and Gluon Interactions 3rd revised and enlarged edition By F. J. Ynduráin

Relativistic Quantum Mechanics and Introduction to Field Theory By F. J. Ynduráin

Quantum Field Theory in Strongly Correlated Electronic Systems By N. Nagaosa

Naoto Nagaosa

Quantum Field Theory in Condensed Matter Physics

Translated by Stefan Heusler
With 33 Figures

Springer

Professor Naoto Nagaosa
Department of Applied Physics
The University of Tokyo
Bunkyo-ku, Tokyo 113, Japan

Translator:
Stefan Heusler
Bothestrasse 106
D-69126 Heidelberg, Germany

Editors

Roger Balian
CEA
Service de Physique Théorique de Saclay
F-91191 Gif-sur-Yvette, France

Wolf Beiglböck
Institut für Angewandte Mathematik
Universität Heidelberg, INF 294
D-69120 Heidelberg, Germany

Harald Grosse
Institut für Theoretische Physik
Universität Wien
Boltzmanngasse 5
A-1090 Wien, Austria

Elliott H. Lieb
Jadwin Hall
Princeton University, P.O. Box 708
Princeton, NJ 08544-0708, USA

Nicolai Reshetikhin
Department of Mathematics
University of California
Berkeley, CA 94720-3840, USA

Herbert Spohn
Zentrum Mathematik
Technische Universität München
D-80290 München, Germany

Walter Thirring
Institut für Theoretische Physik
Universität Wien
Boltzmanngasse 5
A-1090 Wien, Austria

Library of Congress Cataloging-in-Publication Data.
Nagaosa, N. (Naoto), 1958- [Busseiron ni okeru ba no ryōshiron. English] Quantum field theory in condensed matter physics / Naoto Nagaosa ; translated by Stefan Heusler. p. cm. – (Texts and monographs in physics, ISSN 0172-5998) "Originally published in Japanese by Iwanami Shoten, Publishers, Tokyo in 1995"–t.p. verso. Includes bibliographical references and index. ISBN 3-540-65537-9 (hardcover : alk. paper) 1. Quantum field theory. 2. Condensed matter. I. Title. II. Series. QC174.45.N27 1999 530.14'3–dc21
99-35358

Title of the original Japanese edition:
Quantum Field Theory in Strongly Correlated Systems by Naoto Nagaosa © 1995 by Naoto Nagaosa
Originally published in Japanese by Iwanami Shoten, Publishers, Tokyo in 1995

ISSN 0172-5998
ISBN 3-540-65537-9 Springer-Verlag Berlin Heidelberg New York

Typesetting: Data conversion by K. Mattes, Heidelberg
Cover design: *design & production* GmbH, Heidelberg
SPIN: 10683452 55/3144/di - 5 4 3 2 1 0 – Printed on acid-free paper

Preface

Why is quantum field theory of condensed matter physics necessary?

Condensed matter physics deals with a wide variety of topics, ranging from gas to liquids and solids, as well as plasma, where owing to the interplay between the motions of a tremendous number of electrons and nuclei, rich varieties of physical phenomena occur. Quantum field theory is the most appropriate "language", to describe systems with such a large number of degrees of freedom, and therefore its importance for condensed matter physics is obvious. Indeed, up to now, quantum field theory has been succesfully applied to many different topics in condensed matter physics. Recently, quantum field theory has become more and more important in research on the electronic properties of condensed systems, which is the main topic of the present volume.

Up to now, the motion of electrons in solids has been successfully described by focusing on one electron and replacing the Coulomb interaction of all the other electrons by a mean field potential. This method is called mean field theory, which made important contributions to the explananation of the electronic structure in solids, and led to the classification of insulators, semiconductors and metals in terms of the band theory. It might be said that also the present achievements in the field of semiconductor technology rely on these foundations.

In the mean field approximation, effects that arise due to the correlation of the motions of many particles, cannot be described. It has been treated in a perturbative way under the assumption that its effect is small. However, recently, many systems that cannot be described in this standard way have been discovered, and it became clear that a new world opened its doors. Connected to these new aspects of condensed matter physics, the most fundamental problem of quantum theory – the duality between the particle picture and the wave picture – appeared in a very striking way. This particle-wave duality appears in the framework of many-particle quantum field theory as a canonical conjugate relation between the particle number and the quantum mechanical phase.

From this point of view, in systems where the strong repulsive force between the particles fixes the particle number, as is the case, for example, for the Mott insulator and the Wigner crystal, the charge density and the spin density waves are stabilized and the system shows its particle-like face. On the other hand, when the motions of the particles lead to coherence of the quantum mechanical phase, as is the case in superconductors and superfluids, the phase is fixed and becomes visible, and the system shows its wave-like face. The competition between both appears in low-dimensional systems and mesoscopic systems in a very clear manner. Problems like the quantum Hall effect, high-temperature superconductors, organic conductors, metal-insulator transition, superconductor-insulator transition, can all be grasped from this point of view.

The new problems that arise due to this competition are given in the following three points.

(i) Quantum phase transition or quantum critical phenomena – In contrast to the phase transition at finite temperature due to the competition between energy and entropy, these are phase transitions that occur at the absolute zero temperature or low temperature due to quantum fluctuations.

(ii) Novel ground states and low-energy excitations – New types of quantum states have been discovered, such as non-Fermi liquids in relation to high-temperature superconductors, and incompressible quantum liquids in relation to the quantum Hall system. Their elementary excitations, the spinon and the holon, are anyons obeying fractional statistics.

(iii) The quantal phase and its topological properties – The topological aspects of the quantal phase, including the topological defects, show up in the physical properties of solids. Especially, when due to some kind of constraint a gauge field is introduced, phenomena that are also discussed in quantum chromodynamics reappear with some modifications in condensed matter physics.

The present book has been written for graduate students and researchers who are not necessarily specialists in quantum field theory. Starting with a short review of quantum mechanics, the framework of quantum field theory is introduced and applied to problems that are uppermost in the present research in condensed matter physics.

In Chap. 1, most basic principles are reviewed. Topics that are not only important in single-particle quantum mechanics but also in quantum field theory are recalled, namely, canonical conjugate relation, symmetry and the conservation law, and the variation principle. This analogy between single-particle quantum mechanics and quantum field theory can be efficiently applied when quantization is performed using path integral methods, as presented in Chap. 2. The coordinate and momentum in the single-particle problem correspond in the many-particle system to the phase and the amplitude of the quantum field. A similar analogy can be applied for the gauge field and the spin system. In Chap. 3, phase transitions are discussed, being characteristic of field theories, because they cannot occur in systems with only a small number of degrees of freedom. This topic is one fundamental concept of modern condensed matter physics and is developed further to systems where it is difficult to define an order parameter. As examples, the Kosterlitz–Thouless transition, being a topological phase transition that will reappear later on in a different context, and the problem of colour-confinement in gauge theory, are discussed.

After these preliminaries, starting from Chap. 4, explicit applications of quantum field theory to condensed matter physics are discussed. The content of Chap. 4 is a warming up, where representative examples of a fermionic system and a bosonic system are presented, namely the classical RPA theory of an electron gas and the Bogoliubov theory of superfluidity. It is demonstrated that the method of path integrals provides the clearest formulation of the problems. In Chap. 5, many different problems related to superconductors are discussed. Problems that have so far been treated independently, namely the renormalization of the Coulomb interaction, collective modes and gauge invariance in BCS theory, are discussed in a unified approach. In the second part of Chap. 5, the Josephson junction and the two-dimensional superconductor are discussed, being an issue of current interest. In Chap.6, the new quantum state of the (fractional) quantum Hall liquid is discussed within the framework of the Chern–Simons gauge field.

Of course, it is not possible to discuss all the applications of quantum field theory here; therefore, our intention is to reveal their common structure and ideas that provide the tools necessary for further studies.

I owe special thanks to my supervisors and colleagues, especially E. Hanamura, Y. Toyazawa, P. A. Lee, H. Fukuyama, S. Tanaka, M. Imada, K. Ueda, S. Uchida, Y. Tokura, N. Kawakami, A. Furusaki, T. K. Ng, and Y. Kuramoto.

Tokyo, January 1999 *Naoto Nagaosa*

Contents

1. Review of Quantum Mechanics and Basic Principles of Field Theory

The content of this chapter is nothing but a review of the most basic principles. Starting with the quantum mechanics of a single-particle system, then the canonical conjugate relations, the relation between symmetries and conservation laws, the description of multi-particles systems using field theory, finally gauge invariance and the gauge field will be introduced, all being fundamental concepts that built the basis for the whole following discussion. The reader should reconfirm the universality of the quantum mechanical description and get a taste of the efficiency of an analogy.

1.1 Single-Particle Quantum Mechanics

We start by recalling some facts about single-particle quantum mechanics. All points that will be mentioned here will again become important when proceeding to quantum field theory.

The equation of motion of the single-particle system is given by the Schrödinger equation:

$$\mathrm{i}\hbar\frac{\partial\psi(\boldsymbol{r},t)}{\partial t}=\hat{H}\psi(\boldsymbol{r},t)=\left[\frac{\hat{\boldsymbol{p}}}{2m}+V(\hat{\boldsymbol{r}})\right]\psi(\boldsymbol{r},t)\ . \tag{1.1.1}$$

$\psi(\boldsymbol{r},t)$ is the so-called wave function, depending on the space coordinates $\boldsymbol{r}$ and the time t. $\hat{H}$ is the so-called Hamiltonian operator, creating a new wave function $\hat{H}\psi(\boldsymbol{r},t)$ by acting on the wave function $\psi(\boldsymbol{r},t)$. In what follows, operators are assigned by a hat, except for obvious cases where this notation will be omitted. $\hat{\boldsymbol{p}}$ and $\hat{\boldsymbol{r}}$ are three-component vector operators that represent the momentum and space coordinate of the particle, respectively. $\hat{\boldsymbol{p}}^2/2m$ is the kinetic energy, $V(\hat{\boldsymbol{r}})$ the potential energy, and its sum is the total energy of the particle, called the Hamiltonian operator $\hat{H}$. Equation (1.1.1) signifies that the time development of the wave function is determined by the Hamiltonian operator $\hat{H}$. By defining the exponential $\exp(\hat{A})$ of an operator by

$$\exp(\hat{A})=\sum_{n=0}^{\infty}\frac{1}{n!}(\hat{A})^n\ , \tag{1.1.2}$$

its solution can be written as

$$\psi(\boldsymbol{r},t) = \exp\left(-\frac{\mathrm{i}}{\hbar}\hat{H}t\right)\psi(\boldsymbol{r},0) \ . \tag{1.1.3}$$

In quantum mechanics, the wave function is interpreted in terms of probability. The square of the absolute value of the wave function

$$P(\boldsymbol{r},t) = |\psi(\boldsymbol{r},t)|^2 \tag{1.1.4}$$

is interpreted as the probability of detecting the particle at time t at the coordinate $\boldsymbol{r}$. Therefore, because the sum (integral) of the probability over the whole space is 1, we obtain the normalization condition of the wave function:

$$\int \mathrm{d}^3\boldsymbol{r}\, P(\boldsymbol{r},t) = \int \mathrm{d}^3\boldsymbol{r}\, |\psi(\boldsymbol{r},t)|^2 = 1 \ . \tag{1.1.5}$$

We will now explain the matrix formulation of quantum mechanics. We interpret the function $f(\boldsymbol{r})$ as a vector in the Hilbert space (the vector space of functions) and write $|f\rangle$ for the state that the function represents. Doing so, the operator $\hat{A}$ acting on the vectors in this space generates a new vector, which is a linear transformation. Therefore, it corresponds to a matrix. Furthermore, to every vector $|f\rangle$, there exists the conjugate vector $\langle f|$, being specified as the so-called ket- and bra-vector, respectively. Thinking in components, the bra-vector $\langle f|$ can be regarded as the transposed and complex conjugate of the ket-vector $|f\rangle$. The inner product $\langle g \mid f\rangle$ in this vector space is defined by

$$\langle g|f\rangle = \int \mathrm{d}^3\boldsymbol{r}\, g^*(\boldsymbol{r})f(\boldsymbol{r}) = \langle f|g\rangle^* \ . \tag{1.1.6}$$

The matrix element $\langle g|\hat{A}|f\rangle$ of the operator (the matrix) $\hat{A}$ is given by

$$\langle g|\hat{A}|f\rangle = \langle g|\hat{A}f\rangle = \int \mathrm{d}^3\boldsymbol{r}\, g^*(\boldsymbol{r})\hat{A}f(\boldsymbol{r}) \ . \tag{1.1.7}$$

In order to give a more concrete picture of the considerations made so far, we introduce now an orthonormal basis $|i\rangle, i = 1,2,3\ldots$, of the Hilbert space. (We wrote $i = 1,2,3\ldots$; however, the basis is not necessarily a countable set. In general, when the volume of the system is infinite, the set of basis vectors is uncountable. In these cases, the sum $\sum_i$ over the set labelled by i must be replaced by an integral.) Because the basis is orthonormal, the orthonormality condition

$$\langle i|j\rangle = \delta_{i,j} \tag{1.1.8}$$

and the completeness condition

$$\sum_i |i\rangle\langle i| = \hat{1} \tag{1.1.9}$$

hold. Here, the so-called Kronecker delta $\delta_{i,j}$ is defined to equal 1 when $i = j$, and to be zero otherwise. $\hat{1}$ is the identity matrix; in other words, the identity operator. In this basis, the vector $|f\rangle$ can be represented by its components:

$$|f\rangle = \sum_i |i\rangle\langle i|f\rangle \ , \qquad \langle f| = \sum_i \langle f|i\rangle\langle i| \ . \tag{1.1.10}$$

Furthermore, the component representation of $\hat{A}|f\rangle$ is given by

$$\begin{aligned}\hat{A}|f\rangle &= \left(\sum_i |i\rangle\langle i|\right)\hat{A}\left(\sum_j |j\rangle\langle j|\right)|f\rangle \\ &= \sum_{i,j} |i\rangle\langle i|\hat{A}|j\rangle\langle j|f\rangle \end{aligned} \tag{1.1.11}$$

and (1.1.7) can be written as

$$\langle g|\hat{A}|f\rangle = \sum_{i,j} \langle g|i\rangle\langle i|\hat{A}|j\rangle\langle j|f\rangle \ . \tag{1.1.12}$$

We define the Hermitian conjugate $\hat{A}^\dagger$ of $\hat{A}$ by requiring that

$$\langle g|\hat{A}|f\rangle = \langle \hat{A}^\dagger g|f\rangle \tag{1.1.13}$$

holds for every $|f\rangle$ and $|g\rangle$. Comparing the inner product of the conjugate of

$$|\hat{A}^\dagger g\rangle = \sum_{j,i} |j\rangle\langle j|\hat{A}^\dagger|i\rangle\langle i|g\rangle \tag{1.1.14}$$

with $|f\rangle$ and

$$\langle \hat{A}^\dagger g|f\rangle = \sum_{j,i} \langle g|i\rangle\langle j|\hat{A}^\dagger|i\rangle^*\langle j|f\rangle \tag{1.1.15}$$

with (1.1.13), we obtain

$$\langle j|\hat{A}^\dagger|i\rangle = \langle i|\hat{A}|j\rangle^* \ . \tag{1.1.16}$$

This is nothing but the usual definition of the Hermitian conjugation of a matrix. In the case that $\hat{A}$ and $\hat{A}^\dagger$ are equal $\hat{A} = \hat{A}^\dagger$, $\hat{A}$ is called a Hermitian operator. In quantum mechanics, all physical quantities are represented in terms of Hermitian operators.

We now introduce the eigenvalue a and the eigenstate $|a\rangle$ of the Hermitian operator $\hat{A}$:

$$\hat{A}|a\rangle = a|a\rangle \ . \tag{1.1.17}$$

By taking the inner product with $|a\rangle$

$$\langle a|\hat{A}|a\rangle = a\langle a|a\rangle \tag{1.1.18}$$

we can deduce that at the left-hand side due to hermiticity

$$\langle a|\hat{A}|a\rangle = \langle \hat{A}^\dagger a|a\rangle = \langle \hat{A}a|a\rangle = a^*\langle a|a\rangle \tag{1.1.19}$$

holds, and obtain $a = a^*$. Therefore, we conclude that the eigenvalue a is real. Furthermore, for $a \neq a'$ with

$$\langle a'|\hat{A} = a'\langle a'| \tag{1.1.20}$$

from (1.1.19) we can deduce that

$$\langle a'|\hat{A}|a\rangle = a\langle a'|a\rangle = a'\langle a'|a\rangle \tag{1.1.21}$$

and conclude that $\langle a' \mid a\rangle = 0$. This signifies that the eigenstates of a Hermitian operator with different eigenvalues are orthogonal to each other. Therefore, by a suitable normalization it is possible to build an orthonormal basis using the eigenstates of an Hermitian operator by orthogonalizing in eigenspaces belonging to the same eigenvalue.

Naturally, the space coordinate $\hat{\boldsymbol{r}}$ is a Hermitian operator. Every component $\hat{r}_\alpha$ of $\hat{\boldsymbol{r}}$ acts on $f(\hat{\boldsymbol{r}})$

$$\hat{r}_\alpha f(\boldsymbol{r}) = r_\alpha f(\boldsymbol{r}) \tag{1.1.22}$$

creating a new function. Notice that on the right-hand side, r_α is no longer an operator, but the α-component of the function $\boldsymbol{r}$. The generalization of (1.1.22) is

$$V(\hat{\boldsymbol{r}})f(\boldsymbol{r}) = V(\boldsymbol{r})f(\boldsymbol{r}) \tag{1.1.23}$$

with $V(\hat{\boldsymbol{r}})$ being the potential energy of equation (1.1.1). With (1.1.22) we write

$$\begin{aligned}\langle g|\hat{r}_\alpha|f\rangle &= \int \mathrm{d}^3\boldsymbol{r}\, g^*(\boldsymbol{r})\hat{r}_\alpha f(\boldsymbol{r}) = \int \mathrm{d}^3\boldsymbol{r}\, g^*(\boldsymbol{r})r_\alpha f(\boldsymbol{r}) = \int \mathrm{d}^3\boldsymbol{r}\, [r_\alpha g(\boldsymbol{r})]^* f(\boldsymbol{r}) \\ &= \int \mathrm{d}^3\boldsymbol{r}\, [\hat{r}_\alpha g(\boldsymbol{r})]^* f(\boldsymbol{r}) = \langle \hat{r}_\alpha g|f\rangle \ .\end{aligned} \tag{1.1.24}$$

It should be clear from these equations that $\hat{r}_\alpha$ is Hermitian.

We introduce now the state $|\boldsymbol{r}\rangle$ being the eigenstate with eigenvalue $\boldsymbol{r}$ of the operator $\hat{\boldsymbol{r}}$:

$$\hat{\boldsymbol{r}}|\boldsymbol{r}\rangle = \boldsymbol{r}|\boldsymbol{r}\rangle \ . \tag{1.1.25}$$

Because $\langle \boldsymbol{r}' \mid \boldsymbol{r}\rangle = 0$ for $\boldsymbol{r} \neq \boldsymbol{r}'$,

$$\langle \boldsymbol{r}'|\boldsymbol{r}\rangle = \delta(\boldsymbol{r} - \boldsymbol{r}') \ , \tag{1.1.26}$$

with an appropriate choice of the normalization. Here we have introduced the so-called delta function $\delta(\boldsymbol{r}-\boldsymbol{r}')$, defined to be zero for $\boldsymbol{r} \neq \boldsymbol{r}'$, and infinite at $\boldsymbol{r}=\boldsymbol{r}'$, and to give the value 1 when integrated over $\boldsymbol{r}-\boldsymbol{r}'$ in a region containing the origin. Furthermore, $|\boldsymbol{r}\rangle$ and $\langle\boldsymbol{r}|$ fulfil the completeness relation

$$\int \mathrm{d}^3\boldsymbol{r}\, |\boldsymbol{r}\rangle\langle\boldsymbol{r}| = \hat{1} \ . \tag{1.1.27}$$

The reader not familiar with the delta function is referred to the Appendices A and B. As mentioned there, we can introduce a vector space on a discrete lattice. The components of a vector in this space are defined by the values of a function on the discrete lattice points. This vector space approaches the Hilbert space when the number of lattice points N_{L} becomes infinite, that is, when the lattice spacing Δx becomes zero. In this case, the sum $(\Delta x)^3 \sum_{\text{lattice points } i}$ approaches the three-dimensional integral appearing in (1.1.6). As a basis of the N_{L}-dimensional vector space, we define states that are zero at all lattice points except for the coordinate $\boldsymbol{r}_i$, where the value is definded to be $1/(\Delta x)^{3/2}$. Then we have

$$\langle\boldsymbol{r}_i|\boldsymbol{r}_j\rangle = \sum_k \frac{\delta_{\boldsymbol{r}_i,\boldsymbol{r}_k}}{(\Delta x)^{3/2}}\frac{\delta_{\boldsymbol{r}_k,\boldsymbol{r}_j}}{(\Delta x)^{3/2}} = \frac{\delta_{\boldsymbol{r}_i,\boldsymbol{r}_j}}{(\Delta x)^3}$$

and, furthermore,

$$(\Delta x)^3 \sum_i |\boldsymbol{r}_i\rangle\langle\boldsymbol{r}_i| = \hat{1} \ .$$

In the limit as $\Delta x \to 0$, these equations approach the equations of the inner product and the completeness relation of the basis $\boldsymbol{r}$ mentioned above.

Now, owing to the completeness relation of the basis $\boldsymbol{r}$, we can write the inner product (1.1.6) as

$$\langle g|f\rangle = \int \mathrm{d}^3\boldsymbol{r}\, \langle g|\boldsymbol{r}\rangle\langle\boldsymbol{r}|f\rangle \tag{1.1.28}$$

and obtain

$$\begin{aligned} f(\boldsymbol{r}) &= \langle\boldsymbol{r}|f\rangle \ , \\ g^*(\boldsymbol{r}) &= \langle g|\boldsymbol{r}\rangle \ . \end{aligned} \tag{1.1.29}$$

From this point of view, the wave function $\psi(\boldsymbol{r},t)$ is nothing but the $\boldsymbol{r}$-component of the state vector $|\psi(t)\rangle$ of the Hilbert space written in the basis $|\boldsymbol{r}\rangle$.

Now, what about the momentum operator $\hat{\boldsymbol{p}}$? Here, we meet the very first example of the most fundamental relation in quantum mechanics, namely the canonical conjugation relation. A plane wave with wave number vector $\boldsymbol{k}$ can be expressed as $\psi_{\boldsymbol{k}}(\boldsymbol{r}) = (2\pi\hbar)^{-3/2}\,\mathrm{e}^{\mathrm{i}\boldsymbol{k}\boldsymbol{r}}$. Writing the plane wave as a function of $\boldsymbol{r}$, and using

$$\hat{\boldsymbol{p}} = \frac{\hbar}{\mathrm{i}}\frac{\partial}{\partial \boldsymbol{r}} \tag{1.1.30}$$

we obtain

$$\hat{\boldsymbol{p}}\psi_{\boldsymbol{k}}(\boldsymbol{r}) = \hbar\boldsymbol{k}\psi_{\boldsymbol{k}}(\boldsymbol{r}) = \boldsymbol{p}\psi_{\boldsymbol{k}}(\boldsymbol{r}) \tag{1.1.31}$$

and therefore the relation $\boldsymbol{p} = \hbar\boldsymbol{k}$. We now define the following combination of $\hat{\boldsymbol{r}}$ and $\hat{\boldsymbol{p}}$:

$$[\hat{r}_\alpha, \hat{p}_\beta] = \hat{r}_\alpha\hat{p}_\beta - \hat{p}_\beta\hat{r}_\alpha \ . \tag{1.1.32}$$

This is the so-called commutator of $\hat{\boldsymbol{r}}_\alpha$ and $\hat{\boldsymbol{p}}_\alpha$, which is also an operator. Acting with this commutator on an arbitrary function $f(\boldsymbol{r})$, we obtain

$$\begin{aligned}[\hat{r}_\alpha, \hat{p}_\beta]f(\boldsymbol{r}) &= \left(\hat{r}_\alpha\frac{\hbar}{\mathrm{i}}\frac{\partial}{\partial r_\beta} - \frac{\hbar}{\mathrm{i}}\frac{\partial}{\partial r_\beta}\hat{r}_\alpha\right)f(\boldsymbol{r}) \\ &= \frac{\hbar}{\mathrm{i}}\left\{r_\alpha\frac{\partial f(\boldsymbol{r})}{\partial r_\beta} - \frac{\partial}{\partial r_\beta}(r_\alpha f(\boldsymbol{r}))\right\} = \mathrm{i}\hbar\delta_{\alpha,\beta}f(\boldsymbol{r})\end{aligned}$$

and therefore the identity

$$[\hat{r}_\alpha, \hat{r}_\beta] = \mathrm{i}\hbar\delta_{\alpha,\beta} \ . \tag{1.1.33}$$

This is the so-called commutation relation. It follows from (1.1.33) for $\alpha = \beta$ that $[\hat{r}_\alpha, \hat{p}_\alpha] = \mathrm{i}\hbar$. This means that $\hat{r}_\alpha$ and $\hat{p}_\alpha$ are canonical conjugates of each other. This commutation relation, as well as (1.1.30), is the starting point for many very fundamental and wide conceptual developments that will be discussed in what follows. However, we first discuss some aspects of the eigenstates of $\hat{\boldsymbol{p}}$. We can interpret (1.1.31) as

$$\hat{\boldsymbol{p}}|\boldsymbol{p} = \boldsymbol{p}|\boldsymbol{p}\rangle \ , \tag{1.1.34}$$

$$\langle\boldsymbol{r}|\boldsymbol{p}\rangle = \psi_{\boldsymbol{p}/\hbar}(\boldsymbol{r}) = \frac{1}{(2\pi\hbar)^{3/2}}\exp\left(\frac{\mathrm{i}}{\hbar}\boldsymbol{p}\cdot\boldsymbol{r}\right) \ . \tag{1.1.35}$$

$|\boldsymbol{p}\rangle$ also spans a basis; orthogonality can be shown with

$$\begin{aligned}\langle\boldsymbol{p}'|\boldsymbol{p}\rangle &= \int \mathrm{d}^3\boldsymbol{r}\,\langle\boldsymbol{p}'|\boldsymbol{r}\rangle\langle\boldsymbol{r}|\boldsymbol{p}\rangle \\ &= \int\frac{\mathrm{d}^3\boldsymbol{r}}{(2\pi\hbar)^3}\exp\left[\frac{\mathrm{i}}{\hbar}(-\boldsymbol{p}'+\boldsymbol{p})\cdot\boldsymbol{r}\right] = \delta(\boldsymbol{p}-\boldsymbol{p}')\end{aligned} \tag{1.1.36}$$

and, in the same manner, the completeness relation

$$\int \mathrm{d}^3\boldsymbol{p}\,|\boldsymbol{p}\rangle\langle\boldsymbol{p}| = \hat{1} \tag{1.1.37}$$

by acting on it with $\langle \boldsymbol{r}'|$ and $|\boldsymbol{r}\rangle$ on the left- and right-hand sides:

$$\begin{aligned}\int \mathrm{d}^3\boldsymbol{p}\,\langle \boldsymbol{r}'|\boldsymbol{p}\rangle\langle \boldsymbol{p}|\boldsymbol{r}\rangle &= \int \frac{\mathrm{d}^3\boldsymbol{p}}{(2\pi\hbar)^3}\exp\left[\frac{\mathrm{i}}{\hbar}\boldsymbol{p}\cdot(\boldsymbol{r}'-\boldsymbol{r})\right] \\ &= \delta(\boldsymbol{r}-\boldsymbol{r}') = \langle \boldsymbol{r}'|\boldsymbol{r}\rangle\ . \end{aligned} \tag{1.1.38}$$

Equations (1.1.26), (1.1.27), (1.1.36) and (1.1.37) are the basic relations of the Fourier analysis, because

$$f(\boldsymbol{r}) = \langle \boldsymbol{r}|f\rangle = \int \mathrm{d}^3\boldsymbol{p}\,\langle \boldsymbol{r}|\boldsymbol{p}\rangle\langle \boldsymbol{p}|f\rangle = \int \frac{\mathrm{d}^3\boldsymbol{p}}{(2\pi\hbar)^{3/2}}\,\mathrm{e}^{\mathrm{i}\boldsymbol{p}\cdot\boldsymbol{r}/\hbar}\langle \boldsymbol{p}|f\rangle \tag{1.1.39}$$

is the Fourier representation of $f(\boldsymbol{r})$ in terms of $\langle \boldsymbol{p} \mid f\rangle = F(\boldsymbol{p})$, and the inversion of this Fourier transformation can be written as

$$F(\boldsymbol{p}) = \langle \boldsymbol{p}|f\rangle = \int \mathrm{d}^3\boldsymbol{r}\,\langle \boldsymbol{p}|\boldsymbol{r}\rangle\langle \boldsymbol{r}|f\rangle = \int \frac{\mathrm{d}^3\boldsymbol{r}}{(2\pi\hbar)^{3/2}}\,\mathrm{e}^{-\mathrm{i}\boldsymbol{p}\cdot\boldsymbol{r}/\hbar}f(\boldsymbol{r})\ . \tag{1.1.40}$$

We conclude that the Fourier transformation is the basis transformation that links the two basis sets (coordinate sets) $|\boldsymbol{r}\rangle$ and $|\boldsymbol{p}\rangle$ in the Hilbert space. (Explanations about the Fourier transformation can be found in Appendix A.)

We now return to the commutation relation and discuss its meaning in more detail. First, Heisenberg's uncertainty principle can be deduced from (1.1.33). We consider now the expectation values $\hat{r}_\alpha = \langle\psi|\hat{r}_\alpha|\psi\rangle$ and $\hat{p}_\alpha = \langle\psi|\hat{p}_\alpha|\psi\rangle$ of $\hat{r}_\alpha$ and $\hat{p}_\alpha$ in the state $|\psi\rangle$. As mentioned earlier, the interpretation of quantum mechanics is only possible in terms of probabilities, and the observed values of $\hat{p}_\alpha$ and $\hat{r}_\alpha$ should follow a probability distribution around each expectation value. The width of this distribution can in some way be understood as the uncertainty, and in order to make it precise, we define the so-called variation in the following manner:

$$\begin{aligned}\langle(\Delta\hat{r}_\alpha)^2\rangle &= \langle(\hat{r}_\alpha - \langle\hat{r}_\alpha\rangle)^2\rangle = \langle\hat{r}_\alpha^2\rangle - \langle\hat{r}_\alpha\rangle^2\ , \\ \langle(\Delta\hat{p}_\alpha)^2\rangle &= \langle(\hat{p}_\alpha - \langle\hat{p}_\alpha\rangle)^2\rangle = \langle\hat{p}_\alpha^2\rangle - \langle\hat{p}_\alpha\rangle^2\ . \end{aligned} \tag{1.1.41}$$

We now introduce the Schwarz inequality. With λ being an arbitrary complex parameter

$$\begin{aligned}\langle|\Delta\hat{r}_\alpha + \lambda\Delta\hat{p}_\alpha|^2\rangle = {}& \langle(\Delta\hat{r}_\alpha)^2\rangle + \lambda^*\langle\Delta\hat{r}_\alpha\Delta\hat{p}_\alpha\rangle + \lambda\langle\Delta\hat{p}_\alpha\Delta\hat{r}_\alpha\rangle \\ & + |\lambda|^2\langle(\Delta\hat{p}_\alpha)^2\rangle\ , \end{aligned} \tag{1.1.42}$$

we can deduce the Schwarz inequality from the fact that this expression must be positive, therefore

$$\langle(\Delta\hat{r}_\alpha)^2\rangle\langle(\Delta\hat{p}_\alpha)^2\rangle \geqq |\langle\Delta\hat{r}_\alpha\Delta\hat{p}_\alpha\rangle|^2\ . \tag{1.1.43}$$

We make the following decomposition:

$$\Delta\hat{r}_\alpha \Delta\hat{p}_\alpha = \frac{1}{2}\{\Delta\hat{r}_\alpha, \Delta\hat{p}_\alpha\} + \frac{1}{2}[\Delta\hat{r}_\alpha, \Delta\hat{p}_\alpha] \ , \tag{1.1.44}$$

where $\{\hat{A}, \hat{B}\} = \hat{A}\hat{B} + \hat{B}\hat{A}$ is the so-called anti-commutator. Recalling that both $\Delta\hat{r}_\alpha$ and $\Delta\hat{p}_\alpha$ are Hermitian, it follows that the Hermitian conjugate of the first term on the right-hand side of the above equation is

$$\{\Delta\hat{r}_\alpha, \Delta\hat{p}_\alpha\}^\dagger = \{\Delta\hat{r}_\alpha, \Delta\hat{p}_\alpha\} \ . \tag{1.1.45}$$

Therefore, its expectation value is real. On the other hand, the second term equals

$$\frac{1}{2}[\Delta\hat{r}_\alpha, \Delta\hat{p}_\alpha] = \frac{1}{2}[\hat{r}_\alpha, \hat{p}_\alpha] = \frac{i\hbar}{2} \tag{1.1.46}$$

and is therefore complex. Finally, we obtain

$$|\langle\Delta\hat{r}_\alpha \Delta\hat{p}_\alpha\rangle|^2 = \frac{1}{4}\langle\{\Delta\hat{r}_\alpha, \Delta\hat{p}_\alpha\}\rangle^2 + \frac{\hbar^2}{4} \geqq \frac{\hbar^2}{4} \tag{1.1.47}$$

and, in combination with (1.1.43),

$$\langle(\Delta\hat{r}_\alpha)^2\rangle\langle(\Delta\hat{p}_\alpha)^2\rangle \geqq \frac{\hbar}{4} \ . \tag{1.1.48}$$

This is Heisenberg's uncertainty principle. Normally, we forget about the numerical factor and just write

$$\Delta\hat{r}_\alpha \Delta\hat{p}_\alpha \gtrsim \hbar \ . \tag{1.1.49}$$

No state exists that is an eigenstate of both $\hat{r}_\alpha$ and $\hat{p}_\alpha$, which means that it is impossible to determine $\hat{r}_\alpha$ and $\hat{p}_\alpha$ simultaneously, and the product of the uncertainty must be larger than a number of the order of the Planck constant.

We can deduce the following physical picture from the uncertainty principle. As can be seen in (1.1.1), the Hamiltonian is the sum of the kinetic energy $\hat{\boldsymbol{p}}^2/2m$ and the potential energy $V(\hat{\boldsymbol{r}})$. In classical mechanics, because it is possible to determine $\boldsymbol{p}$ and $\boldsymbol{r}$ simultaneously, the ground state is given by $\boldsymbol{p} = 0$ and $\boldsymbol{r} = \boldsymbol{r}_0$ (being the minimum of $V(\boldsymbol{r})$). In quantum mechanics, it follows from (1.1.49) that if we require $\boldsymbol{p} = 0$, then $\boldsymbol{r}$ is totally undetermined, and the gain of the potential energy is lost; on the other hand, if we require $\boldsymbol{r} = \boldsymbol{r}_0$, then $\boldsymbol{p}$ is totally undetermined, and the kinetic term becomes large.

Therefore, owing to the uncertainty principle, $\hat{r}_\alpha$ and $\hat{p}_\alpha$ have a strained relationship with each other. Let us make this more concrete. We start by considering the one-dimensional harmonic oscillator with Hamiltonian

$$\hat{H} = \frac{\hat{p}^2}{2m} + \frac{1}{2}m\omega^2\hat{x}^2 \ . \tag{1.1.50}$$

Writing Δx for the width of the ground state $|0\rangle$ in coordinate space, and Δp in momentum space, we can estimate the expectation value of the Hamiltonian or, in other words, the energy, by

$$E = \langle 0|\hat{H}|0\rangle \sim \frac{(\Delta p)^2}{2m} + \frac{1}{2}m\omega^2(\Delta x)^2 \ . \tag{1.1.51}$$

We now insert the equation $\Delta p \propto \hbar/\Delta x$, obtained from the uncertainty relation:

$$E \sim \frac{\hbar^2}{2m(\Delta x)^2} + \frac{1}{2}m\omega^2(\Delta x)^2 \ . \tag{1.1.52}$$

This is only a function of Δx. Calculating the minimum by $\partial E/\partial(\Delta x) = 0$, we obtain

$$\Delta x_0 \sim \left(\frac{\hbar}{m\omega}\right)^{1/2} \ . \tag{1.1.53}$$

This is the scale that lies behind the Hamiltonian (1.1.50), which can be seen as the compromise point between two competing tendencies, namely the kinetic energy requiring $\Delta p = 0$, and the potential energy requiring $\Delta x = 0$.

Inserting Δx_0 in (1.1.52), it is easy to calculate the zero point energy:

$$E_0 \sim \hbar\omega \ .$$

In much the same manner this calculation can also be performed for the hydrogen atom with the Hamiltonian

$$\hat{H} = \frac{\hat{\boldsymbol{p}}^2}{2m} - \frac{e^2}{|\hat{\boldsymbol{r}}|} \ . \tag{1.1.54}$$

Inserting $|\boldsymbol{p}| \propto \hbar/r$ and $|\boldsymbol{r}| \propto r$, we obtain

$$E \sim \frac{\hbar^2}{2mr^2} - \frac{e^2}{r} \ . \tag{1.1.55}$$

Again, by calculating the minimum $\partial E/\partial r = 0$ we obtain

$$r \sim r_0 = \frac{\hbar^2}{me^2} \tag{1.1.56}$$

and

$$E_0 \sim -R_H = -\frac{me^4}{2\hbar^2} \ . \tag{1.1.57}$$

Here, r_0 is the so-called Bohr radius, and R_{H} is the Rydberg energy. We could argue that the electron of the hydrogen atom does not fall into the nucleus and that the atom does not collapse owing to the uncertainty principle.

Another important aspect that is related to the canonical conjugation relation are symmetry operations. We start with the Taylor expansion:

$$f(\boldsymbol{r}+\boldsymbol{a}) = f(\boldsymbol{r}) + \left(\boldsymbol{a}\cdot\frac{\partial}{\partial \boldsymbol{r}}\right) f(\boldsymbol{r}) + \frac{1}{2}\left(\boldsymbol{a}\cdot\frac{\partial}{\partial \boldsymbol{r}}\right)^2 f(\boldsymbol{r}) + \frac{1}{3!}\left(\boldsymbol{a}\cdot\frac{\partial}{\partial \boldsymbol{r}}\right)^3 f(\boldsymbol{r}) + \cdots \quad . \tag{1.1.58}$$

Using the definition (1.1.2) of the exponential of an operator, we can write this as

$$f(\boldsymbol{r}+\boldsymbol{a}) = \mathrm{e}^{\boldsymbol{a}\cdot\frac{\partial}{\partial \boldsymbol{r}}} f(\boldsymbol{r}) \equiv \hat{U}(\boldsymbol{a}) f(\boldsymbol{r}) \tag{1.1.59}$$

where $\hat{U}(\boldsymbol{a})$ acts like a translation operator about $\boldsymbol{a}$ and, using (1.1.30), can be written as

$$\hat{U}(\boldsymbol{a}) = \mathrm{e}^{\frac{\mathrm{i}}{\hbar}\boldsymbol{a}\cdot\hat{\boldsymbol{p}}} \quad . \tag{1.1.60}$$

It follows that $[\hat{U}(\boldsymbol{a})]^\dagger = \hat{U}(-\boldsymbol{a}) = [\hat{U}(\boldsymbol{a})]^{-1}$, and therefore $\hat{U}(\boldsymbol{a})$ is a unitary operator. An operator written in the exponential, as is the case here for $\hat{\boldsymbol{p}}$, which induces a symmetry operation, is called a generator.

In (1.1.59) we considered a linear operation on the function $f(\boldsymbol{r})$. Next, we consider a linear transformation on the operator $V(\hat{\boldsymbol{r}})$. We start by writing down the conclusion:

$$V(\hat{\boldsymbol{r}}+\boldsymbol{a}) = \hat{U}(\boldsymbol{a})V(\hat{\boldsymbol{r}})[\hat{U}(\boldsymbol{a})]^\dagger = \hat{U}(\boldsymbol{a})V(\hat{\boldsymbol{r}})U(-\boldsymbol{a}) \quad . \tag{1.1.61}$$

The meaning of this equation becomes evident by acting on a function $f(\boldsymbol{r})$

$$\begin{aligned}[\hat{U}(\boldsymbol{a})V(\hat{\boldsymbol{r}})\hat{U}(-\boldsymbol{a})]f(\boldsymbol{r}) &= [\hat{U}(\boldsymbol{a})V(\hat{\boldsymbol{r}})]f(\boldsymbol{r}-\boldsymbol{a}) \\ &= \hat{U}(\boldsymbol{a})[V(\boldsymbol{r})f(\boldsymbol{r}-\boldsymbol{a})] = V(\boldsymbol{r}+\boldsymbol{a})f(\boldsymbol{r}) = V(\hat{\boldsymbol{r}}+\boldsymbol{a})f(\boldsymbol{r}) \quad .\end{aligned} \tag{1.1.62}$$

Since $f(\boldsymbol{r})$ is arbitrary, we obtain (1.1.61). Furthermore, because the operator $\hat{U}(\boldsymbol{a})$ depends only on $\hat{\boldsymbol{p}}$,

$$\hat{U}(\boldsymbol{a})\hat{\boldsymbol{p}}\hat{U}(-\boldsymbol{a}) = \hat{\boldsymbol{p}} \tag{1.1.63}$$

can be proved easily. Therefore, the kinetic energy is invariant:

$$\hat{U}(\boldsymbol{a})\frac{\hat{\boldsymbol{p}}^2}{2m}[\hat{U}(\boldsymbol{a})]^\dagger = \frac{\hat{\boldsymbol{p}}^2}{2m} \quad . \tag{1.1.64}$$

Let us suppose that the system is invariant under translation. In our case, this means that the potential is independent of the position, $V(\hat{\boldsymbol{r}}+\boldsymbol{\alpha}) = V(\hat{\boldsymbol{r}})$. Then, we can write

$$\hat{U}(\boldsymbol{a})\hat{H}[\hat{U}(\boldsymbol{a})]^{-1} = \hat{H} \quad . \tag{1.1.65}$$

Since $\boldsymbol{a}$ is an arbitrary real vector, we can choose it infinitesimally small and write the power series

$$\hat{U}(\pm\boldsymbol{a}) \simeq \hat{1} \pm \frac{\mathrm{i}}{\hbar}\boldsymbol{a}\cdot\hat{\boldsymbol{p}} \ . \tag{1.1.66}$$

Then we obtain from (1.1.65)

$$[\boldsymbol{a}\cdot\hat{\boldsymbol{p}}, \hat{H}] = 0 \ . \tag{1.1.67}$$

Because $\boldsymbol{a}$ is an arbitrary infinitesimal vector, we can conclude

$$[\hat{p}_\alpha, \hat{H}] = 0 \ . \tag{1.1.68}$$

Conversely, it is possible to deduce (1.1.65) from (1.1.68). In order to do so, let λ be an arbitrary real parameter, then we define

$$\hat{H}(\lambda) = \hat{U}(\lambda\boldsymbol{a})\hat{H}\hat{U}(-\lambda\boldsymbol{a}) \ . \tag{1.1.69}$$

Differentiating with respect to λ, with

$$\frac{\partial\hat{U}(\pm\lambda\boldsymbol{a})}{\partial\lambda} = \pm\frac{\mathrm{i}}{\hbar}\boldsymbol{a}\cdot\hat{\boldsymbol{p}}\hat{U}(\pm\lambda\boldsymbol{a}) = \pm\frac{\mathrm{i}}{\hbar}\hat{U}(\pm\lambda\boldsymbol{a})\boldsymbol{a}\cdot\hat{\boldsymbol{p}} \tag{1.1.70}$$

we obtain

$$\frac{\partial\hat{H}(\lambda)}{\partial\lambda} = \frac{\mathrm{i}}{\hbar}\hat{U}(\lambda\boldsymbol{a})[\boldsymbol{a}\cdot\hat{\boldsymbol{p}}, \hat{H}]\hat{U}(-\lambda\boldsymbol{a}) \ . \tag{1.1.71}$$

Owing to (1.1.68), the right-hand side is zero, therefore $H(\lambda)$ does not depend on λ, and we regain (1.1.65) from $\hat{H}(1) = \hat{H}(0)$.

In this way, the symmetry of a system can be interpreted as the fact that the generator of the symmetry operation commutes with the Hamiltonian. And indeed, the commutator with the Hamiltonian has the important meaning of the time development of the physical quantity.

To be complete, we will now review the Heisenberg picture. Equation (1.1.3) describes the time evolution of the wave function ψ. This is the so-called Schrödinger picture. On the other hand, the formal description where the wave function is time independent, and the operators change in time, is called the Heisenberg picture. Explicitly, using (1.1.3), from

$$\begin{aligned}\langle\psi(t)|\hat{A}|\psi(t)\rangle &= \langle\psi(0)|\exp\left(\frac{\mathrm{i}}{\hbar}\hat{H}t\right)\hat{A}\exp\left(-\frac{\mathrm{i}}{\hbar}\hat{H}t\right)|\psi(0)\rangle \\ &= \langle\psi(0)|\hat{A}_{\mathrm{H}}(t)|\psi(0)\rangle\end{aligned} \tag{1.1.72}$$

we obtain the following definition:

$$\hat{A}_{\mathrm{H}}(t) = \exp\left(\frac{\mathrm{i}}{\hbar}\hat{H}t\right)\hat{A}\exp\left(-\frac{\mathrm{i}}{\hbar}\hat{H}t\right) \ . \tag{1.1.73}$$

We obtain the Heisenberg equation of motion as

$$\frac{\mathrm{d}\hat{A}_{\mathrm{H}}(t)}{\mathrm{d}t} = \exp\left(\frac{\mathrm{i}}{\hbar}\hat{H}t\right)\frac{\mathrm{i}}{\hbar}[\hat{H},\hat{A}]\exp\left(-\frac{\mathrm{i}}{\hbar}\hat{H}t\right)$$
$$= \frac{\mathrm{i}}{\hbar}[\hat{H},\hat{A}_{\mathrm{H}}(t)] \ . \tag{1.1.74}$$

Of course, the time evolution of the Hamiltonian itself is given by

$$\hat{H}_{\mathrm{H}}(t) = \hat{H} \tag{1.1.75}$$

and is therefore time independent. This is nothing but the energy conservation law in quantum mechanics.

Returning to the discussion of symmetry, we obtain from (1.1.68)

$$\frac{\mathrm{d}\hat{p}_\alpha(t)}{\mathrm{d}t} = \frac{\mathrm{i}}{\hbar}[\hat{H},\hat{p}_\alpha(t)] = 0 \ . \tag{1.1.76}$$

We see that $\hat{p}_\alpha$ does not change in time and therfore is a conserved quantity. The symmetry operation in the $\hat{\boldsymbol{r}}$ coordinate is written in terms of the canonical conjugate $\hat{\boldsymbol{p}}$ as a symmetry generator, and this symmetry leads to a conservation law for $\hat{\boldsymbol{p}}$. When we proceed to quantum field theory, this will be seen to be related to the Noether theorem.

1.2 Many-Particle Quantum Mechanics: Second Quantization

In this section, we consider the many-particle case. In this case the wave function is a function of the time t and $3N$-dimensional coordinate space (for a moment, we omit the spin dependence)

$$\langle \boldsymbol{r}_1,\ldots,\boldsymbol{r}_N|\psi(t)\rangle = \psi(\boldsymbol{r}_1,\ldots,\boldsymbol{r}_N,t) \ . \tag{1.2.1}$$

Unlike classical mechanics, in many-particle quantum mechanics it is impossible in principle to distinguish particles of the same species. We cannot think about indistinguishable particles as rigid bodies; however, it should be possible to get an idea of it with the following metaphor.

Think about a luminous advertisement screen. By switching the lamps on and off at every point of the surface, it is possible to create a moving picture. Places that are illuminated have more energy than the other places, and therefore there should be a particle. A state with N particles at $\boldsymbol{r}_1\ldots\boldsymbol{r}_\mathrm{N}$ should correspond to the state where N lights are illuminated. In this metaphor, it is clear that it is not possible to distinguish the particles. The particle appears as an illuminated lamp, and it is not possible to trace back the way of it as rigid body.

In mathematical language, this means that exchanging the order of $\boldsymbol{r}_1\ldots\boldsymbol{r}_N$ does not lead to a new state, but should lead to the very same state again. Explicitly, taking care also of the statistics when exchanging $\boldsymbol{r}_i$ with $\boldsymbol{r}_j$, we obtain

$$\psi(\boldsymbol{r}_1, \dots, \boldsymbol{r}_j, \dots, \boldsymbol{r}_i, \dots, \boldsymbol{r}_N) = \begin{cases} +\psi(\boldsymbol{r}_1, \dots, \boldsymbol{r}_i, \dots, \boldsymbol{r}_j, \dots, \boldsymbol{r}_N) & \text{(boson)} \\ -\psi(\boldsymbol{r}_i, \dots, \boldsymbol{r}_i, \dots, \boldsymbol{r}_j, \dots, \boldsymbol{r}_N) & \text{(fermion)} \end{cases} . \tag{1.2.2}$$

However, as the reader might have realized, we now have a little problem with the interpretation of the wave function. Of course,

$$P(\boldsymbol{r}_1, \cdots, \boldsymbol{r}_N; t) = |\psi(\boldsymbol{r}_1, \cdots, \boldsymbol{r}_N; t)|^2 \tag{1.2.3}$$

is the probability of finding at time t the N particles at $\boldsymbol{r}_1 \dots \boldsymbol{r}_\mathrm{N}$. However, the image that we have in mind in the single-particle case, namely that $\psi(\boldsymbol{r}, t)$ is the complex wave amplitude at the position $\boldsymbol{r}$ in the three-dimensional physical space, is ruined because we now have to think mathematically about a $3N$-dimensional space. The answer to the question whether in the many-particle case it is still possible to think about a wave function in the physical three-dimensional space is given by the so-called method of second quantization.

For a detailed discussion of the second quantization the reader is referred to [3]. Here, we proceed in a heuristic way. Let us return to the single-particle case. We decompose the single-particle wave function $\psi(\boldsymbol{r}, t)$ in an orthonormal basis $\phi_n(\boldsymbol{r})$

$$\psi(\boldsymbol{r}, t) = \sum_n a_n(t)\phi_n(\boldsymbol{r}) . \tag{1.2.4}$$

The whole time dependence is given by the expansion coefficients $a_n(t)$. Inserting (1.2.4) into the Schrödinger equation, we obtain

$$\mathrm{i}\hbar \sum_n \frac{\mathrm{d}a_n(t)}{\mathrm{d}t}\phi_n(\boldsymbol{r}) = \sum_n a_n(t)\hat{H}\phi_n(\boldsymbol{r}) . \tag{1.2.5}$$

Multiplying by $\phi_n^*(\boldsymbol{r})$ and integrating over $\boldsymbol{r}$, we obtain

$$\mathrm{i}\hbar \frac{\mathrm{d}a_n(t)}{\mathrm{d}t} = \sum_m \langle\phi_n|\hat{H}|\phi_m\rangle a_m(t) . \tag{1.2.6}$$

The complex conjugate of this equation is given by

$$\mathrm{i}\hbar \frac{\mathrm{d}a_n^*(t)}{\mathrm{d}t} = -\sum_m a_m^*(t)\langle\phi_m|\hat{H}|\phi_n\rangle , \tag{1.2.7}$$

where $\hat{H}^\dagger = \hat{H}$ has been used.

These equations determine the time development of the expansion coefficients $a_n(t)$. We will now modify them a little. In order to do so, we must think about the energy expectation value $\hat{H}$ in the state $|\psi(t)\rangle$:

$$\langle \hat{H} \rangle = \langle \psi(t)|\hat{H}|\psi(t)\rangle = \sum_{n,m} a_n^*(t) a_m(t) \langle \phi_n|\hat{H}|\phi_m\rangle \ . \qquad (1.2.8)$$

Using this expression we can write for (1.2.6) and (1.2.7), respectively

$$\frac{\mathrm{d}a_n(t)}{\mathrm{d}t} = \frac{\partial \langle \hat{H} \rangle}{\partial(\mathrm{i}\hbar a_n^*)} \qquad (1.2.9)$$

and

$$\frac{\mathrm{d}(\mathrm{i}\hbar a_n^*(t))}{\mathrm{d}t} = -\frac{\partial \langle \hat{H} \rangle}{\partial a_n} \ . \qquad (1.2.10)$$

Here, we see that these equations are formally analogous to Hamilton's canonical equations with the correspondences $a_n \leftrightarrow x$ and $\mathrm{i}\hbar a_n^* \leftrightarrow p$. However, of course the expansion coefficients a_n and a_n^* are not dynamical variables of the system.

The idea of second quantization is to promote a_n and a_n^* to operators and to interpret $N_n = N|a_n|^2 = N a_n a_n^*$ as physical quantities of the system. Originally, in single-particle quantum mechanics, N_n is N times the probability of detecting the particle in the state n, and therefore corresponds to the total probability of detecting the particle when the same experiment is performed N times. Then, N is the number of experiments, and therefore in principle this experiment can be performed in single-particle quantum mechanics.

On the other hand, consider a system with N non-interacting particles where only one experiment is performed, and where the number of particles in the single-particle state n is observed to be $\hat{N}_n$. Since both experiments described above are different, N_n and $\hat{N}_n$ are different quantities – repeating number, or observed number of particles. Experience tells us that these two numbers often agree. Admitting this, the number N_n appearing in the single-particle system after performing N experiments will be promoted to the observable (physical quantity) $\hat{N} = \hat{A}_n^\dagger \hat{A}_n$ in the N-particle system. At the same time,

$$\begin{aligned} \sqrt{N} a_n &\to \hat{A}_n \ , \\ \sqrt{N} a_n^* &\to \hat{A}_n^\dagger \end{aligned} \qquad (1.2.11)$$

are promoted to operators. Taking into account also that the energy expectation value $\hat{H}$ is multiplied by N, it is clear from (1.2.9) and (1.2.10) that $\hat{A}_n$ and $\mathrm{i}\hbar\hat{A}_n^*$ are canonical conjugate variables. Therefore, we suppose that they fulfil the same commutation relation as $\hat{\boldsymbol{r}}$ and $\hat{\boldsymbol{p}}$:

$$[\hat{A}_n, \mathrm{i}\hbar\hat{A}_n^\dagger] = \mathrm{i}\hbar \ . \qquad (1.2.12)$$

Further generalizing (1.2.12), we obtain

$$[\hat{A}_n, \mathrm{i}\hbar\hat{A}_m^\dagger] = \mathrm{i}\hbar\delta_{n,m} \quad ,$$
$$[\hat{A}_n, \hat{A}_m] = [\mathrm{i}\hbar\hat{A}_n^\dagger, \mathrm{i}\hbar\hat{A}_m^\dagger] = 0 \quad . \tag{1.2.13}$$

From these commutation relations it can be understood that $\hat{A}_n^\dagger$ and $\hat{A}_n$ are creating and annihilating one particle in the state n, respectively. We write $|N_n\rangle$ for the many-particle state where N_n particles are in the single-particle state n. Then, the following equation holds:

$$\hat{N}_n|N_n\rangle = N_n|N_n\rangle \quad . \tag{1.2.14}$$

Acting with $\hat{N}_n$ on the state $A_n^\dagger|N_n\rangle$, we obtain

$$\begin{aligned}\hat{N}_n\hat{A}_n^\dagger|N_n\rangle &= ([\hat{N}_n, \hat{A}_n^\dagger] + \hat{A}_n^\dagger\hat{N}_n)|N_n\rangle \\ &= (N_n + 1)\hat{A}_n^\dagger|N_n\rangle \quad .\end{aligned} \tag{1.2.15}$$

Here, we used a variation of equation (1.2.13)

$$[\hat{N}_n, \hat{A}_n^\dagger] = [\hat{A}_n^\dagger\hat{A}_n, \hat{A}_n^\dagger] = \hat{A}_n^\dagger[\hat{A}_n, \hat{A}_n^\dagger] = \hat{A}_n^\dagger \quad . \tag{1.2.13$'$}$$

From (1.2.15) it follows that the particle number eigenvalue of the state $A_n^\dagger|N_n\rangle$ is given by $N_n + 1$, and therefore $\hat{A}_n^\dagger$ is an operator that increases the particle number by one. In the same way, it can be seen that $\hat{A}_n$ is an operator that decreases the particle number by one. Starting from the state $|N_n = 0\rangle$, we can create the states $N_n = 0, 1, 2, \ldots$ by acting on it successively with $\hat{A}_n^\dagger$. The particle picture arises because the particle number eigenvalue N_n counts discrete integer numbers.

On the other hand, what might the wave picture be? In order to understand it, we define the field operators $\hat{\psi}(\boldsymbol{r})$ and $\hat{\psi}^\dagger(\boldsymbol{r})$

$$\hat{\psi}(\boldsymbol{r}) = \sum_n \hat{A}_n\phi_n(\boldsymbol{r}) \quad ,$$
$$\hat{\psi}^\dagger(\boldsymbol{r}) = \sum_n \hat{A}_n^\dagger\phi_n^*(\boldsymbol{r}) \quad . \tag{1.2.16}$$

Using the commutator relation (1.2.13) and the fact that $\{\phi_n(\boldsymbol{r})\}$ is an orthonormal basis of single-particle states, we obtain

$$\begin{aligned}[\hat{\psi}(\boldsymbol{r}), \hat{\psi}^\dagger(\boldsymbol{r}')] &= \sum_{n,m}[\hat{A}_n, \hat{A}_m^\dagger]\phi_n(\boldsymbol{r})\phi_m^*(\boldsymbol{r}') = \sum_n \phi_n(\boldsymbol{r})\phi_n^*(\boldsymbol{r}') \\ &= \sum_n \langle\boldsymbol{r}|n\rangle\langle n|\boldsymbol{r}'\rangle = \langle\boldsymbol{r}|\boldsymbol{r}'\rangle = \delta(\boldsymbol{r} - \boldsymbol{r}')\end{aligned} \tag{1.2.17}$$

and

$$[\hat{\psi}(\boldsymbol{r}), \hat{\psi}(\boldsymbol{r}')] = [\hat{\psi}^\dagger(\boldsymbol{r}), \hat{\psi}^\dagger(\boldsymbol{r}')] = 0 \quad . \tag{1.2.18}$$

$n(\boldsymbol{r}) = \hat{\psi}^\dagger(\boldsymbol{r})\hat{\psi}(\boldsymbol{r})$ is the particle density at the position $\boldsymbol{r}$, $\hat{\psi}^\dagger(\boldsymbol{r})$ is the creation operator of a particle at position $\boldsymbol{r}$, and $\hat{\psi}(\boldsymbol{r})$ is the annihilation operator of a particle at position $\boldsymbol{r}$. By promoting the wave functions $\psi(\boldsymbol{r})$ and $\psi^*(\boldsymbol{r})$ to operators $\hat{\psi}(\boldsymbol{r})$ and $\hat{\psi}^\dagger(\boldsymbol{r})$, we regain the picture of wave functions in the three-dimensional physical space, and also the metaphor of the luminous advertisement screen works well. More precisely, the position coordinate $\hat{\boldsymbol{r}}$ is degraded from an operator to a label $\boldsymbol{r}$ defining the position of the light, and instead operators switching the light on ($\hat{\psi}^\dagger(\boldsymbol{r})$) and off ($\hat{\psi}(\boldsymbol{r})$) emerge. A particle is described as the excitation of a field that is created and annihilated.

We now introduce the phase operator $\hat{\theta}_n$ describing the interference of a wave by

$$
\begin{aligned}
\hat{A}_n^\dagger &= (\hat{N}_n)^{1/2}\,\mathrm{e}^{-\frac{\mathrm{i}}{\hbar}\hat{\theta}_n} \quad , \\
\hat{A}_n &= \mathrm{e}^{\frac{\mathrm{i}}{\hbar}\hat{\theta}_n}(\hat{N}_n)^{1/2} \quad .
\end{aligned}
\tag{1.2.19}
$$

We show that by assuming the canonical conjugation relations of $\hat{N}$ and $\hat{\theta}_n$

$$
[\hat{N}_n, \hat{\theta}_n] = \mathrm{i}\hbar \tag{1.2.20}
$$

the commutation relation (1.2.12) is obtained. In order to do so, we define

$$
\hat{N}_n(\lambda) = \exp\left(\frac{\mathrm{i}}{\hbar}\lambda\hat{\theta}_n\right)\hat{N}_n \exp\left(-\frac{\mathrm{i}}{\hbar}\lambda\hat{\theta}_n\right) \quad . \tag{1.2.21}
$$

Then, the following equation holds:

$$
[\hat{A}_n, \hat{A}_n^\dagger] = \hat{N}_n(1) - \hat{N}_n(0) = \int_0^1 \frac{\mathrm{d}\hat{N}_n(\lambda)}{\mathrm{d}\lambda}\,\mathrm{d}\lambda \quad . \tag{1.2.22}
$$

On the other hand, owing to (1.2.20), we obtain

$$
\frac{\mathrm{d}\hat{N}_n(\lambda)}{\mathrm{d}\lambda} = \frac{\mathrm{i}}{\hbar}\exp\left(\frac{\mathrm{i}}{\hbar}\lambda\hat{\theta}_N\right)[\hat{\theta}_n, \hat{N}_n]\exp\left(-\frac{\mathrm{i}}{\hbar}\lambda\hat{\theta}_n\right) = 1 \tag{1.2.23}
$$

and therefore (1.2.12).

Now, having introduced the particle number operator $\hat{N}_n$ and its canonical conjugate, the phase $\hat{\theta}_n$, it is neccessary to stress the following. Obviously, $\hat{N}_n$ is a Hermitian operator; however, exactly speaking, $\hat{\theta}_n$ is not Hermitian. In order to see this, we notice that owing to (1.2.22)

$$
\exp\left(\frac{\mathrm{i}}{\hbar}\hat{\theta}_n\right)\hat{N}_n \exp\left(-\frac{\mathrm{i}}{\hbar}\hat{\theta}_n\right) = \hat{N}_n + 1 \tag{1.2.24}
$$

holds, and for a general integer number m

$$\exp\left(\frac{\mathrm{i}}{\hbar}\hat{\theta}_n\right)(\hat{N}_n)^m \exp\left(-\frac{\mathrm{i}}{\hbar}\hat{\theta}_n\right) = \left[\exp\left(\frac{\mathrm{i}}{\hbar}\hat{\theta}_n\right)\hat{N}_n \exp\left(-\frac{\mathrm{i}}{\hbar}\hat{\theta}_n\right)\right]^m$$
$$= (\hat{N}_n + 1)^m \tag{1.2.25}$$

holds. For a general function $g(N)$ we obtain

$$\exp\left(\frac{\mathrm{i}}{\hbar}\hat{\theta}_n\right) g(\hat{N}_n) \exp\left(-\frac{\mathrm{i}}{\hbar}\hat{\theta}_n\right) = g(\hat{N}_n + 1) \ . \tag{1.2.26}$$

This means that $\hat{U} = \exp(\frac{\mathrm{i}}{\hbar}\hat{\theta}_n)$ is a linear operator acting on $\hat{N}_n$, just like $\hat{U}(\boldsymbol{a})$ in (1.1.60). If $\hat{\theta}_n$ were Hermitian, then $\hat{U}$ would be a unitary operator with $\hat{U}^\dagger\hat{U} = \hat{U}\hat{U}^\dagger = 1$. However, this identity is not true. This can be seen from (1.2.26): $\hat{U}^\dagger$ increases $\hat{N}_n$ by one, $\hat{U}$ decreases $\hat{N}_n$ by one. Acting with $\hat{U}$ on the vacuum state with no particles $|N_n = 0\rangle$ we obtain $\hat{U}|N_n = 0\rangle = 0$. Acting on this equation with $\hat{U}^\dagger$ we obtain of course $\hat{U}^\dagger\hat{U}|N_n = 0\rangle = 0$. However, because of $\hat{U}\hat{U}^\dagger|N_n = 0\rangle = \hat{U}|N_n = 1\rangle = |N_n = 0\rangle \neq 0$ we have just demonstrated that $\hat{U}^\dagger\hat{U} \neq \hat{U}\hat{U}^\dagger$. Therefore, we conclude that because the particle number N_n is bounded from below, $\hat{\theta}$ is not Hermitian. However, when only states with $N_n \gg 1$ are considered, the existence of a lower bound can be neglected, and $\hat{\theta}$ can be regarded to be Hermitian.

Next, we deduce the Hamiltonian occurring after second quantization. Continuing in a heuristic manner as above, we declare in (1.2.8) $\langle\hat{H}\rangle$ to be an operator again and write

$$\hat{H} = \sum_{n,m} \hat{A}_n^\dagger \langle\phi_n|\hat{H}_1|\phi_m\rangle \hat{A}_m \ . \tag{1.2.27}$$

Here, $\hat{H}_1$ is the single-particle Hamiltonian, being an operator in the sense that it acts on single-particle wave functions $\phi_n^*(\boldsymbol{r})$ and $\phi_m(\boldsymbol{r})$. $\hat{H}$ is an operator because $\hat{A}_n^\dagger$ and $\hat{A}_m$ are operators; however, $\langle\phi_n|\hat{H}_1|\phi_m\rangle$ is a simple complex number.

Equation (1.2.27) can also be expressed in terms of the field operators $\hat{\psi}^\dagger(\boldsymbol{r})$ and $\hat{\psi}(\boldsymbol{r})$:

$$\hat{H} = \int \mathrm{d}^3\boldsymbol{r}\, \hat{\psi}^\dagger(\boldsymbol{r})\hat{H}_1\hat{\psi}(\boldsymbol{r}) \ . \tag{1.2.28}$$

The Heisenberg equation of motion of $\hat{\psi}$ is given by

$$\mathrm{i}\hbar\frac{\partial\hat{\psi}(\boldsymbol{r},t)}{\partial t} = [\hat{\psi}(\boldsymbol{r},t), \hat{H}] = \hat{H}_1\hat{\psi}(\boldsymbol{r},t) \ . \tag{1.2.29}$$

If $\hat{\psi}(\boldsymbol{r})$ were a single-particle wave function, then this equation would be the Schrödinger equation (1.1.1); however, again we mention that $\hat{\psi}(\boldsymbol{r})$ is an operator, and the above equation describes the time evolution of this operator in the Heisenberg picture, which leads to a totally different meaning.

In the framework of second quantization, it is also possible to express the interaction between particles in terms of $\hat{\psi}(\boldsymbol{r})$ and $\hat{\psi}^{\dagger}(\boldsymbol{r})$. We mention only the result

$$\sum_{i<j} v(\boldsymbol{r}_i - \boldsymbol{r}_j) \to \frac{1}{2} \int \mathrm{d}^3\boldsymbol{r}\, \mathrm{d}^3\boldsymbol{r}'\, \hat{\psi}^{\dagger}(\boldsymbol{r})\hat{\psi}^{\dagger}(\boldsymbol{r}')v(\boldsymbol{r}-\boldsymbol{r}')\hat{\psi}(\boldsymbol{r}')\hat{\psi}(\boldsymbol{r}) \ . \quad (1.2.30)$$

The Hamiltonian is then the sum of $\hat{H}$ in (1.2.28) and the right-hand side of (1.2.30), and the equation of motion of the field operator is

$$\mathrm{i}\hbar\frac{\partial\hat{\psi}(\boldsymbol{r},t)}{\partial t} = \hat{H}_1\hat{\psi}(\boldsymbol{r},t) + \left[\int \mathrm{d}^3\boldsymbol{r}'\hat{\psi}^{\dagger}(\boldsymbol{r}',t)v(\boldsymbol{r}-\boldsymbol{r}')\hat{\psi}(\boldsymbol{r}',t)\right]\hat{\psi}(\boldsymbol{r},t) \ . \quad (1.2.31)$$

Comparing this expression with (1.2.29), we notice that owing to the interaction, a non-linear term emerges. Because this is not the Schrödinger equation, but the Heisenberg equation, there is no conflict with the superposition principle of quantum mechanics.

Finally, we mention the case of fermions. The whole discussion so far is valid for the case when all particles obey Bose statistics. For fermions, all the commutator relations (1.2.12), (1.2.13), (1.2.17) and (1.2.18) must be replaced by the anti-commutator relations. By doing so, the Hamiltonian (1.2.28), (1.2.30) and the equation of motion of the field (1.2.31) are valid as they stand.

In the case when the particles have a spin degree of freedom, the $\boldsymbol{r}$ coordinate must be extended to $(\boldsymbol{r},\sigma)$ [σ is the spin component, for example the eigenvalue S_z of the spin in the z direction]. The discussion of the phase of the fermions is not that simple compared with the bosonic case. This question will be examined in Chap. 5.

1.3 The Variation Principle and the Noether Theorem

We return to a single-particle system. The Heisenberg equation of motion, describing the time evolution of a particle at position $\hat{\boldsymbol{r}}$ having momentum $\hat{\boldsymbol{p}}$, is given by

$$\mathrm{i}\hbar\frac{\mathrm{d}}{\mathrm{d}t}\hat{\boldsymbol{r}}(t) = [\hat{\boldsymbol{r}}(t),\hat{H}] = \mathrm{i}\hbar\left.\frac{\partial H(\boldsymbol{r},\boldsymbol{p})}{\partial \boldsymbol{p}}\right|_{\substack{\boldsymbol{r}=\hat{\boldsymbol{r}}(t)\\ \boldsymbol{p}=\hat{\boldsymbol{p}}(t)}} , \quad (1.3.1)$$

$$\mathrm{i}\hbar\frac{\mathrm{d}}{\mathrm{d}t}\hat{\boldsymbol{p}}(t) = [\hat{\boldsymbol{p}}(t),\hat{H}] = -\mathrm{i}\hbar\left.\frac{\partial H(\boldsymbol{r},\boldsymbol{p})}{\partial \boldsymbol{r}}\right|_{\substack{\boldsymbol{r}=\hat{\boldsymbol{r}}(t)\\ \boldsymbol{p}=\hat{\boldsymbol{p}}(t)}} . \quad (1.3.2)$$

Here, $H(\boldsymbol{r},\boldsymbol{p})$ is a function of $\boldsymbol{r}$ and $\boldsymbol{p}$, from which the Hamiltonian $\hat{H}$ is obtained by substituting $\boldsymbol{r}\to\hat{\boldsymbol{r}}(t)$ and $\boldsymbol{p}\to\hat{\boldsymbol{p}}(t)$. The above equation has the same structure as the classical canonical equations of the Hamiltonian:

$$\frac{\mathrm{d}}{\mathrm{d}t}\boldsymbol{r}(t) = \frac{\partial H(\boldsymbol{r}(t), \boldsymbol{p}(t))}{\partial \boldsymbol{p}(t)} , \tag{1.3.1'}$$

$$\frac{\mathrm{d}}{\mathrm{d}t}\boldsymbol{p}(t) = -\frac{\partial H(\boldsymbol{r}(t), \boldsymbol{p}(t))}{\partial \boldsymbol{r}(t)} . \tag{1.3.2'}$$

Here, we return for a moment to classical mechanics and use the variation principle of analytical mechanics:

$$L(\boldsymbol{r}, \dot{\boldsymbol{r}}; \boldsymbol{p}) = \boldsymbol{p} \cdot \dot{\boldsymbol{r}} - H(\boldsymbol{r}, \boldsymbol{p}) . \tag{1.3.3}$$

We define the action S as

$$S = \int \mathrm{d}t\, L(\boldsymbol{r}(t), \dot{\boldsymbol{r}}(t); \boldsymbol{p}(t)) . \tag{1.3.4}$$

The variation of S in $\boldsymbol{r}$ and $\boldsymbol{p}$ is given by

$$\begin{aligned} \delta S &= \int \mathrm{d}t \left\{ \delta\boldsymbol{r}(t) \cdot \frac{\partial L}{\partial \boldsymbol{r}(t)} + \delta\dot{\boldsymbol{r}}(t) \cdot \frac{\partial L}{\partial \dot{\boldsymbol{r}}(t)} + \delta\boldsymbol{p}(t) \cdot \frac{\partial L}{\partial \boldsymbol{p}(t)} \right\} \\ &= \int \mathrm{d}t \left\{ \delta\boldsymbol{r}(t) \cdot \left[\frac{\partial L}{\partial \boldsymbol{r}(t)} - \frac{\mathrm{d}}{\mathrm{d}t}\frac{\partial L}{\partial \dot{\boldsymbol{r}}(t)} \right] + \delta\boldsymbol{p}(t) \cdot \frac{\partial L}{\partial \boldsymbol{p}(t)} \right\} . \end{aligned} \tag{1.3.5}$$

(The reader unfamiliar with the variation principle or functional derivative is referred to Appendix B.) By requiring that the variation δS must be zero for arbitrary transformations $\delta\boldsymbol{r}(t)$ and $\delta\boldsymbol{p}(t)$, we obtain directly (1.3.1′) and (1.3.2′).

So, what might be the variation principle corresponding to the field equation (1.2.31)? We know already the analogy $\hat{\boldsymbol{r}} \leftrightarrow \hat{\psi}(\boldsymbol{r})$ and $\hat{\boldsymbol{p}} \leftrightarrow \mathrm{i}\hbar\hat{\psi}^\dagger(\boldsymbol{r})$. From (1.3.3) we can deduce that the Lagrangian must be

$$L(\{\psi(\boldsymbol{r})\}, \{\dot{\psi}(\boldsymbol{r})\}, \{\psi^\dagger(\boldsymbol{r})\}) = \int \mathrm{i}\hbar\psi^\dagger(\boldsymbol{r})\dot{\psi}(\boldsymbol{r})\, \mathrm{d}\boldsymbol{r} - H(\psi(\boldsymbol{r}), \psi^\dagger(\boldsymbol{r})) . \tag{1.3.6}$$

We write for the Lagrange density $\mathcal{L}$

$$\begin{aligned} \mathcal{L}(\{\varphi_A(x)\}, \{\partial_\mu\varphi_A(x)\}) &= \mathrm{i}\hbar\psi^\dagger(\boldsymbol{r},t)\dot{\psi}(\boldsymbol{r},t) \\ &- \frac{\hbar^2}{2m}[\nabla\psi^\dagger(\boldsymbol{r},t)][\nabla\psi(\boldsymbol{r},t)] - V(\boldsymbol{r})\psi^\dagger(\boldsymbol{r},t)\psi(\boldsymbol{r},t) \\ &- \frac{1}{2}\int \mathrm{d}\boldsymbol{r}'\, \psi^\dagger(\boldsymbol{r},t)\psi^\dagger(\boldsymbol{r}',t)v(\boldsymbol{r}-\boldsymbol{r}')\psi(\boldsymbol{r}',t)\psi(\boldsymbol{r},t) . \end{aligned} \tag{1.3.7}$$

We now introduce the combined notation x for the space coordinate $\boldsymbol{r}$ and the time coordinate t, defining the x^μ components to be $(t, \boldsymbol{r})$. The space–time coordinates with lower index x_μ are defined to be $(t, -\boldsymbol{r})$. The partial differential operator ∂_μ corresponding to these coordinate components is given by $\partial_\mu = \partial/\partial x^\mu = (\partial/\partial t, \nabla)$. We wrote the label A to distinguish between different complex fields $\varphi_A(x)$; in the present case we write $\varphi_{A=1} = \psi(\boldsymbol{r},t)$ and $\varphi_{A=2} = \psi^\dagger(\boldsymbol{r},t)$. The action S can be expressed in terms of L or $\mathcal{L}$ as

$$\begin{aligned} S &= \int \mathrm{d}t \, L(\{\varphi_A(x)\}, \{\partial_\mu \varphi_A(x)\}) \\ &= \int \mathrm{d}^4x \, \mathcal{L}(\{\varphi_A(x)\}, \{\partial_\mu \varphi_A(x)\}) \ . \end{aligned} \tag{1.3.8}$$

As in (1.3.5), by taking the variation

$$\begin{aligned} \delta S &= \int \mathrm{d}^4x \left\{ \delta\varphi_A(x) \frac{\delta S}{\delta\varphi_A(x)} + \delta(\delta_\mu \varphi_A(x)) \frac{\delta S}{\delta(\partial_\mu \varphi_A(x))} \right\} \\ &= \int \mathrm{d}^4x \, \delta\varphi_A(x) \left\{ \frac{\delta S}{\delta\varphi_A(x)} - \partial_\mu \left(\frac{\delta S}{\delta(\partial_\mu \varphi_A(x))} \right) \right\} \end{aligned} \tag{1.3.9}$$

and requiring that $\delta S = 0$, we obtain

$$\frac{\delta S}{\delta\varphi_A(x)} - \partial_\mu \left(\frac{\delta S}{\delta(\partial_\mu \varphi_A(x))} \right) = 0 \ . \tag{1.3.10}$$

From (1.3.7), for the case $A = 2$ we obtain

$$\begin{aligned} &\mathrm{i}\hbar\dot{\psi}(\boldsymbol{r},t) - V(\boldsymbol{r})\psi(\boldsymbol{r},t) - \int \mathrm{d}\boldsymbol{r}' \, \psi^\dagger(\boldsymbol{r}',t) v(\boldsymbol{r}-\boldsymbol{r}')\psi(\boldsymbol{r}',t)\psi(\boldsymbol{r},t) \\ &\quad - \nabla \cdot \left\{ -\frac{\hbar^2}{2m} \nabla\psi(\boldsymbol{r},t) \right\} = 0 \ . \end{aligned}$$

By rearranging this equation, we regain equation (1.2.31).

Next, we examine the symmetry operations in quantum field theory. Corresponding to the transformation $\hat{\boldsymbol{r}} \to \hat{\boldsymbol{r}} + \boldsymbol{\alpha}$ in single-particle quantum mechanics, we consider the transformation

$$\varphi_A(x) \to \varphi'_A(x) = \varphi_A(x) + \delta\varphi_A(x) \ , \tag{1.3.11}$$

$$\partial_\mu \varphi_A(x) \to \partial_\mu \varphi'_A(x) = \partial_\mu \varphi_A(x) + \partial_\mu(\delta\varphi_A(x)) \ . \tag{1.3.12}$$

As an explicit example, we consider a phase transformation with constant phase of the field operator

$$\begin{aligned} \varphi_1(x) &= \psi(x) \to \psi'(x) = \mathrm{e}^{-\mathrm{i}a}\psi(x) \simeq \psi(x) - \mathrm{i}a\psi(x) \\ \varphi_2(x) &= \psi^\dagger(x) \to \psi'^\dagger(x) = \mathrm{e}^{+\mathrm{i}a}\psi^\dagger(x) \simeq \psi^\dagger(x) + \mathrm{i}a\psi^\dagger(x) \ . \end{aligned} \tag{1.3.13}$$

Under the transformation (1.3.11) and (1.3.12), the action S that we assume now to be bounded to a space–time region Ω transforms as

$$\begin{aligned} S \to S' &= \int_\Omega \mathrm{d}^4x \, \mathcal{L}(\{\varphi_A(x) + \delta\varphi_A(x)\}, \{\partial_\mu \varphi_A(x) + \partial_\mu(\delta\varphi_A(x))\}) \\ &= S + \int_\Omega \mathrm{d}^4x \left\{ \delta\varphi_A(x) \frac{\delta S}{\delta\varphi_A(x)} + \partial_\mu(\delta\varphi_A(x)) \frac{\delta S}{\delta(\partial_\mu \varphi_A(x))} \right\} \\ &= S + \int_\Omega \mathrm{d}^4x \, \delta\varphi_A(x) \left[\frac{\delta S}{\delta\varphi_A(x)} - \partial_\mu \left(\frac{\delta S}{\delta(\partial_\mu \varphi_A(x))} \right) \right] \\ &\quad + \int_\Omega \mathrm{d}^4x \, \partial_\mu \left[\frac{\delta S}{\delta(\partial_\mu \varphi_A(x))} \delta\varphi_A(x) \right] \ . \end{aligned} \tag{1.3.14}$$

Assuming that φ_A obeys the equation of motion (1.3.10), only the third term of the previous expression contributes to the transformation of S. In the case that the action S is invariant under the transformation (1.3.11) and (1.3.12), in every arbitrary region Ω, we obtain

$$\partial_\mu \left[\frac{\delta S}{\delta(\partial_\mu \varphi_A(x))} \delta\varphi_A(x) \right] = 0 \ . \tag{1.3.15}$$

Defining a current J^μ as

$$J^\mu \propto \frac{\delta S}{\delta(\partial_\mu \varphi_A(x))} \delta\varphi_A(x) \ , \tag{1.3.16}$$

then (1.3.15) becomes the current conservation law

$$\partial_\mu J^\mu = 0 \ . \tag{1.3.17}$$

We just deduced the Noether theorem (in its simplest form). "The invariance of the action S shows up in the symmetry under the transformation, and from this symmetry a current conservation law can be deduced." In the explicit example (1.3.13), the transformation of the action integral S under the phase transformation α can be written via the chain rule with $\psi^\dagger$ and ψ as

$$J^\mu = \frac{\delta S}{\delta(\partial_\mu \psi^\dagger(x))} \frac{\partial \delta\psi^\dagger(x)}{\partial\alpha} + \frac{\delta S}{\delta(\partial_\mu \psi(x))} \frac{\partial \delta\psi(x)}{\partial\alpha} \ . \tag{1.3.18}$$

Because the action is the integral in space and time of the Lagrangian density (1.3.7), we obtain

$$\begin{aligned} J^0 &= \frac{\delta S}{\delta\dot{\psi}^\dagger(x)} \mathrm{i}\psi^\dagger(x) + \frac{\delta S}{\delta\dot{\psi}(x)}(-\mathrm{i}\psi(x)) \\ &= 0 + \mathrm{i}\hbar\psi^\dagger(x)(-\mathrm{i}\psi(x)) \\ &= \hbar\psi^\dagger(x)\psi(x) = \hbar n(x) \qquad (\mu = 0) \ , \end{aligned} \tag{1.3.19}$$

$$\begin{aligned} J^\alpha &= \frac{\delta S}{\delta(\partial_\alpha \psi^\dagger(x))} \mathrm{i}\psi^\dagger(x) + \frac{\delta S}{\delta(\partial_\alpha \psi(x))}(-\mathrm{i}\psi(x)) \\ &= \left(-\frac{\hbar^2}{2m}\partial_\alpha\psi(x)\right)\mathrm{i}\psi^\dagger(x) + \left(-\frac{\hbar^2}{2m}\partial_\alpha\psi^\dagger(x)\right)(-\mathrm{i}\psi(x)) \\ &= \frac{\hbar^2}{2mi}\{\psi^\dagger(x)\partial_\alpha\psi(x) - [\partial_\alpha\psi^\dagger(x)]\psi(x)\} \\ &= \hbar j^\alpha(x) \qquad (\mu = \alpha = 1,2,3) \ . \end{aligned} \tag{1.3.20}$$

Here, $n(x)$ is the particle density, $\boldsymbol{j}(x)$ is the particle current density, and (1.3.17) becomes the well-known continuity equation:

$$\frac{\partial n(x)}{\partial t} + \nabla \cdot \boldsymbol{j}(x) = 0 \ . \tag{1.3.21}$$

The number of particles $N_V(t)$ in a three-dimensional volume V is given by

$$N_V(t) = \int_V n(\boldsymbol{r},t)\,\mathrm{d}^3\boldsymbol{r} \tag{1.3.22}$$

and owing to (1.3.21), its time derivative is given by

$$\frac{\mathrm{d}N_V(t)}{\mathrm{d}t} = \int_V \frac{\partial n(\boldsymbol{r},t)}{\partial t}\,\mathrm{d}^3\boldsymbol{r} = -\int_V \nabla\cdot\boldsymbol{j}(\boldsymbol{r},t)\,\mathrm{d}^3\boldsymbol{r} = -\int_{\partial V} \mathrm{d}\boldsymbol{S}\cdot\boldsymbol{j}(\boldsymbol{r},t)\ . \tag{1.3.23}$$

Here, we re-expressed the volume integral as a surface integral by using Gauss's theorem. For the case when V is the whole space, $N_V(t)$ becomes the total particle number $N(t)$, and ∂V becomes the boundary at infinity, where $\boldsymbol{j}(\boldsymbol{r},t)$ is zero and therefore the surface integral vanishes. We conclude that the total particle number $N(t)$ obeys the conservation law $\mathrm{d}N(t)/\mathrm{d}t = 0$. The conservation law in N that we obtained from the symmetry in the phase is analogous to the conservation law for p that we deduced from the translational symmetry in x.

Last, we discuss the generators of transformations in field theory. Up to now, we have performed functional derivatives of $\psi, \psi^\dagger, N$ and $\boldsymbol{j}$ in a formal manner, ignoring the fact that we are dealing with operators. Now, we need to recall the properties of operators. We define $\hat{U}(\{\delta\psi(x)\})$ as follows:

$$\begin{aligned}\hat{U}(\{\delta\psi\}) &= \exp\left[\frac{\mathrm{i}}{\hbar}\int \mathrm{d}\boldsymbol{r}\ \hat{\Pi}(\boldsymbol{r})\delta\psi(\boldsymbol{r})\right]\\ &\equiv \exp\left[\frac{\mathrm{i}}{\hbar}\hat{Q}(\{\delta\psi\})\right]\ .\end{aligned} \tag{1.3.24}$$

We define $\hat{\Pi}(\boldsymbol{r})$ to be the canonical conjugate of $\hat{\psi}$

$$\hat{\Pi}(\boldsymbol{r}) = \frac{\delta S}{\delta(\partial_t\hat{\psi}(\boldsymbol{r}))}\ , \tag{1.3.25}$$

and assume the commutation relation

$$[\hat{\psi}(\boldsymbol{r}), \hat{\Pi}(\boldsymbol{r}')] = \mathrm{i}\hbar\delta(\boldsymbol{r}-\boldsymbol{r}')\ . \tag{1.3.26}$$

From (1.3.7) we obtain

$$\hat{\Pi}(\boldsymbol{r}) = \mathrm{i}\hbar\hat{\psi}^\dagger(\boldsymbol{r}) \tag{1.3.27}$$

and therefore (1.3.26) is identical to (1.2.17).

It is possible to show that

$$\hat{U}(\{\delta\psi(\boldsymbol{r})\})\hat{\psi}(\boldsymbol{r})\hat{U}(\{-\delta\psi(\boldsymbol{r})\}) = \hat{\psi}(\boldsymbol{r}) + \delta\psi(\boldsymbol{r}) \tag{1.3.28}$$

holds. The proof corresponds to the discussion concerning (1.1.69)–(1.1.71). Starting with the definition

$$\hat{\psi}(\boldsymbol{r};\lambda) = \hat{U}(\{\lambda\delta\psi(\boldsymbol{r})\})\hat{\psi}(\boldsymbol{r})\hat{U}(\{-\lambda\delta\psi(\boldsymbol{r})\}) \tag{1.3.29}$$

we obtain

$$\frac{\partial\hat{\psi}(\boldsymbol{r};\lambda)}{\partial\lambda} = \frac{\mathrm{i}}{\hbar}\exp\left[\frac{\mathrm{i}}{\hbar}\lambda\hat{Q}(\{\delta\psi\})\right][\hat{Q}(\{\delta\psi\}),\hat{\psi}(\boldsymbol{r})] \times \exp\left[-\frac{\mathrm{i}}{\hbar}\lambda\hat{Q}(\{\delta\psi\})\right] \tag{1.3.30}$$

and

$$[\hat{Q}(\{\delta\psi\}),\hat{\psi}(\boldsymbol{r})] = \int \mathrm{d}\boldsymbol{r}'\,[\hat{\Pi}(\boldsymbol{r}'),\hat{\psi}(\boldsymbol{r})]\delta\psi(\boldsymbol{r}') = -\mathrm{i}\hbar\delta\psi(\boldsymbol{r}) \ . \tag{1.3.31}$$

Because $\delta\psi(\boldsymbol{r})$ is simply a function, (1.3.30) becomes

$$\frac{\partial\hat{\psi}(\boldsymbol{r};\lambda)}{\partial\lambda} = \delta\psi(\boldsymbol{r}) \ . \tag{1.3.32}$$

Integrating this equation in λ from zero to one, we obtain (1.3.28). Therefore, (1.3.24) acts on $\hat{\psi}(\boldsymbol{r})$ by shifting it by $\delta\psi(\boldsymbol{r})$, and therefore it is clear that $\hat{\Pi}(\boldsymbol{r})$ given in (1.3.25) is the generator of the transformation.

1.4 Quantization of the Electromagnetic Field

In the previous section we demonstrated that the invariance under the phase transformation (1.3.13) of the operators leads to the conservation law of the particle number $\hat{N}$. We assumed that the angle α is constant and does not depend either on the space coordinate $\boldsymbol{r}$ or on time t. This kind of transformation is called a global gauge transformation. In this chapter we discuss the more general case where invariance under local gauge transformations is required, that is, when α depends on $x = (\boldsymbol{r}, t)$.

Unfortunately, the Lagrange density (1.3.7) is not invariant under this transformation. Problematic is the term with derivative

$$\partial_\mu(\psi \mathrm{e}^{-\mathrm{i}\alpha}) = \mathrm{e}^{-\mathrm{i}\alpha}(\partial_\mu\psi - \mathrm{i}\partial_\mu\alpha\cdot\psi) \ , \tag{1.4.1}$$

where a second term containing $\partial_\mu\alpha$ emerges. Now, a derivative connects the field value at two neighbouring points x and $x + \mathrm{d}x$ in space and time, and (1.4.1) shows that by performing different phase transformations at the two points, then obviously the result will change.

In order to make $\mathcal{L}$ invariant, it would be sufficient to introduce another field that links these two points, and to perform simultaneously to (1.3.13) a transformation of this new field so that the phases of both fields annihilate each other. The field that is introduced in this manner is called a gauge field.

A representative example is the electromagnetic field A_μ; however, also in many other cases gauge fields emerge. In solid state physics, gauge fields play an important role for spin glasses, quantum spin systems, strongly correlated electronic systems, the quantum Hall effect and liquid crystals.

Roughly speaking, gauge fields appear when some constraints or frustration prevent the system from stabilizing in a low energy state. Hopefully, the reader will understand this intuitive picture when dealing with the concrete examples later in this book. Let us now briefly proceed with the mathematical concepts.

We now consider the electromagnetic field A_μ as a gauge field and set the velocity of light c equal to 1. In order to obtain a locally gauge invariant system, the terms $\partial_\mu \psi$ and $\partial_\mu \psi^\dagger$ must be replaced by $(\partial_\mu + \mathrm{i}(e/\hbar)A_\mu)\psi$ and $(\partial_\mu - \mathrm{i}(e\hbar)A_\mu)\psi^\dagger$, respectively, and corresponding to (1.3.13), the transformation of the gauge field must be

$$A_\mu \longrightarrow A_\mu + \frac{\hbar}{e}\partial_\mu \alpha \ . \tag{1.4.2}$$

Here, $-e$ is the charge of the electron. Owing to this replacement, the current $\boldsymbol{J}(x)$ will be defined as

$$\boldsymbol{J}(x) = -\frac{e\hbar}{2m\mathrm{i}}\left(\psi^\dagger\left(\nabla + \frac{\mathrm{i}e}{\hbar}\boldsymbol{A}\right)\psi - \left[\left(\nabla - \frac{\mathrm{i}e}{\hbar}\boldsymbol{A}\right)\psi^\dagger\right]\psi\right) \ . \tag{1.4.3}$$

The conservation law for this current is similar to (1.3.21) with charge density $\rho(x) = -e\psi^\dagger(x)\psi(x)$. Equation (1.3.21) was derived under the assumption that the phase is independent of x, and it can be assumed that a similar equation holds under the more severe local gauge invariance condition. This is indeed the case, all formulas up to (1.3.21) do still hold; however, we will not go into the details.

Now, (1.4.2) is the gauge transformation that follows from the theory of electromagnetism, and we will now review the dynamics and the quantization of the electromagnetic field including this transformation. The Lagrangian of the electromagnetic field is given by

$$\mathcal{L}_{\mathrm{em}} = -\frac{1}{16\pi}F_{\mu\nu}F^{\mu\nu} \ . \tag{1.4.4}$$

It is simple to see that $F_{\mu\nu} = \partial A_\mu/\partial x^\nu - \partial A_\nu/\partial x^\mu$ and $F^{\mu\nu} = \partial A^\mu/\partial x_\nu - \partial A^\nu/\partial x_\mu$ are invariant under the gauge transformation (1.4.4). Therefore, (1.4.4) is invariant under local gauge transformtions.

Now, we write the full Lagrangian including (1.4.4) and the matter fields as

$$\begin{aligned}\mathcal{L}_{\mathrm{total}} = {} & \mathrm{i}\hbar\psi^\dagger\left(\dot{\psi} + \mathrm{i}\frac{e}{\hbar}A_0\psi\right) - \frac{\hbar^2}{2m}\left[\left(\nabla - \frac{ie}{\hbar}\boldsymbol{A}\right)\psi^\dagger\right]\left[\left(\nabla + \frac{ie}{\hbar}\boldsymbol{A}\right)\psi\right] \\ & -\frac{1}{2}\int \mathrm{d}\boldsymbol{r}'\, \psi^\dagger(\boldsymbol{r})\psi^\dagger(\boldsymbol{r}')v(\boldsymbol{r}-\boldsymbol{r}')\psi(\boldsymbol{r}')\psi(\boldsymbol{r}) - \frac{1}{16\pi}F_{\mu\nu}F^{\mu\nu} \ . \end{aligned} \tag{1.4.5}$$

Varying this equation with respect to A_μ and requiring the result to vanish, we obtain (the derivation is left as an exercise for the reader)

$$-\nabla^2 A_0(x) + \frac{\partial}{\partial t}\nabla\cdot\boldsymbol{A}(x) = -4\pi\rho(x) \ , \tag{1.4.6}$$

$$\left(\nabla^2 - \frac{\partial^2}{\partial t^2}\right)\boldsymbol{A}(x) + \frac{\partial}{\partial t}\nabla A_0(x) - \nabla(\nabla\cdot\boldsymbol{A}(x)) = -4\pi\boldsymbol{J}(x) \ . \tag{1.4.7}$$

These equations are the field equations. By setting $\boldsymbol{H} = \nabla\times\boldsymbol{A}$ and $\boldsymbol{E} = \nabla A_0 - \partial\boldsymbol{A}/\partial t$, two of the four Maxwell equations are automatically fulfilled, and inserting these expressions into the other two Maxwell equations, the above equations are regained. The fields $\boldsymbol{E}$ and $\boldsymbol{H}$ have six components; however, at this step the components are reduced to four.

Equations (1.4.6) and (1.4.7) are complicated, so let us try to simplify them by using the freedom of gauge invariance (1.4.2) in an appropriate manner. Here, we have many different possibilities, and because in the Lagrangian used for solid state physics, Lorentz invariance is already broken, in many cases the so-called Coulomb (transverse) gauge is chosen, where the degrees of freedom of the electromagnetic field appear very clearly. By choosing α in (1.4.2) in such a way that $(\hbar/e)\nabla^2\alpha = -\nabla\boldsymbol{A}$ holds, owing to such a gauge transformation, we obtain

$$\nabla\cdot\boldsymbol{A} = 0 \ . \tag{1.4.8}$$

Here, α is determined uniquely when boundary conditions are imposed because of the uniqueness of the solution of the Poisson equation. Therefore, notice that in the Coulomb gauge, the gauge is totally fixed. With (1.4.8), equations (1.4.6) and (1.4.7) become

$$-\nabla^2 A_0(x) = -4\pi\rho(x) \ , \tag{1.4.9}$$

$$\left(\nabla^2 - \frac{\partial^2}{\partial t^2}\right)\boldsymbol{A}(x) = -4\pi\boldsymbol{J}(x) - \frac{\partial}{\partial t}\nabla A_0(x) \ . \tag{1.4.10}$$

Under the boundary condition $A_0(x) \to 0$ for $\boldsymbol{r} \to 0$, the solution of (1.4.9) is

$$-A_0(x) = \int \mathrm{d}^3\boldsymbol{r}' \, \frac{\rho(\boldsymbol{r}', t)}{|\boldsymbol{r} - \boldsymbol{r}'|} \ . \tag{1.4.11}$$

A_0 is not an independent field, but can be expressed in terms of the charge distribution of the matter field. Notice that no retardation effect occurs in (1.4.11); ρ at time t determines A_0 at time t. This is not an approximation, but an exact result in the Coulomb gauge. Finally, owing to the condition $\boldsymbol{\nabla}\cdot\boldsymbol{A} = 0$ that lowers the degrees of freedom once more, we are left with two degrees of freedom.

Now, using (1.4.5), we define the "momentum" conjugated to A_μ:

$$\Pi^0 = \frac{\partial \mathcal{L}}{\partial \dot{A}_0} = 0 \ , \tag{1.4.12a}$$

$$\boldsymbol{\Pi} = \frac{\partial \mathcal{L}}{\partial \dot{\boldsymbol{A}}} = +\frac{1}{4\pi}\left[-\nabla A^0 + \frac{\partial \boldsymbol{A}}{\partial t}\right] = -\frac{\boldsymbol{E}}{4\pi} \ . \tag{1.4.12b}$$

For simplicity, we now ignore the matter field and discuss only (1.4.4) $\mathcal{L}_{\text{em}}$. Then, in the Coulomb gauge, with $\rho(\boldsymbol{r}', t) = 0$ due to (1.4.11), it follows that $A_0 = 0$ and

$$\mathcal{L}_{\text{em}} = \frac{1}{8\pi}\{\dot{\boldsymbol{A}}^2 - (\nabla \times \boldsymbol{A})^2\} \ . \tag{1.4.13}$$

The corresponding Hamiltonian density is given by

$$\begin{aligned} \mathcal{H}_{\text{em}} &= \boldsymbol{\Pi} \cdot \dot{\boldsymbol{A}} - \mathcal{L}_{\text{em}} \\ &= \frac{1}{8\pi}\{\dot{\boldsymbol{A}}^2 + (\nabla \times \boldsymbol{A})^2\} \\ &= \frac{1}{8\pi}\{\boldsymbol{E}^2 + \boldsymbol{H}^2\} \ . \end{aligned} \tag{1.4.14}$$

Here, the analogy $\boldsymbol{\Pi} \leftrightarrow \boldsymbol{p}$ and $\boldsymbol{A} \leftrightarrow \boldsymbol{r}$ is obvious, and this suggests that for quantization, the commutator between $\boldsymbol{A}$ and $\boldsymbol{\Pi}$ at the equal time t should be introduced as

$$[A_i, (\boldsymbol{r}, t), \Pi_j(\boldsymbol{r}', t)] = \mathrm{i}\hbar\delta_{i,j}\delta^3(\boldsymbol{r} - \boldsymbol{r}') \ . \tag{1.4.15}$$

However, we have a problem with (1.4.15). By acting with nabla ∇ with respect to $\boldsymbol{r}$ on both sides, we obtain

$$[\nabla \cdot \boldsymbol{A}(\boldsymbol{r}, t), \Pi_j(\boldsymbol{r}', t)] = \mathrm{i}\hbar\partial_j\delta^3(\boldsymbol{r} - \boldsymbol{r}') \ . \tag{1.4.16}$$

The left-hand side vanishes due to $\boldsymbol{\nabla} \cdot \boldsymbol{A} = 0$; however, the right-hand side is not zero. In order to resolve the problem, we write instead of (1.4.15)

$$[A_i(\boldsymbol{r}, t), \Pi_j(\boldsymbol{r}', t)] = \mathrm{i}\hbar\left(\delta_{i,j} - \frac{\partial_i\partial_j}{\nabla^2}\right)\delta^3(\boldsymbol{r} - \boldsymbol{r}') \ . \tag{1.4.17}$$

If you are confused about a Laplacian appearing in the denominator, perform a Fourier transformation and interpret the right-hand side as $\mathrm{i}\hbar(\delta_{i,j} - k_ik_j/k^2)$. The nabla operator in the k-space leads to $\mathrm{i}k_iA_i$, therefore

$$\sum_i \mathrm{i}k_i(\delta_{i,j} - k_ik_j/k^2) = \mathrm{i}k_j - \mathrm{i}k_j = 0$$

and the inconsistency is removed. As an exercise, you may show that with the commutator (1.4.17), the Heisenberg equation deduced from the Hamiltonian (1.4.14) agrees with the field equation.

2. Quantization with Path Integral Methods

In the previous chapter we learned about operators occurring in quantum mechanics. In this chapter we demonstrate that it is also possible to describe quantum mechanics using integrals in the space of functions, namely functional integrals and path integrals, instead of using operators.

2.1 Single-Particle Quantum Mechanics and Path Integrals

Everybody who learnt quantum mechanics should have heard about the interference experiment of electron waves through slits. Figure 2.1 shows the principle. Electrons emitted from an electron gun pass either through slit A or slit B of a shield and finally illuminate a screen. Because it is possible to avoid more than one electron reaching the screen at the same time by adjusting the intensity of the electron gun to be small enough, it is clear that electrons can indeed be interpreted as propagating "particles". However, when this kind of experiment is performed over a long period, the observed distribution of electrons at all points of the screen becomes the interference pattern of a wave.

In quantum mechanics, this interference pattern is explained as follows. We call the paths from the electron gun through slit A or slit B to the point P on the screen $P_{\rm A}$ or $P_{\rm B}$, respectively. $P_{\rm A}$ and $P_{\rm B}$ each corresponds to a complex amplitude $a_{\rm A}$ and $a_{\rm B}$ of a quantum mechanical wave. Then, the phase difference φ of the complex functions $a_{\rm A}$ and $a_{\rm B}$ ($a_{\rm A}/a_{\rm B} \propto {\rm e}^{{\rm i}\varphi}$) equals the phase difference of the waves. The complex amplitude corresponding to the process that an electron started from the electron gun and reached the point P, without asking whether the electron passed through slit A or slit B is given by the sum of the amplitudes $P_{\rm A}$ and $P_{\rm B}$. The intensity of the wave reaching P (in quantum mechanics the probability of reaching P) is given by the absolute value of the square of the complex amplitude

$$|a_{\rm A} + a_{\rm B}|^2 = |a_{\rm A}|^2 + |a_{\rm B}|^2 + 2|a_{\rm A}||a_{\rm B}|\cos\varphi \ .$$

This expression varies periodically depending on φ.

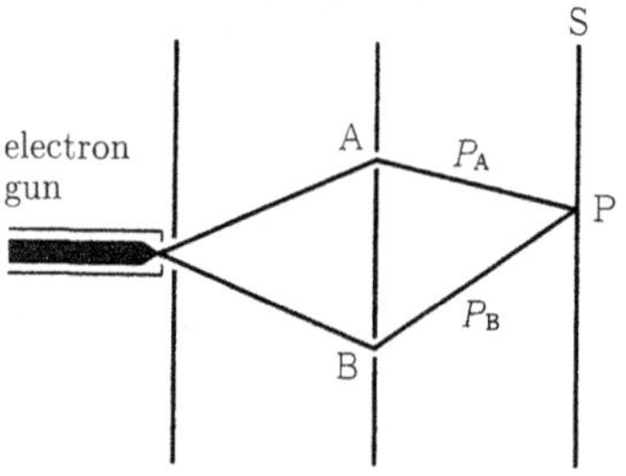

Fig. 2.1. Interference experiment with electron waves. The waves that passed through slits A and B interfere on the screen S

Next, we generalize this approach as shown in Fig. 2.2. Different from Fig. 2.1, there is not just one screening shield, but many, and there are not two slits in each, but a large number (Fig. 2.2a).

Depending on the way at each screen, many different combinations c for passing through the slits are possible, each corresponding to an amplitude a_c, the total amplitude a is given by $\sum_c a_c$. By increasing the number of screening layers and slits more and more, and finally reaching an infinite number, each screening layer will have many slits and finally will disappear. The interval between the gun and the screen will become a continuous space (Fig. 2.2b), and depending on the "combination of slits" the path will mutate to an arbitrary path from the gun to the screen, and "the sum of all amplitudes of possible ways" will mutate to an integral over the paths, that is, the path integral. That is to say, the principle that "the amplitude corresponding to a transition from a starting point to an end point corresponds to an integral over the amplitudes of all possible paths linking these two points" can be regarded as the principle of quantum mechanics.

When considering quantum mechanics, one might have in mind the wave function, and the Schrödinger equation of the wave function, all describing wave properties of the particle. It should be possible to link this description with the picture of the path integral described above by using the superposition principle of waves. However, it required the genius of Feynman to make this discovery.

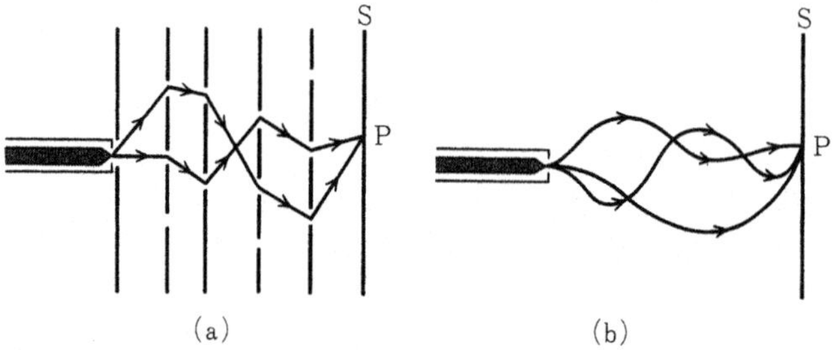

Fig. 2.2a, b. Generalization of the interference experiment of Fig. 2.1. The number of screens and the number of slits in the screen is increased

We now turn to exact mathematics. We start from the Schrödinger equation

$$\begin{aligned} \mathrm{i}\hbar \frac{\partial \psi(x,t)}{\partial t} &= -\frac{\hbar^2}{2m}\frac{\mathrm{d}^2\psi(x,t)}{\mathrm{d}x^2} + V(x)\psi(x,t) \\ &= H\psi(x,t) \ . \end{aligned} \tag{2.1.1}$$

For a moment, we restrict the motion of the particle to a one-dimensional space with space coordinate x.

As shown in (1.1.3), it is possible to integrate (2.1.1) formally

$$|\psi(t')\rangle = U(t',t)|\psi(t)\rangle = \mathrm{e}^{-(\mathrm{i}/\hbar)H(t'-t)}|\psi(t)\rangle \ . \tag{2.1.2}$$

We choose $t' > t$. Equation (2.1.2) describes the time evolution of the state $|\psi(t)\rangle$ occurring at t through $U(t',t)$ during the time $t'-t$, leading to the state $|\psi(t')\rangle$. Not the wave function can be expressed with the path integral mentioned before, but the operator $U(t,t')$ describing the time evolution. In (2.1.2), the label x has been omitted intentionally, because $|\psi(t)\rangle$ should be regarded as a continuous infinite dimensional vector with components $\psi(x,t)$ labelled by x, and $U(t',t)$ should be regarded as a matrix with components $U(x',t';x,t)$.

In x-representation, we obtain

$$\psi(x',t') = \int \mathrm{d}x\, U(x',t';x,t)\psi(x,t) \ . \tag{2.1.3}$$

We decompose $t'-t$ into small time intervals. We write $t'-t = N\Delta t$ and $t_k = t + k\Delta t$. The time evolution corresponding to these steps can be expressed as

$$U(t',t) = U(t',t_{N-1})U(t_{N-1},t_{N-2})\dots U(t_2,t_1)U(t_1,t) \tag{2.1.4}$$

using matrix multiplication of $U(t_{k+1},t_k)$. In components, we obtain

$$\begin{aligned} U(x',t';x,t) = \int & \mathrm{d}x\,(N-1)\,\mathrm{d}x\,(N-2)\dots\mathrm{d}x\,(1) \\ & \times U(x',t';x(N-1),t_{N-1}) \\ & \times U(x(N-1),t_{N-1};x(N-2),t_{N-2}) \\ & \times \dots U(x(1),t_1;x,t) \ . \end{aligned} \tag{2.1.5}$$

Here, the matrix components are given by the "bra- and ket-sandwich"

$$U(x(k+1),t_{k+1};x(k),t_k) = \langle x(k+1)|\,\mathrm{e}^{-(\mathrm{i}/\hbar)H\Delta t}|x(k)\rangle \ . \tag{2.1.6}$$

When Δt is small enough, the exponential can be expanded:

$$\langle x(k+1)|\,\mathrm{e}^{-(\mathrm{i}/\hbar)H\Delta t}|x(k)\rangle \cong \langle x(k+1)|\left(1 - \frac{\mathrm{i}}{\hbar}\Delta t H\right)|x(k)\rangle \ . \tag{2.1.7}$$

Using the momentum eigenstates introduced in (1.1.34) and below, by inserting the completeness relations (1.1.37) we obtain

$$\langle x(k+1)|H|x(k)\rangle = \int \mathrm{d}p\,(k)\langle x(k+1)|p(k)\rangle\langle p(k)|H|x(k)\rangle \ . \tag{2.1.8}$$

Because the Hamiltonian (2.1.1) is expressed in terms of $\hat{p}$ and $\hat{x}$, we obtain

$$\langle p(k)|H(\hat{p},\hat{x})|x(k)\rangle = H(p(k),x(k))\langle p(k)|x(k)\rangle \ .$$

Here, $\hat{p} = (\hbar/\mathrm{i})\,\mathrm{d}/\mathrm{d}x$, and for $\hat{x}$ the hat has been written to stress that it is an operator.

Using this, (2.1.6) and (2.1.7) can be written as

$$\begin{aligned} &U(x(k+1),t_{k+1};x(k),t_k) \\ &\cong \int \mathrm{d}p\,(k)\langle x(k+1)|p(k)\rangle\langle p(k)|x(k)\rangle\left(1-\frac{\mathrm{i}}{\hbar}\Delta t H(p(k),x(k))\right) \\ &\cong \int \frac{\mathrm{d}p\,(k)}{2\pi\hbar}\exp\left[\mathrm{i}\frac{p(k)}{\hbar}(x(k+1)-x(k))-\frac{\mathrm{i}}{\hbar}\Delta t H(p(k),x(k))\right] . \end{aligned} \tag{2.1.9}$$

Inserting this for each term in (2.1.5), we obtain for $U(x',t';x,t)$

$$\begin{aligned} U(x',t';x,t) &= \int \frac{\mathrm{d}p\,(N-1)}{2\pi\hbar}\cdots\int\frac{\mathrm{d}p\,(0)}{2\pi\hbar}\int \mathrm{d}x\,(N-1)\cdots\int \mathrm{d}x\,(1) \\ &\times \exp\left[\frac{\mathrm{i}}{\hbar}\sum_{k=0}^{N-1}\left[p(k)(x(k+1)-x(k))-\Delta t H(p(k),x(k))\right]\right] \ . \end{aligned} \tag{2.1.10}$$

By performing the limit $N\to\infty$ and $\Delta t\to 0$, we obtain

$$x(k+1)-x(k)\to \dot{x}(t)\Delta t$$

$$\sum_{k=0}^{N-1}\Delta t \to \int_t^{t'} \mathrm{d}t \ ,$$

and writing $\mathcal{D}p(t)$ and $\mathcal{D}x(t)$ for the multiple integrals over $p(k)$ and $x(k)$ in (2.1.10), respectively, the result is

$$\begin{aligned} U(x',t';x,t) &= \int_{\substack{x(t)=x \\ x(t')=x'}} \mathcal{D}p(t'')\mathcal{D}x(t'') \\ &\times \exp\left[\frac{\mathrm{i}}{\hbar}\int_t^{t'}\left[p(t'')\dot{x}(t'')-H(p(t''),x(t''))\right]\mathrm{d}t''\right] \ . \end{aligned} \tag{2.1.11}$$

Recalling the definition (1.3.3) and (1.3.4) of the action S, we notice that the expression in the exponent of (2.1.11) is exactly $(\mathrm{i}/\hbar)S$. We conclude that the amplitude corresponding to the path $x(t'')$ and $p(t'')$ $(t\le t''\le t')$ is given by $\exp[\mathrm{i}S(x(t''),p(t''))]$. The action is defined for every arbitrary path $x(t'')$ and $p(t'')$, and the relation $p(t'')=m\dot{x}(t'')$ need not necessarily be valid.

Because the dependence on $p(t'')$ in S is given by

$$\int_t^{t'} \left[p(t'')\dot{x}(t'') - \frac{p(t'')^2}{2m} \right] \mathrm{d}t'' = \int_t^{t'} \left\{ -\frac{1}{2m} \left(p(t'') - m\dot{x}(t'') \right)^2 + \frac{m\dot{x}(t'')^2}{2} \right\} \mathrm{d}t'' \; , \tag{2.1.12}$$

it is possible to perform the integral in $p(t'')$ obtaining

$$\begin{aligned} U(x',t';x,t) &= \int_{\substack{x(t)=x \\ x(t')=x'}} \mathcal{D}x(t'') \exp\left[\frac{\mathrm{i}}{\hbar} S(\{x(t'')\}) \right] \\ &= \int_{\substack{x(t)=x \\ x(t')=x'}} \mathcal{D}x(t'') \exp\left[\frac{\mathrm{i}}{\hbar} \int_t^{t'} \left\{ \frac{m\dot{x}(t')^2}{2} - V(x(t'')) \right\} \mathrm{d}t'' \right] . \end{aligned} \tag{2.1.13}$$

The equation (2.1.13) corresponds to the Lagrangian in the sense that L is given as a function of $\dot{x}$ and x, and (2.1.11) corresponds to the canonical (Hamiltonian) formalism. In textbooks, equation (2.1.13) is often presented; however, (2.1.11) is the more basic equation. This is due to the fact that both canonical coordinates x and p appear in the equation, and that the term $\mathrm{i}p\dot{x}$ has the important interpretation of the Berry phase, as will be explained in Sect. 2.5. In the above case, it has been possible to integrate out the momentum; however, in general this is not always the case.

Next, as a repetition of (1.3.5) of the previous chapter, let us take the variation of $S(x(t''), p(t''))$ appearing in (2.1.11) and $S(x(t''))$ in (2.1.13):

$$\begin{aligned} \delta S(\{x(t'')\},\{p(t'')\}) &= \delta \int_t^{t'} \mathrm{d}t'' \left[p(t'')\dot{x}(t'') - \frac{p(t'')^2}{2m} - V(x(t'')) \right] \\ &= \int_t^{t'} \mathrm{d}t'' \left\{ \delta p(t'') \left[\dot{x}(t'') - \frac{p(t'')}{m} \right] \right. \\ &\qquad \left. + \delta x(t'') \left[-\dot{p}(t'') - V'(x(t'')) \right] \right\} \; , \end{aligned} \tag{2.1.14}$$

$$\begin{aligned} \delta S(\{x(t'')\}) &= \delta \int_t^{t'} \mathrm{d}t'' \left[\frac{m\dot{x}(t'')^2}{2} - V(x(t'')) \right] \\ &= \int_t^{t'} \mathrm{d}t'' \, \delta x(t'') \left[-m\ddot{x}(t'') - V'(x(t'')) \right] \; . \end{aligned} \tag{2.1.15}$$

Here we used $\delta x(t) = \delta x(t') = 0$ for partial integration.

$\delta S = 0$ in (2.1.14) and (2.1.15) leads to the classical equations of motion. From the point of view of classical dynamics, this is the well-known variation principle, but what will be the meaning from the point of view of the path integral? The meaning of equations (2.1.11) and (2.1.13) is that the transition amplitude from x, t to x', t' is given by integrating over all amplitudes:

$$\exp\left[\frac{\mathrm{i}}{\hbar}S(\{x(t'')\},\{p(t'')\})\right] , \qquad \exp\left[\frac{\mathrm{i}}{\hbar}S(\{x(t'')\})\right]$$

of every single path $x(t'')$ and $p(t'')$. However, even when all paths have to be taken into account, not every path is equally important. Especially, considering the fact that $\hbar$ is "small" (in the classical limit, this is a good approximation), we see that even a small change of the path will alter drastically the phase in $\exp[(\mathrm{i}/\hbar)S]$. Taking the sum, the contribution of the phase average will therefore be extremely small. An exception is a path where a small change does no affect S, that is, the path satisfying $\delta S = 0$.

In summary, we have just learnt that quantum mechanics can be formulated without the use of operators, only with c-numbers. The price we pay is that we have to perform an infinite dimensional integral in paths (being in the present case functions of time). The largest contribution to this integral is given by the classical path.

Next, we discuss the term $\mathrm{i}p\dot{x}$. We consider the following path integral of the functional $F(x(t'), p(t'))$ of $x(t'), p(t')$:

$$\begin{aligned}&\langle F(\{x(t')\},\{p(t')\})\rangle \\ &\quad = \int \mathcal{D}x(t')\mathcal{D}p(t')F(\{x(t')\},\{p(t')\})\,\mathrm{e}^{\frac{\mathrm{i}}{\hbar}S(\{x(t')\},\{p(t')\})} \ . \end{aligned} \tag{2.1.16}$$

We split the time intervals as done starting from (2.1.4):

$$\begin{aligned}F(\{x(t')\},\{p(t')\}) &\to F(x(1),\dots,x(N-1);p(0),p(1),\dots,p(N-1)) \\ &= F(x(k),p(k)) \ ,\end{aligned} \tag{2.1.17}$$

$$\int \mathcal{D}x(t')\mathcal{D}p(t') = \prod_k \int \mathrm{d}x\,(k)\mathrm{d}p\,(k) \ . \tag{2.1.18}$$

We perform now a transformation of the integration measure by introducing the infinitesimal variables $\eta(k)$ and $\xi(k)$:

$$x(k) = \tilde{x}(k) + \eta(k) \quad \text{and} \quad p(k) = \tilde{p}(k) + \zeta(k) \tag{2.1.19}$$

and write the integral in the $\tilde{x}(k)$ and $\tilde{p}(k)$ variables introduced in this manner. Then, the integral becomes

$$\begin{aligned}&\langle F(x(k)),p(k))\rangle \\ &\quad = \int \prod_{k'} \mathrm{d}\tilde{x}(k')\,\mathrm{d}\tilde{p}(k')\,F(\tilde{x}(k)+\eta(k),\tilde{p}(k)+\zeta(k)) \\ &\qquad \times \exp\left\{\frac{\mathrm{i}}{\hbar}S[\tilde{x}(k)+\eta(k),\tilde{p}(k)+\zeta(k)]\right\}\end{aligned}$$

$$= \int \prod_{k'} \mathrm{d}x(k')\,\mathrm{d}p(k') \left(F + \sum_k \eta(k) \frac{\partial F}{\partial x(k)} + \sum_k \zeta(k) \frac{\partial F}{\partial p(k)} \right)$$
$$\times \left(1 + \frac{\mathrm{i}}{\hbar} \sum_k \eta(k) \frac{\partial S}{\partial x(k)} + \frac{\mathrm{i}}{\hbar} \sum \zeta(k) \frac{\partial S}{\partial p(k)} \right) \mathrm{e}^{\mathrm{i}S/\hbar}$$
$$= \langle F(x(k), p(k)) \rangle + \sum_k \eta(k) \left\langle \frac{\partial F}{\partial x(k)} + \frac{\mathrm{i}}{\hbar} F \frac{\partial S}{\partial x(k)} \right\rangle$$
$$+ \sum_k \zeta(k) \left\langle \frac{\partial F}{\partial p(k)} + \frac{\mathrm{i}}{\hbar} F \frac{\partial S}{\partial p(k)} \right\rangle \ . \tag{2.1.20}$$

We expanded the expression up to first order in $\eta(k)$ and $\xi(k)$. Equation (2.1.10) holds for every $\eta(k)$ and $\xi(k)$, therefore

$$\begin{aligned} \left\langle \frac{\partial F}{\partial x(k)} \right\rangle &= -\frac{\mathrm{i}}{\hbar} \left\langle F \frac{\partial S}{\partial x(k)} \right\rangle \ , \\ \left\langle \frac{\partial F}{\partial p(k)} \right\rangle &= -\frac{\mathrm{i}}{\hbar} \left\langle F \frac{\partial S}{\partial p(k)} \right\rangle \end{aligned} \tag{2.1.21}$$

hold. By setting $F(x(k), p(k)) = p(k_0)$, (2.1.21) reads

$$1 = -\frac{\mathrm{i}}{\hbar} \left\langle p(k_0) \frac{\partial S}{\partial p(k_0)} \right\rangle \ . \tag{2.1.22}$$

On the other hand, because of

$$S = \sum_k p(k)(x(k+1) - x(k)) - \Delta t \sum_k H(p(k), x(k)) \tag{2.1.23}$$

we obtain

$$\frac{\partial S}{\partial p(k_0)} = x(k_0 + 1) - x(k_0) - \Delta t \frac{\partial H(p(k_0), x(k_0))}{\partial p(k_0)} \tag{2.1.24}$$

and for $\Delta t \to 0$ the third term disappears. Therefore, (2.1.22) becomes

$$\mathrm{i}\hbar = \langle p(k_0)[x(k_0 + 1) - x(k_0)] \rangle \ . \tag{2.1.25}$$

To the expectation value of which operator does the right-hand side of this equation correspond? As is clear from the discussion in (2.1.7)–(2.1.10), the order of the operators is important:

$$\begin{aligned} \langle \hat{x}\hat{p} \rangle_{t=t_0} &\to \langle x(k_0 + 1) p(k_0) \rangle \ , \\ \langle \hat{p}\hat{x} \rangle_{t=t_0} &\to \langle p(k_0) x(k_0) \rangle \ . \end{aligned} \tag{2.1.26}$$

Therefore, the right-hand side of (2.1.25) corresponds to the expectation value of the equal time commutator of $\hat{x}$ and $\hat{p}$:

$$[\hat{x}, \hat{p}] = \mathrm{i}\hbar \ . \tag{2.1.27}$$

In this manner, $ip\dot{x}$ shows that $\hat{x}$ and $\hat{p}$ are canonical conjugate operators, and the ordering of these operators is represented in the path integral as time ordering.

Next, we consider the formalism in imaginary time. As is clear from (2.1.2), we used the operator $\exp[(-\mathrm{i}/\hbar)Ht]$ for the discussion. Writing $-\mathrm{i}\tau$ instead of t, and redoing the steps as before, we obtain

$$\begin{aligned}\langle x'|\,\mathrm{e}^{-H\tau/\hbar}|x\rangle &= \int_{\substack{x(0)=x\\ x(\tau)=x'}} \mathcal{D}x(\tau')\int \mathcal{D}p(\tau')\,\exp\left[-\frac{1}{\hbar}S(\{x(\tau')\},\{p(\tau')\})\right]\\ &= \int_{\substack{x(0)=x\\ x(\tau)=x'}} \mathcal{D}x(\tau')\int \mathcal{D}p(\tau')\\ &\quad\times\exp\left[-\frac{1}{\hbar}\int_0^\tau\left\{-\mathrm{i}p(\tau')\dot{x}(\tau')+\frac{p(\tau')^2}{2m}+V(x(\tau'))\right\}\mathrm{d}\tau'\right], \end{aligned} \tag{2.1.28}$$

$$\begin{aligned}\langle x'|\,\mathrm{e}^{-H\tau/\hbar}|x\rangle &= \int_{\substack{x(0)=x\\ x(\tau)=x'}} \mathcal{D}x(\tau')\,\exp\left[-\frac{1}{\hbar}S(\{x(\tau')\})\right]\\ &= \int_{\substack{x(0)=x\\ x(\tau)=x'}} \mathcal{D}x(\tau')\exp\left[-\frac{1}{\hbar}\int_0^\tau\left\{\frac{m\dot{x}(\tau')^2}{2}+V(x(\tau'))\right\}\mathrm{d}\tau'\right]. \end{aligned} \tag{2.1.29}$$

Here, we replaced

$$\int \mathrm{d}t'' \to -\mathrm{i}\int \mathrm{d}\tau' \quad \text{and} \quad \dot{x}(t'') \to \mathrm{i}\dot{x}(\tau') \ .$$

In particular, when setting $\tau = \beta\hbar$ and $x' = x$ and integrating in x (see Appendix C)

$$Z = \mathrm{Tr}\,\mathrm{e}^{-\beta H} = \int \mathrm{d}x \langle x|\,\mathrm{e}^{-\beta H}|x\rangle \ , \tag{2.1.30}$$

we obtain the partition function of the system. Using the path integral formalism in imaginary time, it is therefore also possible to apply it to statistical physics.

Notice that the factor i in (2.1.28) does not disappear in the term $\mathrm{i}p(\tau')\dot{x}(\tau')$ in the complex-time formalism. This term indicates the phase (Berry phase). On the other hand, after the $p(\tau')$-integration, no complex term is present in (2.1.29), and the exponent is positive as usual. In such a case, the partition function (2.1.30) corresponds to that of one string $x(\tau')$ in classical statistical mechanics. In general, a d-dimensional system in quantum physics can be associated with a $(d+1)$-dimensional classical system in such a manner. However, for the case that the phase factor mentioned above remains, no equivalent classical model exists.

Next, we determine the path with the largest contribution to (2.1.29), as was done for (2.1.13). Taking the variation, we obtain

$$\delta \int_0^\beta \left\{ \frac{m\dot{x}(\tau')^2}{2} + V(x(\tau')) \right\} \mathrm{d}\tau' = \int_0^\beta [-m\ddot{x}(\tau') + V'(x(\tau'))]\delta x(\tau')\mathrm{d}\tau'$$
$$= 0 \ . \qquad (2.1.31)$$

This path obeys the equation $m\ddot{x}(\tau') = V'(x(\tau'))$. This is a classical equation of motion, corresponding to the classical motion of a particle in a potential with reversed sign, that is, where up and down are reversed. For example, a potential $V(x)$ as shown in Fig. 2.3 with two valleys becomes the potential $-V(x)$ with two mountains. In this case, possible motions are the way from the top of one mountain to the other, falling down in the valley, or climbing up in the valley.

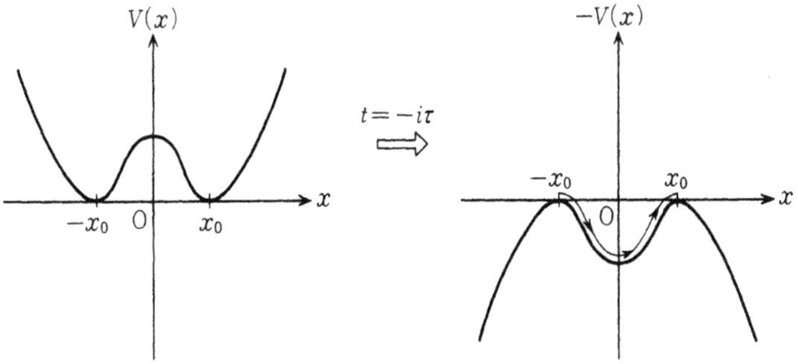

Fig. 2.3. Classical movement in the complex-time formalism. When the real time is changed to complex time, the potential is reversed, and therefore classical movement in regions that has been forbidden before becomes possible

The classical solution for a path in the imaginary-time formalism is called the instanton, and has the physical meaning of tunnelling within the framework of quantum mechanics. Because the potential is reversed, motion in classical forbidden regions becomes possible, and in contrast to the normal description of tunnelling, where the wave number becomes imaginary and the wave is damped, in this case the imaginary time represents the damping of the wave.

In order to confirm that the instanton is indeed responsible for the tunnelling, we perform the following calculation. We set $\hbar = 1$, and for T large enough, we want to calculate in (2.1.29) the amplitude of the transition from $\pm x_0$ to $\pm x_0$:

$$\langle x_0 | \mathrm{e}^{-HT} | x_0 \rangle = \langle -x_0 | \mathrm{e}^{-HT} | -x_0 \rangle \ , \qquad (2.1.32)$$

and from $\pm x_0$ to $\mp x_0$:

$$\langle x_0 | \mathrm{e}^{-HT} | -x_0 \rangle = \langle -x_0 | \mathrm{e}^{-HT} | x_0 \rangle \ . \qquad (2.1.33)$$

Writing τ $(0 \le \tau \le T)$ on the x axis and $x(\tau)$ on the y axis, one type of path rests at $x(\tau) = \pm x_0$ and the second type connects the points x_0 and $-x_0$, corresponding to the instanton (anti-instanton) described above. With the boundary condition $x(0) = \pm x(T)$, the number of instantons and anti-instantons in (2.1.32) is equal; in (2.1.33) the numbers differ by one.

In the vicinity of the top of the mountain, $V(x)$ can be written approximately as

$$-V(x) = -V(x_0) - \frac{m\omega^2}{2}(x - x_0)^2$$

and we obtain the equation of motion $m\ddot{x}(\tau) = \omega^2(x - x_0)$. The instanton solution behaves as $x - x_0 \propto \mathrm{e}^{-\omega\tau}$. In the case when the distance between the instanton and anti-instanton is large compared with the width of the instanton solution itself, the instantons can be regarded as a dilute free gas living on the τ axis, assigned to distinct places $0 \le \tau_1 \le \tau_2 \ldots \le \tau_n \le T$ (the dilute gas approximation). If no instanton is present, then we obtain

$$V(x) = \frac{m\omega^2}{2}(x \pm x_0)^2$$

as the solution of the harmonic oscillator problem

$$\begin{aligned}\langle x_0|\,\mathrm{e}^{-HT}|x_0\rangle &= \langle -x_0|\,\mathrm{e}^{-HT}| - x_0\rangle \\ &= \left(\frac{m\omega}{\pi}\right)^{1/2} \mathrm{e}^{-\omega T/2} \ .\end{aligned} \tag{2.1.34}$$

Recall that the ground state of the harmonic oscillator with energy $\omega/2$ is given by

$$\varphi_0(x) = \left(\frac{m\omega}{\pi}\right)^{1/4} \mathrm{e}^{-m\omega(x \pm x_0)^2/2} \ .$$

For the case when n instantons and anti-instantons are present, the answer will become

$$\int_0^T \mathrm{d}\tau_n \int_0^{\tau_n} \mathrm{d}\tau_{n-1} \ldots \int_0^{\tau_2} \mathrm{d}\tau_1 \left(\frac{m\omega}{\pi}\right)^{1/2} \mathrm{e}^{-\omega T/2} K^n\, \mathrm{e}^{-nS_0} \ . \tag{2.1.35}$$

Here, S_0 is the action of the one-instanton solution $S(x(\tau))$, and K is the ratio of the Gauss integral with and without the instanton. More detailed discussions can be found in the literature [G.6]. Here, it is sufficient to assume that K is just a number. It is important to notice that the term $(K\,\mathrm{e}^{-S_0})^n$ appears in the dilute gas approximation. The imaginary time integral in (2.1.35) can be performed easily, with the result $T^n/n!$, and taking the sum over n, we obtain

$$\begin{aligned}\langle x_0|\,\mathrm{e}^{-HT}|x_0\rangle &= \langle -x_0|\,\mathrm{e}^{-HT}|-x_0\rangle \\ &= \left(\frac{m\omega}{\pi}\right)^{1/2} \mathrm{e}^{-\omega T/2} \sum_{n=0\ (\mathrm{even})}^{\infty} \frac{(TK\,\mathrm{e}^{-S_0})^n}{n!} \\ &= \left(\frac{m\omega}{\pi}\right)^{1/2} \mathrm{e}^{-\omega T/2} \cosh(TK\,\mathrm{e}^{-S_0}) \ , \end{aligned} \tag{2.1.36}$$

$$\begin{aligned}\langle x_0|\,\mathrm{e}^{-HT}|-x_0\rangle &= \langle -x_0|\,\mathrm{e}^{-HT}|x_0\rangle \\ &= \left(\frac{m\omega}{\pi}\right)^{1/2} \mathrm{e}^{-\omega T/2} \sum_{n=1\ (\mathrm{odd})}^{\infty} \frac{(TK\,\mathrm{e}^{-S_0})^n}{n!} \\ &= \left(\frac{m\omega}{\pi}\right)^{1/2} \mathrm{e}^{-\omega T/2} \sinh(TK\,\mathrm{e}^{-S_0}) \ . \end{aligned} \tag{2.1.37}$$

On the other hand, using the eigenstates $|n\rangle$ and eigenvalues of the energy E_n in (2.1.29), we obtain

$$\begin{aligned}\langle x'|\,\mathrm{e}^{-HT}|x\rangle &= \sum_n \langle x'|n\rangle\langle n|x\rangle\,\mathrm{e}^{-E_nT} \\ &= \sum_n \phi_n(x')\phi_n^*(x)\,\mathrm{e}^{-E_nT} \ . \end{aligned} \tag{2.1.38}$$

Comparing (2.1.36) with (2.1.37), we obtain two energy levels:

$$E_\pm = \frac{\omega}{2} \pm K\,\mathrm{e}^{-S_0} \ . \tag{2.1.39}$$

This can be interpreted as splitting of the ground state energies around $\pm x_0$ due to the tunnelling contribution $K\,\mathrm{e}^{-S_0}$.

The reader might wonder why such a long-winded method has been presented, whereas the normal WKB method would have been sufficient for the calculation. Indeed, for this simple case of a single particle in one dimension, this is the case; however, when proceeding to many-particle systems, that is, quantum field theory, the generalization of the instanton calculation can be done easily. On the other hand, the WKB method, based on fitting boundary conditions of solutions in different regions of a differential equation, becomes very complicated in a more generalized case.

2.2 The Path Integral for Bosons

In the previous section, we introduced the path integral for a single-particle system. In the present and the following section, the path integral of the many-particle system using second quantization is presented. A more precise discussion can be found in the literature [G.16].

As first step in this section, the bosonic system will be discussed, which is quite analogous to the classical case, so the method will be almost parallel to the previous section. Two points have to be noticed: first observe the correspondence of the operators

$$\hat{x} \leftrightarrow \hat{\psi}(\boldsymbol{r}) \quad \text{and} \quad \hat{p} \leftrightarrow \mathrm{i}\hbar\hat{\psi}^\dagger(\boldsymbol{r}) \ , \tag{2.2.1}$$

and next notice that it becomes necessary to perform the sum (integral) in $\boldsymbol{r}$. For simplicity, we set $\hbar = 1$ in this section. Doing so, the partition function in the imaginary time formalism becomes by analogy to (2.1.28)

$$Z = \int \mathcal{D}\bar{\psi}(\boldsymbol{r},\tau)\mathcal{D}\psi(\boldsymbol{r},\tau)\,\mathrm{e}^{-S(\{\bar{\psi}\},\{\psi\})} \ , \tag{2.2.2}$$

$$S = \int_0^\beta \mathrm{d}\tau \int \mathrm{d}\boldsymbol{r}\,\bar{\psi}(\boldsymbol{r},\tau)\partial_t\psi(\boldsymbol{r},\tau) + \int_0^\beta \mathrm{d}\tau\, H(\tau) \ , \tag{2.2.3}$$

with

$$\begin{aligned} H(\tau) = & \int \mathrm{d}\boldsymbol{r} \left(\frac{1}{2m}\nabla\bar{\psi}(\boldsymbol{r},\tau)\nabla\psi(\boldsymbol{r},\tau) - \mu\bar{\psi}(\boldsymbol{r},\tau)\psi(\boldsymbol{r},\tau) \right) \\ & + \frac{1}{2}\int \mathrm{d}\boldsymbol{r}\,\mathrm{d}\boldsymbol{r}'\,\bar{\psi}(\boldsymbol{r})\bar{\psi}(\boldsymbol{r}')v(\boldsymbol{r}-\boldsymbol{r}')\psi(\boldsymbol{r}')\psi(\boldsymbol{r}) \ . \end{aligned} \tag{2.2.4}$$

In (2.2.3) and (2.2.4), $\bar{\psi}(\boldsymbol{r},\tau)$ and $\psi(\boldsymbol{r},\tau)$ correspond to the operators $\hat{\psi}^\dagger(\boldsymbol{r},\tau)$ and $\hat{\psi}(\boldsymbol{r},\tau)$. However, they are not operators, but *c*-number functions of $\boldsymbol{r}$ and τ.

Notice that we have chosen the grand canonical ensemble, where instead of a constant particle number, the chemical potential μ has been introduced.

The reader might think that ψ and $\bar{\psi}$ are complex conjugate; however, this must not be the case. When restricting to one $\boldsymbol{r}, t$, we have the line integrals

$$\int \mathrm{d}\bar{\psi}\,(\boldsymbol{r},\tau) \ , \qquad \int \mathrm{d}\psi\,(\boldsymbol{r},\tau) \ ,$$

and since the path on the $\bar{\psi}$-plane and the ψ-plane can be altered independently, in general $\bar{\psi}$ can be independent of ψ. Starting from Chap. 4, the path of this complex contour integral will be chosen by the saddle-point method, where S becomes larger in every direction when the path is deviating from the saddle-point solution.

Next, we introduce by the following equation another important physical function besides the partition function, namely the thermal Green function $\mathcal{G}(\boldsymbol{r},\boldsymbol{r}';\tau,\tau')$:

$$\begin{aligned} \mathcal{G}(\boldsymbol{r},\boldsymbol{r}';\tau,\tau') &= -\langle \mathrm{T}_\tau \psi(\boldsymbol{r},\tau)\psi^\dagger(\boldsymbol{r}',\tau')\rangle \\ &= \begin{cases} -\langle \psi(\boldsymbol{r},\tau)\psi^\dagger(\boldsymbol{r}',\tau')\rangle & (\text{for } \tau > \tau') \\ -\langle \psi^\dagger(\boldsymbol{r}',\tau')\psi(\boldsymbol{r},\tau)\rangle & (\text{for } \tau < \tau') \end{cases} \ . \end{aligned} \tag{2.2.5}$$

Here, T_τ is the time ordering operator. In terms of the functional integral, the Green function is given by

$$\mathcal{G}(\boldsymbol{r},\boldsymbol{r}';\tau,\tau') = -\frac{1}{Z}\int \mathcal{D}\bar{\psi}\mathcal{D}\psi\, \psi(\boldsymbol{r},\tau)\bar{\psi}(\boldsymbol{r}',\tau)\,\mathrm{e}^{-S(\{\bar{\psi}\},\{\psi\})} \ . \tag{2.2.6}$$

The proof proceeds in the same manner as in Sect. 2.1. Notice that in the path integral, the Green function is automatically time ordered. Now, let us consider the Green function explicitly in the case where the interaction is set to zero ($v(\boldsymbol{r}-\boldsymbol{r}')=0$). We perform the Fourier transformation

$$\psi(\boldsymbol{r},\tau) = (\beta V)^{-1/2}\sum_{\boldsymbol{k},\omega}\mathrm{e}^{-\mathrm{i}\omega\tau+\mathrm{i}\boldsymbol{k}\cdot\boldsymbol{r}}a(\boldsymbol{k},\omega) \ , \tag{2.2.7}$$

$$\bar{\psi}(\boldsymbol{r},\tau) = (\beta V)^{-1/2}\sum_{\boldsymbol{k},\omega}\mathrm{e}^{\mathrm{i}\omega\tau-\mathrm{i}\boldsymbol{k}\cdot\boldsymbol{r}}\bar{a}(\boldsymbol{k},\omega) \ . \tag{2.2.8}$$

Imposing the boundary conditions $\psi(\boldsymbol{r},\beta)=\psi(\boldsymbol{r},0)$ and $\bar{\psi}(\boldsymbol{r},\beta)=\bar{\psi}(\boldsymbol{r},0)$, only discrete values $\omega = 2\pi n/\beta$ for ω ($n=$ integer) become possible, which are the so-called Matsubara frequencies. S becomes

$$S_0 = \sum_{\boldsymbol{k},\omega}\left(-\mathrm{i}\omega + \frac{\boldsymbol{k}^2}{2m} - \mu\right)\bar{a}(\boldsymbol{k},\omega)a(\boldsymbol{k},\omega) \tag{2.2.9}$$

and (2.2.5) is equal to

$$\begin{aligned}\mathcal{G}(\boldsymbol{r},\boldsymbol{r}';\tau,\tau') = &-\frac{1}{\beta}\sum_{\boldsymbol{k}_1,\omega_1}\sum_{\boldsymbol{k}_2,\omega_2}\frac{1}{Z}\int\prod_{\boldsymbol{k},\omega}\mathrm{d}a\,(\boldsymbol{k},\omega)\,\mathrm{d}\bar{a}\,(\boldsymbol{k},\omega)\\ &\times a(\boldsymbol{k}_1,\omega_1)\bar{a}(\boldsymbol{k}_2,\omega_2)\\ &\times\exp\left[\sum_{\boldsymbol{k},\omega}\left(\mathrm{i}\omega - \frac{\boldsymbol{k}^2}{2m}+\mu\right)\bar{a}(\boldsymbol{k},\omega)a(\boldsymbol{k},\omega)\right]\\ &\times\exp[\mathrm{i}\,(-\omega_1\tau+\boldsymbol{k}_1\cdot\boldsymbol{r})]\exp[\mathrm{i}\,(\omega_2\tau'-\boldsymbol{k}_2\cdot\boldsymbol{r}')] \ .\end{aligned} \tag{2.2.10}$$

Essentially, by setting $\bar{a}(\boldsymbol{k},\omega) = [a(\boldsymbol{k},\omega)]^* = a'(\boldsymbol{k},\omega) - \mathrm{i}a''(\boldsymbol{k},\omega)$ and performing the Gauss integral in a' and a'' from $-\infty$ to $+\infty$, only the term with $\boldsymbol{k_1}=\boldsymbol{k_2}$ and $\omega_1=\omega_2$ contributes. Then, (2.2.10) becomes

$$\mathcal{G}(\boldsymbol{r},\boldsymbol{r}';\tau,\tau') = \frac{1}{\beta V}\sum_{\boldsymbol{k}_1,\omega_1}\frac{1}{\mathrm{i}\omega_1 - \boldsymbol{k}_1^2/2m+\mu}\,\mathrm{e}^{-\mathrm{i}\omega_1(\tau-\tau')+\mathrm{i}\boldsymbol{k}_1\cdot(\boldsymbol{r}-\boldsymbol{r}')} \ . \tag{2.2.11}$$

Writing $\mathcal{G}(\boldsymbol{k},\mathrm{i}\omega_n)$ $[\omega_n = 2\pi kTn]$ for the Fourier transformation of $\mathcal{G}(\boldsymbol{r},\boldsymbol{r}';\tau,\tau')$, we obtain $\mathcal{G}(\boldsymbol{k},\mathrm{i}\omega_n) = (\mathrm{i}\omega_n - \boldsymbol{k}^2/2m+\mu)^{-1}$.

The reader might already know the following direct proof for $\mathcal{G}(\boldsymbol{k},\mathrm{i}\omega_n) = (\mathrm{i}\omega_n - \boldsymbol{k}^2/2m+\mu)^{-1}$ starting from (2.2.5). The Hamiltonian is given by

$$\begin{aligned} H &= \sum_{\boldsymbol{k}} \left(\frac{\boldsymbol{k}^2}{2m} - \mu \right) a^\dagger(\boldsymbol{k}) a(\boldsymbol{k}) \\ &\equiv \sum_{\boldsymbol{k}} \xi_{\boldsymbol{k}} a^\dagger(\boldsymbol{k}) a(\boldsymbol{k}) \end{aligned} \tag{2.2.12}$$

and with

$$a(\boldsymbol{k},\tau) = \mathrm{e}^{-\xi_{\boldsymbol{k}}\tau} a(\boldsymbol{k}) \ , \qquad a^\dagger(\boldsymbol{k},\tau) = \mathrm{e}^{\xi_{\boldsymbol{k}}\tau} a^\dagger(\boldsymbol{k})$$

for $0 < \tau < \beta$, we obtain

$$\begin{aligned} \mathcal{G}(\boldsymbol{r},\boldsymbol{r}';\tau,0) &= -\frac{1}{V} \sum_{\boldsymbol{k}_1,\boldsymbol{k}_2} \langle a(\boldsymbol{k}_1) a^\dagger(\boldsymbol{k}_2) \rangle \, \mathrm{e}^{-\xi_{\boldsymbol{k}_1}\tau} \, \mathrm{e}^{\mathrm{i}(\boldsymbol{k}_1\cdot\boldsymbol{r} - \boldsymbol{k}_2\cdot\boldsymbol{r}')} \\ &= -\frac{1}{V} \sum_{\boldsymbol{k}_i} [1 + n(\boldsymbol{k}_1)] \, \mathrm{e}^{-\xi_{\boldsymbol{k}_1}\tau} \, \mathrm{e}^{\mathrm{i}\boldsymbol{k}_1\cdot(\boldsymbol{r}-\boldsymbol{r}')} \ . \end{aligned} \tag{2.2.13}$$

Performing the Fourier transformation, we obtain

$$\begin{aligned} \mathcal{G}(\boldsymbol{k},\mathrm{i}\omega_n) &= \int_0^\beta \mathrm{d}\tau \, \mathrm{e}^{\mathrm{i}\omega_n\tau} \int \mathrm{d}(\boldsymbol{r}-\boldsymbol{r}') \, \mathrm{e}^{-\mathrm{i}\boldsymbol{k}\cdot(\boldsymbol{r}-\boldsymbol{r}')} \mathcal{G}(\boldsymbol{r},\boldsymbol{r}';\tau,0) \\ &= -[1+n(\boldsymbol{k})] \int_0^\beta \mathrm{d}\tau \exp[(\mathrm{i}\omega_n - \xi_{\boldsymbol{k}})\tau] \\ &= -[1+n(\boldsymbol{k})] \frac{\exp[-\beta\xi_{\boldsymbol{k}}] - 1}{\mathrm{i}\omega_n - \xi_{\boldsymbol{k}}} = \frac{1}{\mathrm{i}\omega_n - \xi_{\boldsymbol{k}}} \ . \end{aligned} \tag{2.2.14}$$

This expression agrees with the result we obtained before. More generally, using the energy eigenvalues and eigenfunctions of the system in (2.2.5), we obtain the spectral decomposition

$$\begin{aligned} \mathcal{G}(\boldsymbol{k},\mathrm{i}\omega_n) &= -\int_0^\beta \mathrm{d}\tau \, \mathrm{e}^{\mathrm{i}\omega_n\tau} \langle a(\boldsymbol{k},\tau) a^\dagger(\boldsymbol{k}) \rangle \\ &= -\frac{1}{Z} \int_0^\beta \mathrm{d}\tau \, \mathrm{e}^{\mathrm{i}\omega_n\tau} \sum_{n,m} \mathrm{e}^{-\beta E_n} \langle n | a(\boldsymbol{k}) | m \rangle \, \mathrm{e}^{\tau(E_n - E_m)} \langle m | a^\dagger(\boldsymbol{k}) | n \rangle \\ &= -\frac{1}{Z} \sum_{n,m} |\langle n | a(\boldsymbol{k}) | m \rangle|^2 \int_0^\beta \mathrm{d}\tau \, \mathrm{e}^{\tau(\mathrm{i}\omega_n + E_n - E_m)} \, \mathrm{e}^{-\beta E_n} \\ &= \frac{1}{Z} \sum_{n,m} \frac{|\langle n | a(\boldsymbol{k}) | m \rangle|^2}{\mathrm{i}\omega_n - (E_m - E_n)} (\mathrm{e}^{-\beta E_n} - \mathrm{e}^{-\beta E_m}) \ . \end{aligned} \tag{2.2.15}$$

Next, we want to link the imaginary time formalism developed so far with the real-time formalism. For this purpose, we introduce the advanced Green function and the retarded Green function $G^{\mathrm{A}}(\boldsymbol{k},\omega)$, $G^{\mathrm{R}}(\boldsymbol{k},\omega)$

$$G^{\mathrm{A}}(\boldsymbol{k},\omega) = +\mathrm{i} \int_{-\infty}^0 \langle [a(\boldsymbol{k},t), a^\dagger(\boldsymbol{k})] \rangle \, \mathrm{e}^{\mathrm{i}\omega t + \delta t} \mathrm{d}t \ , \tag{2.2.16}$$

$$G^{\mathrm{R}}(\boldsymbol{k},\omega) = -\mathrm{i}\int_0^\infty \langle[a(\boldsymbol{k},t),a^\dagger(\boldsymbol{k})]\rangle\,\mathrm{e}^{\mathrm{i}\omega t-\delta t}\mathrm{d}t \ , \tag{2.2.17}$$

where δ is an infinitesimal positive constant introduced to make the t integral converge. The spectral decomposition corresponding to (2.2.15) now reads

$$G^{\mathrm{A,R}}(\boldsymbol{k},\omega) = \frac{1}{Z}\sum_{n,m}\frac{|\langle n|a(\boldsymbol{k})|m\rangle|^2}{\omega \mp \mathrm{i}\delta - (E_m - E_n)}(\mathrm{e}^{-\beta E_n} - \mathrm{e}^{-\beta E_m}) \ . \tag{2.2.18}$$

Comparing this with (2.2.15), (2.2.16) and (2.2.17), we see that by analytic continuation of $\mathrm{i}\omega_n$ in the thermal Green function to $\omega - \mathrm{i}\delta$ $(\omega + \mathrm{i}\delta)$, the advanced Green function (retarded Green function) can be obtained. For a system at finite temperature, this suggests the method to calculate first the thermal Green function in the complex time formalism, and then to obtain by analytic continuation $\mathrm{i}\omega_n \to \pm\omega \mp \mathrm{i}\delta$ the Green functions G^A and G^R in the real-time formalism. This formalism is named after its inventor and is called the Matsubara formalism.

Next, we want to compare the advantages of the path integral method with those of the operator formalism. The advantage is that no operators are present, but only c-numbers (commutators are trivial). However, the price we pay is an infinite-dimensional integral.

In the case when an interaction $v(\boldsymbol{r} - \boldsymbol{r}')$ is present, the most popular technique might be perturbation theory. This means splitting S in (2.2.6) into $S_0 + S_1$, and expanding e^{-S_1}, and because S_1 is a simple c-number function containing no operators, the expansion is possible. In this case the cumulant analysis of the Gauss-integral can be applied; for example,

$$\langle\psi_1\psi_2\psi_3^\dagger\psi_4^\dagger\rangle = \langle\psi_1\psi_3^\dagger\rangle\langle\psi_2\psi_4^\dagger\rangle + \langle\psi_1\psi_4^\dagger\rangle\langle\psi_2\psi_3^\dagger\rangle \ . \tag{2.2.19}$$

Now, because we are dealing with c-numbers, the variational method can be introduced easily. Splitting S in (2.2.2) into the sum of the "trial action" S_0 and $S - S_0$, we can write

$$Z = Z_0\langle\,\mathrm{e}^{-S+S_0}\rangle_{S_0} \ . \tag{2.2.20}$$

Here, Z_0 is the partition function of the action S_0 and $\langle\ \rangle_{S_0}$ means averaging with respect to e^{-S_0}. Because S and S_0 are in general real, with $\exp[(x_1 + x_2)/2] \le (\mathrm{e}^{x_1} + \mathrm{e}^{x_2})/2$, we obtain $\exp[-\langle S - S_0\rangle_{S_0}] \le \langle\exp[-S + S_0]\rangle_{S_0}$ and therefore

$$J = -k_{\mathrm{B}}T\ln Z \leqq -k_{\mathrm{B}}T\ln Z_0 + k_{\mathrm{B}}T\langle S - S_0\rangle_{S_0} \ . \tag{2.2.21}$$

Here, k_{B} is the Boltzmann's constant and $J = -pV$ is the thermal potential of the grand canonical ensemble. The right-hand side can be calculated when S_0 is determined, and by optimizing the variation parameter in it, it is possible to determine the best S_0 for this framework. In practise, because almost only Gauss integrals can be performed, S_0 is often quadratic in ψ and $\bar{\psi}$, and often this approximation agrees with the mean field approximation.

So far we have discussed the bosonic path integral. Starting from Chap. 1, ψ appeared on the stage and "evolved". This flow is shown in Fig. 2.4, and when we meet ψ, it is important to remember to which step ψ corresponds.

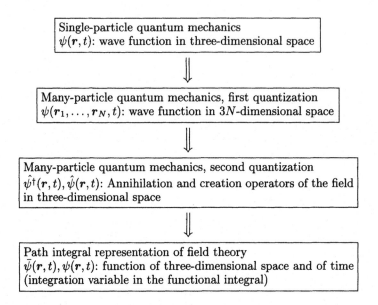

Single-particle quantum mechanics
$\psi(\boldsymbol{r}, t)$: wave function in three-dimensional space

$\Downarrow$

Many-particle quantum mechanics, first quantization
$\psi(\boldsymbol{r}_1, \ldots, \boldsymbol{r}_N, t)$: wave function in $3N$-dimensional space

$\Downarrow$

Many-particle quantum mechanics, second quantization
$\hat{\psi}^\dagger(\boldsymbol{r}, t), \hat{\psi}(\boldsymbol{r}, t)$: Annihilation and creation operators of the field in three-dimensional space

$\Downarrow$

Path integral representation of field theory
$\bar{\psi}(\boldsymbol{r}, t), \psi(\boldsymbol{r}, t)$: function of three-dimensional space and of time (integration variable in the functional integral)

Fig. 2.4

2.3 The Path Integral for Fermions

In this section we discuss the case of fermions. Different from the bosonic case, a correspondence to a classical system does not exist. The reader might think that in this case the path integral method is useless; however, this is not the case. At least mathematically, by introducing Grassmann numbers, the description can be given in a manner totally similar to the bosonic case.

Corresponding to the anti-commutation relations of fermions, Grassmann numbers are defined to be anti-commuting, that is $x_i x_j + x_j x_i = 0$. Therefore, $x_i^2 = 0$ holds, and we conclude that every function in $x_1, \ldots, x_n$ can be written as

$$f(x_1 \ldots, x_N) = \sum_{n=0}^{N} \sum_{i_1 < i_2 < \cdots < i_n} C_n(i_1, \ldots, i_n)\, x_{i_1} \ldots x_{i_n} \ . \tag{2.3.1}$$

Here $i_1, \ldots, i_n$ is a disjunct set of indices $1 - N$, and we define the order to be $i_1 < i_2 < \cdots < i_n$. The coefficient $C_n(i_1, \ldots, i_n)$ is a simple complex function. We define the "integral" of this function as follows:

$$\int x_{i_1} \dots x_{i_n} \, \mathrm{d}x_{j_n} \dots \mathrm{d}x_{j_1} = \varepsilon \begin{pmatrix} i_1, \dots, i_n \\ j_1, \dots, j_n \end{pmatrix} \quad . \tag{2.3.2}$$

Here, $\varepsilon(\cdots)$ is different from zero only for the case when $(i_1, \dots, i_n)$ equals $(j_1, \dots, j_n)$. Then, it is $+1$ when both sets become identical under an even number of permutations and -1 when both set become identical under an odd number of permutations. We choose N even $(= 2n)$ and split $x_1, \dots, x_N$ into $\psi_1, \dots, \psi_n$ and $\bar{\psi}_1, \dots, \bar{\psi}_n$, with ψ and $\bar{\psi}$ corresponding to the fermionic operators $\hat{\psi}_i$ and $\hat{\psi}_i^\dagger$, respectively. The index i then indicates the space coordinates $\boldsymbol{r}$ and the complex time τ $(i = (\boldsymbol{r}, \tau))$. From (2.3.2), we deduce that for a c-number $n \times n$ matrix A with components A_{ij}

$$\int \exp\left[-\sum_{i,j} \bar{\psi}_i A_{ij} \psi_j \right] \prod_{i=1}^{n} \mathrm{d}\bar{\psi}_i \, \mathrm{d}\psi_i = \det A \tag{2.3.3}$$

holds. (The proof is left as an exercise.) If ψ and $\bar{\psi}$ had been c-numbers, then the result would have been $(\det A)^{-1}$.

As for the case of the bosons, we define the coherent state of one fermion as follows:

$$|\{\psi(\boldsymbol{r})\}\rangle = \prod_{\boldsymbol{r}} (|0_{\boldsymbol{r}}\rangle + |1_{\boldsymbol{r}}\rangle \psi(\boldsymbol{r})) \quad , \tag{2.3.4}$$

$$\langle\{\bar{\psi}(\boldsymbol{r})\}| = \prod_{\boldsymbol{r}} (\langle 0_{\boldsymbol{r}}| + \bar{\psi}(\boldsymbol{r}) \langle 1_{\boldsymbol{r}}|) \quad . \tag{2.3.5}$$

Here, $|0_{\boldsymbol{r}}\rangle, |1_{\boldsymbol{r}}\rangle$ are the states where at $\boldsymbol{r}$ no fermion or one fermion is present, respectively. The completeness relation

$$\int |\{\psi(\boldsymbol{r})\}\rangle\langle\{\bar{\psi}(\boldsymbol{r})\}| \exp\left[-\sum_{\boldsymbol{r}} \bar{\psi}(\boldsymbol{r}) \psi(\boldsymbol{r}) \right] \prod_{\boldsymbol{r}} \mathrm{d}\bar{\psi}(\boldsymbol{r}) \, \mathrm{d}\psi(\boldsymbol{r}) = 1 \tag{2.3.6}$$

holds because at every $\boldsymbol{r}$ the equation

$$\begin{aligned} &\int (|0\rangle + |1\rangle\psi)(\langle 0| + \bar{\psi}\langle 1|) \, \mathrm{e}^{-\bar{\psi}\psi} \, \mathrm{d}\bar{\psi} \, \mathrm{d}\psi \\ &= \int (|0\rangle\langle 0|(1 + \psi\bar{\psi}) + |0\rangle\langle 1|\bar{\psi} + |1\rangle\langle 0|\psi + |1\rangle\langle 1|\psi\bar{\psi}) \, \mathrm{d}\bar{\psi} \, \mathrm{d}\psi \\ &= |0\rangle\langle 0| + |1\rangle\langle 1| = 1 \end{aligned} \tag{2.3.7}$$

holds.

Similarly to (2.3.7), for every operator A we obtain

$$\begin{aligned} \int \langle -\bar{\psi}|A|\psi\rangle \, \mathrm{e}^{-\bar{\psi}\psi} \, \mathrm{d}\bar{\psi} \, \mathrm{d}\psi &= \int (\langle 0| - \bar{\psi}\langle 1|) A (|0\rangle + \psi|1\rangle)(1 - \bar{\psi}\psi) \, \mathrm{d}\bar{\psi} \, \mathrm{d}\psi \\ &= \langle 0|A|0\rangle + \langle 1|A|1\rangle = \mathrm{Tr}\, A \quad . \end{aligned} \tag{2.3.8}$$

Using (2.3.7) and (2.3.8), the partition function becomes

$$
\begin{aligned}
Z &= \operatorname{Tr} \mathrm{e}^{-\beta H} \\
&= \int \langle -\bar{\psi}_0 | \, \mathrm{e}^{-\beta H/N} | \psi_{N-1} \rangle \langle \bar{\psi}_{N-1} | \, \mathrm{e}^{-\beta H/N} | \psi_{N-2} \rangle \cdots \langle \bar{\psi}_1 | \, \mathrm{e}^{-\beta H/N} | \psi_0 \rangle \\
&\quad \times \exp\left[-\sum_{i=0}^{N-1} \bar{\psi}_i \psi_i \right] \prod_{i=0}^{N-1} \mathrm{d}\bar{\psi}_i \, \mathrm{d}\psi_i \\
&= \int \exp\left[-\bar{\psi}_0 \psi_{N-1} - \frac{\beta}{N} H(-\bar{\psi}_0, \psi_{N-1}) \right] \\
&\quad \times \exp\left[\bar{\psi}_{N-1} \psi_{N-2} - \frac{\beta}{N} H(\bar{\psi}_{N-1}, \psi_{N-2}) \right] \\
&\quad \times \cdots \times \exp\left[\bar{\psi}_1 \psi_0 - \frac{\beta}{N} H(\bar{\psi}_1, \psi_0) \right] \exp\left[-\sum_{i=0}^{N-1} \bar{\psi}_i \psi_i \right] \prod_{i=0}^{N-1} \mathrm{d}\bar{\psi}_i \, \mathrm{d}\psi_i \\
&= \int \exp\Bigg[-\bar{\psi}_0 \psi_{N-1} - \bar{\psi}_0 \psi_0 - \frac{\beta}{N} H(-\bar{\psi}_0, \psi_{N-1}) \\
&\quad - \sum_{i=1}^{N-1} \left\{ \bar{\psi}_i (\psi_i - \psi_{i-1}) + \frac{\beta}{N} H(\bar{\psi}_i, \psi_{i-1}) \right\} \Bigg] \prod_{i=0}^{N-1} \mathrm{d}\bar{\psi}_i \, \mathrm{d}\psi_i \ .
\end{aligned}
\tag{2.3.9}
$$

Requiring for $\bar{\psi}_N$ and ψ_N the anti-periodic boundary conditions

$$
\psi_N = -\psi_0 \ , \qquad \bar{\psi}_N = -\bar{\psi}_0 \ , \tag{2.3.10}
$$

the sign in the exponential of (2.3.9) changes, and the action S becomes

$$
S = \sum_{i=1}^{N} \left[\bar{\psi}_i (\psi_i - \psi_{i-1}) + \frac{\beta}{N} H(\bar{\psi}_i, \psi_{i-1}) \right] \ , \tag{2.3.11}
$$

and formally taking the limit $N \to \infty$, we can write

$$
S = \int_0^\beta \mathrm{d}\tau \, [\bar{\psi}(\tau) \partial_\tau \psi(\tau) + H(\bar{\psi}(\tau), \psi(\tau))] \ . \tag{2.3.12}
$$

In this case, corresponding to (2.3.10), we write

$$
\psi(\beta) = -\psi(0) \ , \qquad \bar{\psi}(\beta) = -\bar{\psi}(0) \ . \tag{2.3.13}
$$

Therefore, the Fourier transformation of $\bar{\psi}(\tau)$ and $\psi(\tau)$ is given by

$$
\begin{aligned}
\psi(\tau) &= \frac{1}{\beta} \sum_{\mathrm{i}\omega_n} \mathrm{e}^{-\mathrm{i}\omega_n \tau} \psi(\omega_n) \ , \\
\bar{\psi}(\tau) &= \frac{1}{\beta} \sum_{\mathrm{i}\omega_n} \mathrm{e}^{\mathrm{i}\omega_n \tau} \bar{\psi}(\omega_n) \ .
\end{aligned}
\tag{2.3.14}
$$

The Matsubara frequencies ω_n become $\omega_n = \pi k_{\mathrm{B}} T(2n+1)$, and the Green function $\mathcal{G}(\tau) = -\langle T\psi(\tau)\psi^\dagger(0)\rangle$ is anti-periodic $\mathcal{G}(\tau+\beta) = -\mathcal{G}(\tau)$.

2.4 The Path Integral for the Gauge Field

Up to now, we have discussed the quantization of the bosons and fermions. Next, we will discuss the quantization of the gauge field with path integral methods.

For the case of a vanishing interaction, the action is quadratic, and the Green function is the inverse of the coefficient matrix of this quadratic form. Let us apply this fact to the electromagnetic field. In the imaginary time formalism, we set $x_\mu = (\tau, x, y, z)$ and $k_\mu = (\omega, k_x, k_y, k_z)$. Because in this formalism the space–time is Euclidian, we do not need to distinguish upper and lower indices. At the zero temperature, the action is given by

$$\begin{aligned} S &= \sum_{\mu,\nu} \frac{1}{16\pi} \int \mathrm{d}^4 x F_{\mu\nu}^2(x) \\ &= \sum_{\mu,\nu} \frac{1}{16\pi} \int \mathrm{d}^4 x (\partial_\mu A_\nu(x) - \partial_\nu A_\mu(x))^2 \\ &= \frac{1}{16\pi} \sum_{\mu,\nu,k} (k_\mu A_\nu(k) - k_\nu A_\mu(k))(k_\mu A_\nu(-k) - k_\nu A_\mu(-k)) \ . \end{aligned} \quad (2.4.1)$$

Splitting $A_\mu(k)$ into its longitudinal component $A_\mu^{\mathrm{L}}(k) = (k_\mu k_\nu / k^2) A_\nu(k)$ and its transversal component $A_\mu^{\mathrm{T}}(k) = A_\mu(k) - A_\mu^{\mathrm{L}}$, we can write (2.4.1) as

$$\begin{aligned} S &= \frac{1}{8\pi} \sum_{k,\mu,\nu} (k^2 \delta_{\mu,\nu} - k_\mu k_\nu) A_\mu(k) A_\nu(-k) \\ &= \frac{1}{8\pi} \sum_{k,\mu,\nu} (k^2 \delta_{\mu,\nu} - k_\mu k_\nu) A_\mu^{\mathrm{T}}(k) A_\nu^{\mathrm{T}}(-k) \ . \end{aligned} \quad (2.4.2)$$

This equation does not contain A_μ^{L}. Therefore, it is possible to define a Green function for the transversal component, but, curiously, not for the longitudinal component.

In Sect. 1.4, we performed the canonical quantization of the electromagnetic field using the commutation relations. So, what is the problem with the path integral description? The answer is that we fixed a gauge ($\operatorname{div} \boldsymbol{A} = 0$) in the previous case, and in (2.4.1) we did not say anything about the gauge. Therefore, when exponentiating (2.4.1) and performing the path integral $\int \mathcal{D}A_\mu$, paths that are in reality equal and differ only by a different choice of the gauge pile up in the calculation. Because there exist infinitely many choices for the gauge, there is nothing strange about it when the path integral diverges, and indeed because (2.4.2) does not contain A_μ^{L}, the functional integral in A_μ^{L} diverges.

Therefore, the question is: How do we implement the gauge fixing conditions in the path integral? The answer is given by the Faddeev–Popov technique. In what follows, we will explain this method.

Let us choose for A_μ one fixed function $\bar{A}_\mu(\boldsymbol{r},\tau)$. Performing a gauge transformation with $\Lambda(\boldsymbol{r},\tau)$, then $\bar{A}_\mu(\boldsymbol{r},\tau)$ is transformed to $A_\mu = \bar{A}_\mu(\boldsymbol{r},\tau)+\partial_\mu\Lambda$. Physically, this A_μ is identical to $\bar{A}_\mu$, and because the action is gauge invariant, $S(A_\mu) = S(\bar{A}_\mu)$. When performing the path integral $\int \mathcal{D}A_\mu$, for all possible functions Λ the physically identical field A_μ reappears again and again. We write

$$
\begin{aligned}
Z &= \int \mathcal{D}A_\mu \, \mathrm{e}^{-S(\{A_\mu\})} \\
&= \int \mathcal{D}\bar{A}_\nu \int \mathcal{D}\Lambda \, \mathrm{e}^{-S(\{\bar{A}_\mu\})} \quad . \qquad (2.4.3)
\end{aligned}
$$

Here, $\int \mathcal{D}\Lambda$ is responsible for the divergence.

In order to absorb this divergent integral, we introduce a functional $F(A_\mu)$ of A_μ and write instead of $\int \mathcal{D}\Lambda$

$$
\begin{aligned}
\int \mathcal{D}\Lambda &\to \int \mathcal{D}F(A_\mu) \exp\left[-\int \mathrm{d}\boldsymbol{r}\, \mathrm{d}\tau \frac{F^2}{2\alpha}\right] \\
&= \int \mathcal{D}\Lambda \det\left(\frac{\delta F}{\delta \Lambda}\right) \exp\left[-\int \mathrm{d}\boldsymbol{r}\, \mathrm{d}\tau \frac{F^2}{2\alpha}\right] \quad . \qquad (2.4.4)
\end{aligned}
$$

This rewriting contains arbitrariness, because every function leading to a convergent result can be used. Because we want to recover the integral $\int \mathcal{D}\bar{A}_\mu \int \mathcal{D}\Lambda = \int \mathcal{D}A_\mu$ (this means that we want to perform the functional integration in A_μ without restrictions), we choose (2.4.4) in such a way that it is independent of $\bar{A}_\mu$ and contains a functional integral in Λ. Concerning F, from the requirement that $-F^2/2\alpha$ should be quadratic in A_μ; normally F is set to be

$$
F = \partial_\mu A_\mu \quad . \qquad (2.4.5)
$$

$F = 0$ leads to the Lorentz gauge. In the above case we obtain $\delta F = F(\Lambda + \delta\Lambda) - F(\Lambda) = \partial_\mu^2 \delta\Lambda$ and therefore $\det[\delta F/\delta\Lambda] = \det[\partial_\mu^2]$. Because this is an A_μ-independent constant, we finally obtain

$$
Z = \int \mathcal{D}A_\mu \exp\left[-S - \frac{1}{2\alpha}\int (\partial_\mu A_\mu)^2 \, \mathrm{d}\boldsymbol{r}\, \mathrm{d}\tau\right] \quad . \qquad (2.4.6)
$$

Here, instead of (2.4.2), in the exponent we obtain

$$
-\frac{1}{8\pi}\sum_{k,\mu,\nu}(k^2\delta_{\mu,\nu} - k_\mu k_\nu)A_\mu^{\mathrm{T}}(k)A_\nu^{\mathrm{T}}(-k) - \frac{1}{2\alpha}\sum_{k,\mu,\nu} k_\mu k_\nu A_\mu^{\mathrm{L}}(k)A_\nu^{\mathrm{L}}(-k) \qquad (2.4.7)
$$

and a Green function for A_μ^{L} can be defined.

As demonstrated, for the case when the gauge is not totally fixed, the functional integral over the remaining degrees of freedom is made to converge by adding a new term to the action (in the above example this is the gauge fixing term $F^2/2\alpha$). This is the so-called the Faddeev–Popov technique.

2.5 The Path Integral for the Spin System

Quantization using path integral techniques is possible also for a spin system. For simplicity, we consider a system with only one spin $I = 1/2$. In this case, the space of states can be written as a two-vector with up-spin component $\uparrow$ and down-spin component $\downarrow$:

$$|\psi\rangle = \alpha|\uparrow\rangle + \beta|\downarrow\rangle = \begin{bmatrix} \alpha \\ \beta \end{bmatrix} . \tag{2.5.1}$$

Here, α and β are complex numbers, and the normalization is $|\alpha|^2 + |\beta|^2 = 1$. The system has three real degrees of freedom. Using the variables b, θ and φ we can write

$$|\psi\rangle = |b, \theta, \varphi\rangle = \mathrm{e}^{ib}\left(\mathrm{e}^{-\mathrm{i}\varphi/2}\cos\frac{\theta}{2}|\uparrow\rangle + \mathrm{e}^{\mathrm{i}\varphi/2}\sin\frac{\theta}{2}|\downarrow\rangle\right) , \tag{2.5.2}$$

where e^{ib} is an overall phase factor and therefore has no influence on the physics. More precisely, b corresponds to the degree of freedom of the gauge transformation. It can be proved as follows that $|b, \theta, \varphi\rangle$ is a complete set with b fixed and θ and φ variable:

$$\begin{aligned} \int_0^{\pi} \sin\theta\, \mathrm{d}\theta \int_0^{2\pi} \frac{\mathrm{d}\varphi}{2\pi} |b, \theta, \varphi\rangle\langle b, \theta, \varphi| &= |\uparrow\rangle\langle\uparrow| + |\downarrow\rangle\langle\downarrow| \\ &= \hat{1} . \end{aligned} \tag{2.5.3}$$

Furthermore, the expectation value of the spin operator $\hat{\boldsymbol{I}} = \frac{1}{2}\boldsymbol{\sigma}$ is given by

$$\begin{aligned} \langle b, \theta, \varphi|\hat{\boldsymbol{I}}|b, \theta, \varphi\rangle &= \tfrac{1}{2}(\sin\theta\cos\varphi, \sin\theta\sin\varphi, \cos\theta) \\ &= \tfrac{1}{2}\boldsymbol{n} . \end{aligned} \tag{2.5.4}$$

$\boldsymbol{\sigma}$ are the so-called Pauli matrices

$$\sigma^x = \begin{bmatrix} 0 & 1 \\ 1 & 0 \end{bmatrix} , \quad \sigma^y = \begin{bmatrix} 0 & -\mathrm{i} \\ \mathrm{i} & 0 \end{bmatrix} , \quad \sigma^z = \begin{bmatrix} 1 & 0 \\ 0 & -1 \end{bmatrix} , \tag{2.5.5}$$

written as a three-component vector.

As shown in (2.5.3), $|b, \theta, \varphi\rangle$ is a complete set; however, for different (θ, φ), the states $|b, \theta, \varphi\rangle$ are not orthogonal. However, when returning to the discussion of Sect. 2.1, it is clear that for the insertion of intermediate states into the path integral, (2.5.3) is sufficient. An important step for the calculation of the sum of states in $|\tau\rangle = |b(\tau), \theta(\tau), \varphi(\tau)\rangle$, is the following estimation of the time evolution during the infinitesimally small complex time $\Delta\tau$:

$$
\begin{aligned}
\langle\tau+\Delta\tau|\,\mathrm{e}^{-\Delta\tau H}|\tau\rangle &\cong \langle\tau+\Delta\tau|(1-\Delta\tau H)|\tau\rangle \\
&= \langle\tau+\Delta\tau|\tau\rangle - \Delta\tau\langle\tau+\Delta\tau|H|\tau\rangle \\
&\cong \left[\langle\tau| + \Delta\tau\left(\frac{\mathrm{d}\langle\tau|}{\mathrm{d}\tau}\right)\right]|\tau\rangle - \Delta\tau\langle\tau|H|\tau\rangle \\
&= 1 + \Delta\tau[\langle\dot{\tau}|\tau\rangle - \langle\tau|H|\tau\rangle] \\
&\cong \mathrm{e}^{\Delta\tau[\langle\dot{\tau}|\tau\rangle - \langle\tau|H|\tau\rangle]} \quad .
\end{aligned} \tag{2.5.6}
$$

Because of

$$
\begin{aligned}
0 = \frac{\mathrm{d}}{\mathrm{d}\tau}\langle\tau|\tau\rangle &= \langle\dot{\tau}|\tau\rangle + \langle\tau|\dot{\tau}\rangle \\
&= 2\,\mathrm{Re}\langle\tau|\dot{\tau}\rangle
\end{aligned} \tag{2.5.7}
$$

the first term in the exponential of (2.5.6) is imaginary and therefore leads to a phase factor. Because this phase was originally written as $\langle\tau+\Delta\tau \mid \tau\rangle$, in the time evolution of the system it has the meaning of the "overlap integral" of the wave functions at infinitesimally separated times, or in mathematical language is has the meaning of a "connection". For the single-particle path integral (Sect. 2.1), the corresponding factor is given by

$$
\begin{aligned}
\langle x(\tau+\Delta\tau)|x(\tau)\rangle &= \int \mathrm{d}p(\tau)\,\langle x(\tau+\Delta\tau)|p(\tau)\rangle\langle p(r)|x(\tau)\rangle \\
&= \int \frac{\mathrm{d}p(\tau)}{2\pi h}\,\mathrm{e}^{\mathrm{i}p(\tau)[x(\tau+\Delta\tau)-x(\tau)]} \\
&= \int \frac{\mathrm{d}p(\tau)}{2\pi h}\,\mathrm{e}^{\mathrm{i}\Delta\tau p(\tau)\dot{x}(\tau)} \quad .
\end{aligned} \tag{2.5.8}
$$

Notice that this is just the factor $\mathrm{i}p\dot{x}$.

Now, let us determine $\langle\dot{\tau} \mid \tau\rangle$ explicitely. With (2.5.2), we obtain

$$
\begin{aligned}
\frac{\mathrm{d}}{\mathrm{d}\tau}|\tau\rangle = \mathrm{i}\dot{b}|\tau\rangle + \mathrm{e}^{\mathrm{i}b}\bigg\{&\left(-\frac{\mathrm{i}\dot{\varphi}}{2}\cos\frac{\theta}{2} - \frac{\dot{\theta}}{2}\sin\frac{\theta}{2}\right)\mathrm{e}^{-\mathrm{i}\varphi/2}|\uparrow\rangle \\
&+\left(\frac{\mathrm{i}\dot{\varphi}}{2}\sin\frac{\theta}{2} + \frac{\dot{\theta}}{2}\cos\frac{\theta}{2}\right)\mathrm{e}^{\mathrm{i}\varphi/2}|\downarrow\rangle\bigg\} \quad .
\end{aligned} \tag{2.5.9}
$$

Taking the inner product with $\langle\tau|$, we obtain

$$
\begin{aligned}
\langle\tau|\dot{\tau}\rangle &= -\langle\dot{\tau}|\tau\rangle \\
&= \mathrm{i}\dot{b} + \cos\frac{\theta}{2}\left(-\frac{\mathrm{i}\dot{\varphi}}{2}\cos\frac{\theta}{2} - \frac{\dot{\theta}}{2}\sin\frac{\theta}{2}\right) + \sin\frac{\theta}{2}\left(\frac{\mathrm{i}\dot{\varphi}}{2}\sin\frac{\theta}{2} + \frac{\dot{\theta}}{2}\cos\frac{\theta}{2}\right) \\
&= \mathrm{i}\left(\dot{b} - \frac{1}{2}\dot{\varphi}\cos\theta\right) \quad .
\end{aligned} \tag{2.5.10}
$$

Here, we will fix the gauge b with the following requirement. As is clear from equation (2.5.4), the physical meaning of θ and φ is the direction of the spin, and we require that $|\theta, \varphi\rangle$ should be unambiguously defined by the direction $\boldsymbol{n}$ of the spin. Then, the boundary condition $\boldsymbol{n}(\beta) = \boldsymbol{n}(0)$ in the path integral equals $|\theta(\beta), \varphi(\beta)\rangle = |\theta(0), \varphi(0)\rangle$, the periodicity of the wave function, which is convenient. In order to obtain an unambiguous expression, the factor $\mathrm{e}^{\pm \mathrm{i}\varphi/2}$ in (2.5.2) is problematic. This is due to the fact that when shifting $\varphi \to \varphi \pm 2\pi$, the direction of $\boldsymbol{n}$ is the same; however, $\mathrm{e}^{\pm \mathrm{i}\varphi/2}$ changes its sign. In order to resolve this problem, we could set $b = \pm\varphi/2$, so that $\mathrm{e}^{\pm \mathrm{i}\varphi/2}$ becomes either 1 or $\mathrm{e}^{\pm \mathrm{i}\varphi}$.

Now, let us choose $b = \varphi/2$. In order to calculate the sum of states, we have to perform the path integral in $\theta(\tau)$ and $\varphi(\tau)$ under the boundary condition $|\tau = \beta\rangle = |\tau = 0\rangle$. Every path corresponds to a closed path on the unit sphere, described by the vector $\boldsymbol{n}$ of (2.5.4). Then, the integrand of the integral in (2.5.6) becomes e^{-S}, and the action S for $I = 1/2$ becomes

$$\begin{aligned} S &= -\int_0^\infty \langle \dot{\tau}|\tau\rangle \, \mathrm{d}\tau + \int_0^\beta \langle \tau|\hat{H}|\tau\rangle \, \mathrm{d}\tau \\ &= \mathrm{i}I \int_0^\beta (1 - \cos\theta)\dot{\varphi} \, \mathrm{d}\tau + \int_0^\beta H(I\boldsymbol{n}(\tau)) \, \mathrm{d}\tau \ . \end{aligned} \qquad (2.5.11)$$

The above equation also holds for general spin I.

The first term in (2.5.11) is called the Berry phase and has the following geometrical meaning, as shown in Fig. 2.5. Noting that $\Delta\tau\dot{\varphi}(1 - \cos\theta)$ describes the solid angle between the z axis, $\boldsymbol{n}(\tau)$ and $\boldsymbol{n}(\tau + \Delta\tau)$, we may also express the Berry phase by the solid angle ω subtended by the closed path described by $\boldsymbol{n}$:

$$\mathrm{i}I\omega \ .$$

The solid angle is determined moduli 4π, the surface of the unit ball. However, because $2I$ is an integer, $\mathrm{e}^{4\pi \mathrm{i}I} = 1$ and therefore this ambiguity does not affect physics. Saying it the other way round, we conclude that the spin I cannot reach any value, but is quantized in such a way that $2I$ is an integer.

Furthermore, the canonical conjugate relations of p and x in $\mathrm{i}p\dot{x}$ are also reflected in the commutation relations of the components of the spin. The

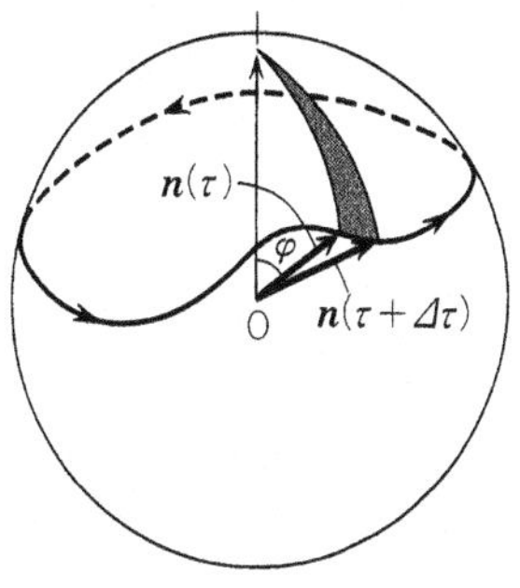

Fig. 2.5. The topological meaning of the Berry phase

term $\mathrm{i}I\cos\theta\dot{\varphi}$ signals that in the angle φ of the spin in the xy plane and $I^z = I\cos\theta$ are canonical conjugate. Indeed, from (2.5.4) for $I = 1/2$ we obtain

$$\hat{I}^{\pm} = \hat{I}^x \pm \mathrm{i}\hat{I}^y = I\,\mathrm{e}^{\pm\mathrm{i}\hat{\varphi}}\sin\hat{\theta} \ , \tag{2.5.12}$$

the definition of the operators $\hat{\varphi}$ and $\hat{\theta}$. The commutation relation of the spin

$$[\hat{I}^z, \hat{I}^{\pm}] = \pm\hat{I}^{\pm} \tag{2.5.13}$$

equals

$$[\hat{I}^z, \mathrm{e}^{\pm\mathrm{i}\hat{\varphi}}] = \pm\mathrm{e}^{\pm\mathrm{i}\hat{\varphi}} \tag{2.5.14}$$

and this induces

$$[\hat{\varphi}, \hat{I}^z] = \mathrm{i} \ . \tag{2.5.15}$$

The second term in (2.5.11) is the expectation value of the Hamiltonian $\hat{H} = H(\hat{\boldsymbol{I}})$ in the state $|\tau\rangle$ expressed with the spin variable $\boldsymbol{I} = I\boldsymbol{n}$. Let us ensure that the classical equations of motion can be obtained from the action (2.5.11). First, the variation with respect to the solid angle leads to

$$\begin{aligned} \delta\omega &= \int_0^\beta \mathrm{d}\tau\,[\dot{\varphi}\delta\theta\sin\theta - \delta\dot{\varphi}\cos\theta] \\ &= \int_0^\beta \mathrm{d}\tau\,[\dot{\varphi}\delta\theta - \dot{\theta}\delta\varphi]\sin\theta \\ &= \int_0^\beta \mathrm{d}\tau\,\delta\boldsymbol{n}(\tau)\cdot\left(\frac{\mathrm{d}\boldsymbol{n}(\tau)}{\mathrm{d}\tau}\times\boldsymbol{n}(\tau)\right) \ . \end{aligned} \tag{2.5.16}$$

Therefore, $\delta S = 0$ leads to the equation

$$\mathrm{i}I\frac{\mathrm{d}\boldsymbol{n}(\tau)}{\mathrm{d}\tau}\times\boldsymbol{n}(\tau) = -\frac{\partial H(I\boldsymbol{n}(\tau))}{\partial\boldsymbol{n}(\tau)} \ . \tag{2.5.17}$$

We now transform from $\tau = it$ to real time and take the cross product with $\boldsymbol{n}(\tau)$ on both sides:

$$I\frac{\mathrm{d}\boldsymbol{n}(t)}{\mathrm{d}t} = \boldsymbol{n}(t)\times\frac{\partial H(I\boldsymbol{n}(t))}{\partial\boldsymbol{n}(t)} \ . \tag{2.5.18}$$

This equation equals the classical equation of motion of the spin. The torque acting on the spin is on the right-hand side.

3. Symmetry Breaking and Phase Transition

Qualitatively new features arise for the case when the system has infinitely many degrees of freedom, namely spontaneous symmetry breaking and phase transition. These concepts are of great interest in solid state physics, dealing with many-particle systems with about 10^{23} particles. In this chapter, starting with the concept of the order parameter and the basics of the Landau theory, we proceed to a discussion of the Goldstone mode and the absence of order in low-dimensional systems. In the second half of the chapter we will focus on phase transitions, where the topological defects play important roles and it is not easy to find the order parameters, as is the case for the Kosterlitz–Thouless transition, and the confinement in gauge theories on a lattice.

3.1 Spontaneous Symmetry Breaking

One of the most important aspects of field theory is symmetry breaking. We now introduce the XY model as an example. We define a d-dimensional cubic lattice and put on every lattice point a spin $\boldsymbol{I}_i = I(\cos\theta_i, \sin\theta_i)$. For a while, let us consider a classical spin system. Quantum properties will be discussed next. The Hamiltonian of this system can be written as

$$H = -J\sum_{\langle ij\rangle} \boldsymbol{I}_i \cdot \boldsymbol{I}_j = -JI^2 \sum_{\langle ij\rangle} \cos(\theta_i - \theta_j) \ , \tag{3.1.1}$$

where $\langle ij\rangle$ runs over all neighbouring lattice points. The partition function of the system is given by

$$Z(\{\boldsymbol{h}_i\}) = \operatorname{Tr}\exp\left\{-\beta\left(H - \sum_{i=1}^{N} \boldsymbol{h}_i \cdot \boldsymbol{I}_i\right)\right\} \ . \tag{3.1.2}$$

Here, we have added a "magnetic field" $\boldsymbol{h}_i$ and expressed Z as a function of $\boldsymbol{h}_i$. The average of $\boldsymbol{I}_i$, that is, the magnetization $\boldsymbol{M}_i$, can be calculated as

$$\begin{aligned} \boldsymbol{M}_i &= \langle \boldsymbol{I}_i \rangle \\ &= \operatorname{Tr}\left[\boldsymbol{I}_i \exp\left\{-\beta\left(H - \sum_{i=1}^{N} \boldsymbol{h}_i \cdot \boldsymbol{I}_i\right)\right\}\right] \Big/ Z(\{\boldsymbol{h}_i\}) \end{aligned}$$

$$= \frac{1}{\beta}\frac{\partial}{\partial \boldsymbol{h}_i} \ln Z(\{\boldsymbol{h}_i\}) = -\frac{\partial F}{\partial \boldsymbol{h}_i} \ , \tag{3.1.3}$$

where N is the number of lattice points, and F is the free energy of the system.

Now, let us first consider the case $\boldsymbol{h}_i = \boldsymbol{0}$. Then, because $\theta_i - \theta_j$ remains unchanged when all spins are shifted simultaneously from θ_i to $\theta_i + \pi$, the Hamiltonian is invariant under this transformation. Therefore, we obtain

$$\begin{aligned}\langle \boldsymbol{I}_i \rangle &= (I\langle \cos\theta_i\rangle, I\langle \sin\theta_i\rangle) \\ &= (I\langle \cos(\theta_i + \pi)\rangle, I\langle \sin(\theta_i + \pi)\rangle) \\ &= -(I\langle \cos\theta_i\rangle, I\langle \sin\theta_i\rangle) \\ &= -\langle \boldsymbol{I}_i \rangle \ , \end{aligned} \tag{3.1.4}$$

and we conclude that $\langle \boldsymbol{I}_i \rangle$ vanishes. In this manner, when the Hamiltonian is invariant under a transformation, a physical quantity that is not invariant under this transformation is zero on average. However, following this argument, everything in this world would become (in a thermal equilibrium state) homogeneous and extended owing to translational symmetry, or everything would become round owing to rotational symmetry. The reason why so many structures are present in the real world is that we are dealing with the macroscopic world, and that qualitatively new features arise because of the large number of $N \sim 10^{23}$ particles. Mathematically speaking, only in the limit $N \to \infty$ does a singularity in the partition function Z or the free energy arise; however, we will not give a precise mathematical discussion, but focus on the description of the physical picture.

As can be seen in (3.1.2), the ratio of the interaction JI^2 and the temperature $T = 1/\beta$ determine the degree of disorder of the θ_i. That is, for the case when two neighbouring spins are parallel ($\theta_i - \theta_j = \pm\pi$), the energy loss is $2JI^2$, and as a result the probability is as small as $e^{-2JI^2/T}$. Writing the trace in (3.1.2) explicitly, we obtain the N-dimensional integral

$$\mathrm{Tr} = \int_0^{2\pi} d\theta_1 \cdots \int_0^{2\pi} d\theta_N \ . \tag{3.1.5}$$

In the N-dimensional integral of (3.1.2), when the spins are in a parallel arrangement, the energy H becomes small and the integrand $e^{-\beta H}$ large; however, the number of such arrangements is limited, that is, the volume in this N-dimensional space.

The entropy S is given by the logarithm of the volume in the N-dimensional space, and as a result of the competition between the entropy trying to attain a maximum and the energy trying to attain a minimum, at some temperature T a phase transition occurs. Because in thermal equilibrium at finite temperature, not the energy E, but the free energy $F = E - TS$ must be minimized, at high temperature the second term becomes important and the disorder of the system increases; therefore above a temperature T_c

the systems passes from an ordered state to a disordered state. However, also at temperatures higher than T_c the spins $\boldsymbol{I}_i$ and $\boldsymbol{I}_j$ are not totally independent, but the effect of J is still present. Therefore, nearby spins have the tendency to become parallel.

The function that measures this correlation explicitly is defined by

$$C(\boldsymbol{R}_i, \boldsymbol{R}_j) = C(\boldsymbol{R}_i - \boldsymbol{R}_j) = \langle \boldsymbol{I}(\boldsymbol{R}_i) \cdot \boldsymbol{I}(\boldsymbol{R}_j) \rangle \ , \tag{3.1.6}$$

where $\boldsymbol{R}_i$ is the coordinate vector of site i, and the first equality in (3.1.6) is a result of translational invariance. With (3.1.2), we obtain

$$C(\boldsymbol{R}_i - \boldsymbol{R}_j) = -\frac{1}{\beta} \sum_\alpha \frac{\partial^2 F}{\partial h_\alpha(\boldsymbol{R}_i) \partial h_\alpha(\boldsymbol{R}_j)} + \langle \boldsymbol{I}(\boldsymbol{R}_i) \rangle \cdot \langle \boldsymbol{I}(\boldsymbol{R}_j) \rangle \ . \tag{3.1.7}$$

The second term vanishes for temperatures higher than T_c.

When $\boldsymbol{R}_i$ and $\boldsymbol{R}_j$ are sufficiently far from each other, the correlation function decays for $T > T_c$ as

$$C(\boldsymbol{R}_i - \boldsymbol{R}_j) \sim \exp\left[-\frac{|\boldsymbol{R}_i - \boldsymbol{R}_j|}{\xi} \right] \ . \tag{3.1.8}$$

This fact can be proved in the limit of high temperature as follows. Starting from the definition of the correlation function,

$$\begin{aligned} &C(\boldsymbol{R}_1 - \boldsymbol{R}_2) \\ &= I^2 \langle \cos(\theta(\boldsymbol{R}_1) - \theta(\boldsymbol{R}_2)) \rangle \\ &= \frac{1}{Z} \mathrm{Tr} \left[\exp\left\{ \beta J I^2 \sum_{\langle ij \rangle} \cos(\theta(\boldsymbol{R}_i) - \theta(\boldsymbol{R}_j)) \right\} I^2 \cos(\theta(\boldsymbol{R}_1) - \theta(\boldsymbol{R}_2)) \right] \ , \end{aligned} \tag{3.1.9}$$

we expand the exponential for the case when $\beta J I^2$ is small compared with 1 ($T \gg J I^2$):

$$C(\boldsymbol{R}_1 - \boldsymbol{R}_2) = \frac{\mathrm{Tr} \left[I^2 \cos(\theta_1 - \theta_2) \sum_{m=1}^\infty \frac{1}{m!} (\beta J I^2)^m \left\{ \sum_{\langle ij \rangle} \cos(\theta_i - \theta_j) \right\}^m \right]}{\mathrm{Tr} \left[\sum_{m=1}^\infty \frac{1}{m!} (\beta J I^2)^m \left\{ \sum_{\langle ij \rangle} \cos(\theta_i - \theta_j) \right\}^m \right]} \ . \tag{3.1.10}$$

This kind of expansion is called a high-temperature expansion.

In every term

$$\left\{ \sum_{\langle ij \rangle} \cos(\theta_i - \theta_j) \right\}^m = \frac{1}{2^m} \left\{ \sum_{\langle ij \rangle} \left(\mathrm{e}^{\mathrm{i}(\theta_i - \theta_j)} + \mathrm{e}^{\mathrm{i}(\theta_j - \theta_i)} \right) \right\}^m \tag{3.1.11}$$

of the expansion, we think of $\mathrm{e}^{\mathrm{i}(\theta_i - \theta_j)}$ as a bond with a direction arrow from the site $\boldsymbol{R}_j$ to $\boldsymbol{R}_i$ and represent them by diagrams as demonstrated in Fig. 3.1. Let us take the trace Tr of (3.1.11). As written explicitly in (3.1.5), the Tr is

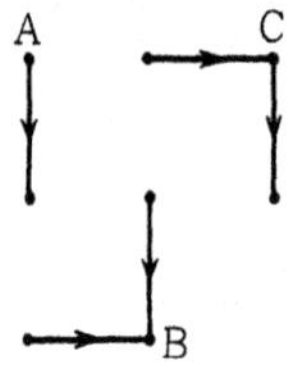

Fig. 3.1. Diagrams corresponding to the different types of terms arising in the expansion (3.1.11)

an integral over all angels θ_i at the lattice sites. For the point A, where only one site with an outgoing arrow is present, we obtain

$$\int_0^{2\pi} d\theta_A \, e^{-i\theta_A} = 0 \ . \tag{3.1.12}$$

For the point B with two incoming sites, we obtain

$$\int_0^{2\pi} d\theta_B \, e^{2i\theta_B} = 0 \ . \tag{3.1.13}$$

Both integrals vanish. Only for the integral of the point C, where the number of ingoing and outgoing arrows is equal, do we obtain a finite contribution

$$\int_0^{2\pi} d\theta_C \, e^{i(\theta_C - \theta_C)} = 2\pi \tag{3.1.14}$$

leading to an effective contribution in the integral.

Therefore, only diagrams that are constructed of points similar to C make a contribution, as shown for example in Fig. 3.2.

Above, we discussed the denominator of the expansion in (3.1.10). Concerning the numerator

$$\cos(\theta_1 - \theta_2) = \left(e^{i(\theta_1 - \theta_2)} + e^{i(\theta_2 - \theta_1)}\right)/2 \ ,$$

the same considerations as for (3.1.11) lead to the conclusion that only connected paths from point 1 to point 2 (or vice versa) lead to a finite contribution. As shown in Fig. 3.3, a separated closed loop may also be present, leading to a higher power contribution in $\beta J I^2 (\ll 1)$. The dominant contribution of the numerator of (3.1.10) comes from the shortest path linking point 1 and point 2. For the case when $\boldsymbol{R}_1$ and $\boldsymbol{R}_2$ are far enough apart, the

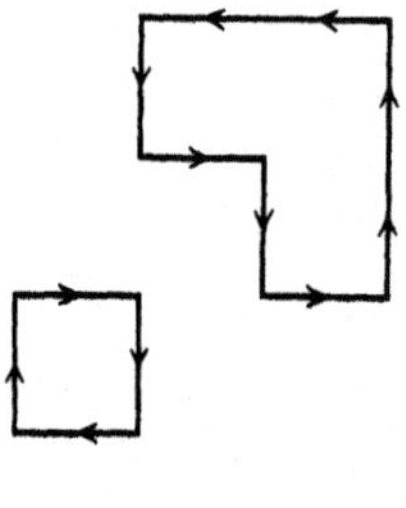

Fig. 3.2. Diagrams leading to a finite contribution in (3.1.11) after taking the trace

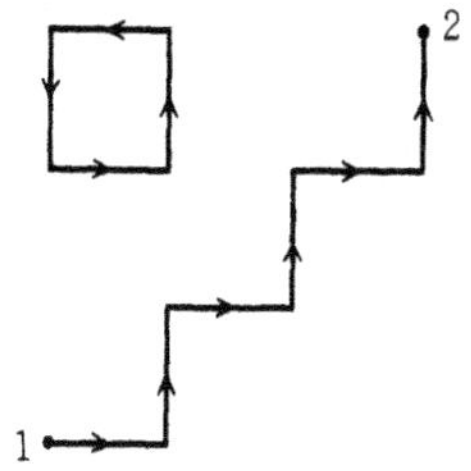

Fig. 3.3. Diagrams leading to a finite contribution in the numerator of (3.1.10)

step number m is about the order of $|\boldsymbol{R}_1 - \boldsymbol{R}_2|$. Because in the expansion (3.1.11) the number of similar diagrams is given by $m!$, the factor $1/m!$ from the expansion of the exponential cancels, and we obtain finally

$$
\begin{aligned}
C(\boldsymbol{R}_1 - \boldsymbol{R}_2) &\sim \left(\frac{\beta J I^2}{2}\right)^m \sim \left(\frac{\beta J I^2}{2}\right)^{|\boldsymbol{R}_1 - \boldsymbol{R}_2|} \\
&= \exp\left[-\ln\left(\frac{2}{\beta J I^2}\right) \cdot |\boldsymbol{R}_1 - \boldsymbol{R}_2|\right] . \qquad (3.1.15)
\end{aligned}
$$

Comparing this equation with (3.1.8), for ξ we obtain

$$
\xi = \left[\ln\left(\frac{2}{\beta J I^2}\right)\right]^{-1} \qquad (3.1.16)
$$

and, as expected, this number becomes smaller for higher temperatures.

ξ is called the correlation length. In the distance ξ, the spins reverse their direction. In other words, in a region of size ξ (called a domain) most of the spins are aligned in the same direction. With lower and lower temperature, the size ξ of the domain becomes larger and larger, and finally at T_c it becomes infinite.

In this manner, when the size of the domains becomes larger, when an external magnetic field is applied, large numbers of spins are aligned at once in its direction, and therefore also the susceptibility χ becomes larger. At T_c, the correlation length ξ diverges and χ also becomes infinite, and below T_c the macroscopic system as a whole becomes a domain. An infinitely small magnetic field leads to a reversion of the macroscopic magnetization $\boldsymbol{M}$ from $-\boldsymbol{M}_0$ to $\boldsymbol{M}_0$. The occurrence of a magnetization $\boldsymbol{M}$ in the ordered phase can be explained by assuming the presence of such an initial infinitesimal magnetic field. Mathematically speaking, this corresponds to first taking the limit $N \to \infty$ and then approaching $\boldsymbol{h} \to \boldsymbol{0}$ ($\lim_{\boldsymbol{h}\to\boldsymbol{0}} \lim_{N\to\infty}$). This corresponds to the "Bogoliubov quasi-average". In this way, it is possible to avoid the dilemma of (3.1.4). Physical quantities like this magnetization, which are not zero in an ordered phase with broken symmetry, are called order parameters.

In order to make this discussion more precise, let us consider the magnetization curve in Fig. 3.4. In what follows we discuss the isotropic system, where for example the magnetic field points in the x direction $\boldsymbol{h} = (h, 0, 0)$

and the magnetization is therefore given by $\boldsymbol{M} = (M, 0, 0)$. The susceptibility at vanishing external magnetization is given by $\mathrm{d}M/\mathrm{d}h\,|_{h=0}$. At $T = T_\mathrm{c}$, for small h the magnetization is proportional to $M \propto h^{1/\delta}$ with $\delta > 1$, and therefore $\chi \to \infty$.

For $T < T_\mathrm{c}$, at $h = 0$ the magnetization curve is not continuous. Physically speaking, as described above, this is due to the fact that a macroscopic number of spins have a fixed relation relative to each other. If we try to obtain this result mathematically from (3.1.2), we might think that it is quite a difficult problem to determine a discontinuous curve.

The answer to this problem has been given by Landau, and it is astonishingly simple and clear. In a word, exchange the x axis and the y axis in Fig. 3.4 (or look at the figure from the side). Mathematically speaking, perform a Legendre transformation from the free energy $F(h)$ to the Gibbs free energy

$$G(M) = F(h) + M \cdot h \ . \tag{3.1.17}$$

The above equation changes to

$$\begin{aligned} \mathrm{d}G(M) &= \mathrm{d}F(h) + M \cdot \mathrm{d}h + h \cdot \mathrm{d}M \\ &= \frac{\partial F}{\partial T}\mathrm{d}T + \left(\frac{\partial F}{\partial h} + M\right) \cdot \mathrm{d}h + h \cdot \mathrm{d}M \\ &= -S\,\mathrm{d}T + h \cdot \mathrm{d}M \ , \end{aligned} \tag{3.1.18}$$

where we have used (3.1.3). From (3.1.18) we obtain

$$h = \frac{\partial G(M)}{\partial M} \ . \tag{3.1.19}$$

Equation (3.1.19) expresses h in terms of M. Looking at Fig. 3.4 from the side, we see the derivative of the function $G(M)$. Here, we assume that the function $G(M)$ is smooth and differentiable with respect to M, so that the derivative of $G(M)$ between M_0 and $-M_0$ is given by the dotted line.

For example, for

$$G(M) = aM^2 + bM^4 \qquad (b > 0) \tag{3.1.20}$$

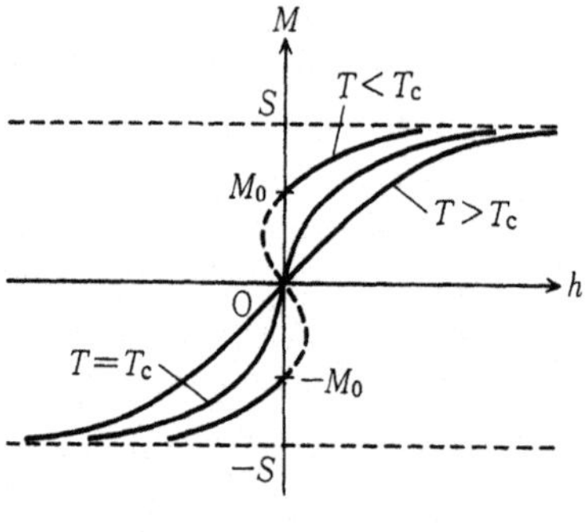

Fig. 3.4. Magnetization as function of the external magnetic field

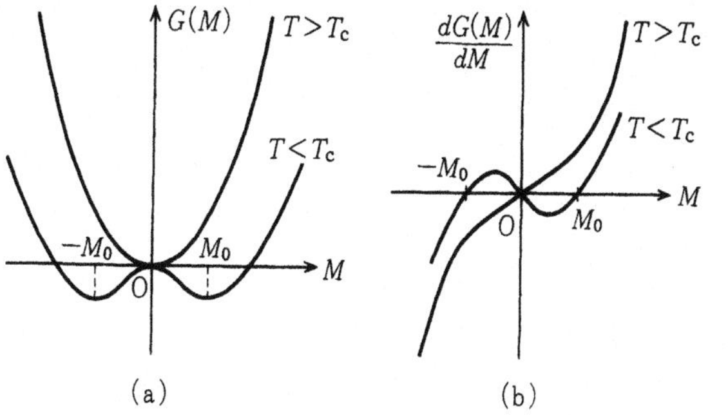

Fig. 3.5a, b

the corresponding curves for $a > 0$, that is $T > T_c$, and $a < 0$, that is $T < T_c$, are shown in Fig. 3.5a. For $T < T_c$, the origin $M = 0$ becomes a maximum, and two new minima arise at $\pm M_0$. Differentiating these curves, we regain Fig. 3.4 for $T < T_c$, as shown in Fig. 3.5b. In this case, the region between $-M_0$ and $+M_0$ is the mountain in Fig. 3.5a and therefore unstable, and will not be realized in a thermal equilibrium state. What physically happens is that for $h = -0$ the extremum $-M_0$, and for $h = +0$ the extremum $+M_0$, is energetically stable, and therefore the magnetization M is discontinuous at $h = 0$. In this manner, the Landau theory describes a scenario where $G(M)$ is a smooth function of temperature and magnetization without discontinuities and singularities; however, by minimizing it with respect to M, singularities and discontinuities arise in $F(h)$ and $\mathrm{d}F(h)/\mathrm{d}h$, respectively.

The careful reader will have recognized that the equations below (3.1.17) are independent of the position, and that the assumptions $h_i = h$ and $M_i = M$ have been made. At the very beginning, the free energy deduced from the partition function (3.1.2) has been written as a function of $h_1, \ldots, h_N$. Consider $h(\boldsymbol{r})$ as a function of the position $\boldsymbol{r}$ instead of h_i, then F becomes a "functional" of $h(\boldsymbol{r})$, that is, a function of a function. We write for this functional $F(\{h(\boldsymbol{r})\})$. Analogously, we generalize (3.1.17) by writing

$$G(\{M(\boldsymbol{r})\}) = F(\{h(\boldsymbol{r})\}) + \int \mathrm{d}\boldsymbol{r}\, M(\boldsymbol{r})h(\boldsymbol{r}) \tag{3.1.21}$$

and G also becomes a functional of $M(\boldsymbol{r})$. Here, we state implicitly that $h(\boldsymbol{r})$ on the right-hand side of (3.1.21) has been expressed by using $M(\boldsymbol{r})$ in

$$M(\boldsymbol{r}) = -\delta F(\{h(\boldsymbol{r})\})/\delta h(\boldsymbol{r}) \ . \tag{3.1.22}$$

Equation (3.1.21) is, of course, the generalization $G(\{M(\boldsymbol{r})\})$ of the Gibbs free energy, but in quantum field theory this function is called the "vertex function" (with reversed sign), being of central importance for the discussion of phase transitions and the renormalization group. By applying a functional derivative with respect to $M(\boldsymbol{r}')$ in (3.1.22), we obtain

$$\frac{\delta M(\boldsymbol{r})}{\partial M(\boldsymbol{r}')} = \delta(\boldsymbol{r}-\boldsymbol{r}') = -\frac{\delta^2 F(\{h(\boldsymbol{r})\})}{\delta h(\boldsymbol{r})\delta M(\boldsymbol{r}')}$$
$$= -\int \mathrm{d}\boldsymbol{r}'' \frac{\delta^2 F(\{h(\boldsymbol{r})\})}{\delta h(\boldsymbol{r})\delta h(\boldsymbol{r}'')} \cdot \frac{\delta h(\boldsymbol{r}'')}{\delta M(\boldsymbol{r}')}$$
$$= -\int \mathrm{d}\boldsymbol{r}'' \frac{\delta^2 F(\{h(\boldsymbol{r})\})}{\delta h(\boldsymbol{r})\delta h(\boldsymbol{r}'')} \cdot \frac{\delta^2 G(\{M(\boldsymbol{r})\})}{\delta M(\boldsymbol{r}'')\delta M(\boldsymbol{r}')} \ . \tag{3.1.23}$$

Setting $h(\boldsymbol{r}) = 0$, we obtain with (3.1.7)

$$\delta(\boldsymbol{r}-\boldsymbol{r}') = \int \mathrm{d}\boldsymbol{r}'' G_{\mathrm{c}}(\boldsymbol{r}-\boldsymbol{r}'')\Gamma^{(2)}(\boldsymbol{r}''-\boldsymbol{r}') \ . \tag{3.1.24}$$

Here,

$$G_{\mathrm{c}}(\boldsymbol{r}-\boldsymbol{r}') \equiv \langle I(\boldsymbol{r})I(\boldsymbol{r}')\rangle_{h=0} - \langle I(\boldsymbol{r})\rangle_{h=0}\langle I(\boldsymbol{r}')\rangle_{h=0} \tag{3.1.25}$$

is the Green function and

$$\Gamma^{(2)}(\boldsymbol{r}-\boldsymbol{r}') \equiv \beta \left.\frac{\delta^2 G(\{M(\boldsymbol{r})\})}{\delta M(\boldsymbol{r})\delta M(\boldsymbol{r}')}\right|_{h=0} \tag{3.1.26}$$

is called the vertex function. Consider a matrix G_{c}, $\Gamma^{(2)}$, with $G_{\mathrm{c}}(\boldsymbol{r}-\boldsymbol{r}')$ ($\Gamma^{(2)}(\boldsymbol{r}-\boldsymbol{r}')$) as its $\boldsymbol{r}, \boldsymbol{r}'$-component. Then we can express (3.1.24) with $\hat{1}$ being the unit matrix as

$$\hat{1} = G_{\mathrm{c}}\Gamma^{(2)} \ . \tag{3.1.27}$$

G_{c} is the inverse matrix of $\Gamma^{(2)}$. Taking the Fourier transformation on both sides of (3.1.24), owing to the theorem of convolution integrals we obtain the following algebraic relation:

$$G_{\mathrm{c}}(\boldsymbol{k})\Gamma^{(2)}(\boldsymbol{k}) = 1 \ . \tag{3.1.28}$$

with $\boldsymbol{k}$ being the wave vector.

Now, what will be the explicit form of $G_{\mathrm{c}}(\boldsymbol{k})$ and $\Gamma^{(2)}(\boldsymbol{k})$? For $\boldsymbol{k} = 0$, that is, for a uniform field, $G(M(\boldsymbol{r}))$ can be assumed to be given by $G(M)$ in (3.1.20). When a space dependence is present, $G(M(\boldsymbol{r}))$ can be expanded in terms of $\nabla M(\boldsymbol{r})$, and to lowest order we obtain

$$\Gamma^{(2)}(\boldsymbol{k}) = 2\beta(a + c|\boldsymbol{k}|^2) \ . \tag{3.1.29}$$

For $T > T_{\mathrm{c}}$, the Fourier transform of $G_{\mathrm{c}}(\boldsymbol{k}) = [\Gamma^{(2)}(\boldsymbol{k})]^{-1} = [2\beta(a+c\,|\boldsymbol{k}^2|)]^{-1}$ behaves like (3.1.8), and therefore we obtain

$$\xi = \sqrt{\frac{c}{a}} \ . \tag{3.1.30}$$

In the Landau theory, a is a function of temperature that passes through $T = T_{\mathrm{c}}$ without singularity, and therefore in the vicinity of T_{c} can be written as $a = a'(T - T_{\mathrm{c}})$. We conclude therefore from (3.1.30) that $\xi \propto |T - T_{\mathrm{c}}|^{-1/2}$.

Furthermore, because the magnetic susceptibility of the wave vector $\boldsymbol{k}$ is proportional to $G_c(\boldsymbol{k})$, we obtain

$$\chi(\boldsymbol{k}) \propto \frac{1}{|\boldsymbol{k}|^2 + \xi^{-2}} \ , \tag{3.1.31}$$

and therefore

$$\chi(\boldsymbol{k} = 0) \propto \xi^2 \propto |T - T_c|^{-1} \ , \tag{3.1.32}$$

$$\chi(\boldsymbol{k}, T = T_c) \propto |\boldsymbol{k}|^{-2} \ . \tag{3.1.33}$$

Equation (3.1.32) is the mathematical expression for "increasing the magnetic susceptibility due to the growth of domains", which was pointed out below (3.1.16).

Above, we gave an incomplete overview of the so-called classical theory of phase transitions. In the vicinity of T_c, the above theory is not exactly valid (for the case when the dimension is large enough, for example in $d = 10$, the theory is correct). In the classical theory it is assumed that "the functional $G(\{M(\boldsymbol{r})\})$ has no singularities and can be expanded in $M(\boldsymbol{r})$", and it is this point that becomes problematic. This is the starting point of Wilson's theory of the renormalization group. We will not go into detail, but refer to the literature [G.4], [G.5], [8].

Next, we will explain another point that is related to symmetry breaking, namely rigidity. As explained earlier, symmetry breaking means that an infinitesimal magnetic field h initiates a macroscopic magnetization M, and we considered an infinitesimal magnetic field being homogeneous in space. Now, we want to change the point of view and consider a quite strong magnetic field; however, the space region where it acts is extremely small, and can be ignored in the limit $N \to \infty$. As an explicit example, we consider a d-dimensional sample where at one surface (for example on the right-hand side) a strong magnetic field aligns the spins in its direction. The number of spins at one surface is $N^{(d-1)/d}$ and can be ignored for $N \to \infty$ compared with N. However, in the ordered phase, starting from this small magnetization, the whole sample will align in this direction, and also the spins at the left-hand side will align in the same direction as the spins at the right-hand side. When the spins at one side are reversed, this reversing will go through the sample to the other side. In this sense, the spin system in the ordered phase is quite similar to a rigid stick; we could say that the spins became "solid".

Furthermore, when the spins at the right-hand side and the left-hand side have a slightly different alignment (just like the twist of a stick), calling this angle θ, then for small θ the increase in free energy is proportional to $\Delta F = \kappa\theta^2$. This κ is called the coefficient of rigidity, and is a generalization of the coefficient of elasticity. The scenario described above can be summarized as: "owing to symmetry breaking, the systems gain rigidity". The Meissner effect in superconductivity can be understood in exactly the same manner, with κ replaced by the superfluidity density ρ_s. This will we explained in Sect. 5.1.

3.2 The Goldstone Mode

In this chapter we discuss the low energy excitations in the ordered phase. However, the Hamiltonian (3.1.1) does not contain any dynamics with respect to θ. The reason why (3.1.1) is sufficient for the description of the classical statistical mechanics can be explained as follows from the point of view of path integrals.

As was explained in Sect. 2.1, the partition function Z of a single-particle system can be written in the imaginary time formalism as

$$Z = \int \mathcal{D}p(\tau)\mathcal{D}x(\tau) \exp\left\{-\int_0^{\beta\hbar} \left[-\mathrm{i}p(\tau)\dot{x}(\tau) + \frac{p^2}{2m} + V(x)\right] \mathrm{d}\tau\right\} \quad (3.2.1)$$

$$= \int_{x(0)=x(\beta)} \mathcal{D}x(\tau) \exp\left\{-\int_0^{\beta\hbar} \left[\frac{m}{2}\dot{x}^2(\tau) + V(x)\right] \mathrm{d}\tau\right\} . \quad (3.2.2)$$

At high temperatures, when β is small, it is important to notice that there is almost no τ-dependence in $x(\tau)$. As shown in Fig. 3.6, when the margin in the direction of the imaginary time becomes "small", the freedom to oscillate decreases. Of course, the importance of the τ-dependence is determined by the influence of $V(x)$.

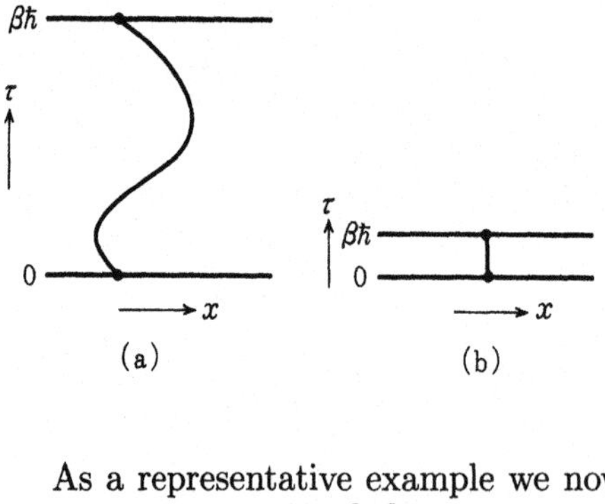

Fig. 3.6a, b. Characteristic path for the low-temperature case (**a**) and the high-temperature case (**b**)

As a representative example we now discuss the harmonic oscillator potential $V(x) = (m/2)\omega^2 x^2$. The larger ω is, the more the contribution of $x(\tau)$ is concentrated in the area around $x = 0$. Therefore, the detailed structure of the path also becomes important. Thus for small β the quantum effects remain important. Writing $x(\tau)$ in its Fourier components

$$x(\tau) = (\beta\hbar)^{-1/2} \sum_n \mathrm{e}^{-\mathrm{i}\omega_n \tau} x_n \qquad \left(\omega_n = \frac{2\pi}{\beta\hbar} n\right) \quad (3.2.3)$$

we obtain

$$Z = \int \prod_n \mathrm{d}x_n \exp\left[-\sum_n \frac{m}{2}\left(\omega_n^2 + \omega^2\right) |x_n|^2\right] . \quad (3.2.4)$$

In the sum of the above equation only the contribution $\omega_n = 0$ is important in the case when $|\omega_{\pm 1}| = 2\pi/\beta\hbar \gg \omega$, that is, for $k_B T \gg \hbar\omega/2\pi$. In this manner, the quantum effects arise in the path integral owing to the (imaginary) time dependence. Having this in mind, it is plausible that when describing a d-dimensional quantum field, adding one dimension for the time direction, there exists a corresponding description in terms of a $d+1$-dimensional classical field. For simplicity, we will set $k_B = \hbar = 1$ below.

So, what will the quantum dynamics be in our problem? Of course, the answer will be model dependent, and thinking in this paragraph of the Hamiltonian (3.1.1) as the description of the Josephson contact, θ_i becomes the Josephson phase of the ith microscopic superconductor. Then, using the relation between the phase and the particle number, as described in Chap. 1, we obtain the following relation to the number of Cooper pairs n_i:

$$n_i = \mathrm{i}\frac{\partial}{\partial \theta_i} \tag{3.2.5}$$

and we can write a generalization of (3.1.1) as

$$\begin{aligned} \mathcal{H} &= \sum_i \frac{n_i^2}{2C} - J\sum_{\langle ij\rangle} \cos(\theta_i - \theta_j) \\ &= -\sum_i \frac{1}{2C}\left(\frac{\partial}{\partial \theta_i}\right)^2 - J\sum_{\langle ij\rangle} \cos(\theta_i - \theta_j) \ . \end{aligned} \tag{3.2.6}$$

Here, C is the capacity of every microscopic superconductor, and J is the Josephson coupling energy.

Quantization of this Hamiltonian with path integral methods leads to the partition function

$$Z = \int \mathcal{D}\theta_i(\tau) \exp\left[-\int_0^\beta \mathrm{d}\tau \left\{\frac{C}{2}\sum_i (\dot{\theta}_i(\tau))^2 - J\sum_{\langle ij\rangle} \cos(\theta_i(\tau) - \theta_j(\tau))\right\}\right]. \tag{3.2.7}$$

First, we consider the perfectly ordered case $\theta_i = \langle\theta\rangle$ and expand around this point the action up to second order in the variation $\delta\theta_i$:

$$S_0 = \int_0^\beta \mathrm{d}\tau \left[\frac{C}{2}\sum_i (\delta\dot{\theta}_i(\tau))^2 + \frac{J}{2}\sum_{\langle ij\rangle} (\delta\theta_i(\tau) - \delta\theta_j(\tau))^2\right] . \tag{3.2.8}$$

Performing a Fourier transformation

$$\delta\theta_i(\tau) = \frac{1}{\sqrt{\beta}}\frac{1}{\sqrt{N}}\sum_n \sum_{\boldsymbol{k}} \mathrm{e}^{-\mathrm{i}\omega_n \tau}\, \mathrm{e}^{\mathrm{i}\boldsymbol{k}\cdot\boldsymbol{R}_i}\, \theta(\mathrm{i}\omega_n, \boldsymbol{k}) \tag{3.2.9}$$

the action becomes

$$S_0 = \sum_n \sum_k \left[\frac{C}{2}\omega_n^2 + J\sum_{l=1}^{d}(1-\cos k_l a) \right] \theta(\mathrm{i}\omega_n, \boldsymbol{k})\theta(-\mathrm{i}\omega_n, -\boldsymbol{k}) \ . \quad (3.2.10)$$

Here, a is the lattice spacing. Owing to the correspondence with the harmonic oscillator (3.2.4) for frequency ω, S_0 describes an ensemble of harmonic oscillators with mass C and frequency

$$\omega_{\boldsymbol{k}}^2 = \frac{2J}{C}\sum_{l=1}^{d}(1-\cos k_l a) \ .$$

In particular, for small $|\boldsymbol{k}|$, $\cos k_l a$ can be expanded, and as result

$$\omega_{\boldsymbol{k}}^2 \cong \frac{Ja^2}{C}|\boldsymbol{k}|^2 \quad (3.2.11)$$

and we obtain acoustic waves with velocity $(Ja^2/C)^{1/2}$.

In the same manner, it is generally known that in the phase where a continuous symmetry is broken (in this case the symmetry of θ rotation), in the limit $|\boldsymbol{k}| \to 0$, a mode with $\omega_{\boldsymbol{k}} \to 0$ exists. This theorem is called the Goldstone theorem, and the mode is called the Goldstone mode. Other examples are phonons in a solid body occurring due to broken translational invariance ($\omega_{\boldsymbol{k}} \simeq |\boldsymbol{k}|$), or ferromagnetic spin waves ($\omega_{\boldsymbol{k}} \propto |\boldsymbol{k}^2|$) and antiferromagnetic spin waves ($\omega_{\boldsymbol{k}} \propto |\boldsymbol{k}|$), occurring due to rotational symmetry breaking in the spin space.

Intuitively, Goldstone's theorem can be explained as follows. First, each direction of magnetization $\langle\theta\rangle$ is as good as every other, and even when a special direction is chosen due to symmetry breaking, there exist an infinite number of energetically degenerate states. Because $\boldsymbol{k} \to \boldsymbol{0}$ acts exactly in the same way on all θ_i, it can be seen as the excitation that tries to shift the system to one of the other possible ground states. Of course, the excitation energy for the transition into a state with the same energy is zero.

In this way, we can say that the Goldstone mode tries to "walk around" in the other possible states and at the same time tries to "restore the symmetry". At the absolute zero point, the zero point oscillation of the Goldstone mode, and at finite temperature, thermal excitations also are intending to break the order.

The zero point oscillation is not present when the order parameter O commutes with the Hamiltonian H. Both operators can be diagonalized, and therefore it is possible that the ground state is an eigenstate of both H and O, and in this case a finite value for the eigenvalue of O is possible. The spin waves of ferromagnetism are such an example; the total magnetization $\boldsymbol{M} = \sum_i \boldsymbol{I}_i$ and H do commute, and for the ground state, no problem occurs with a state where all spins are aligned in the same direction. On the other hand, for example in the above case, $M_x = \sum_i I\cos\theta_i$ and $M_y = \sum_i I\sin\theta_i$ do not commute with the Hamiltonian (3.2.6). Also in the case of Heisenberg antiferromagnetism, the alternating magnetization $\boldsymbol{M}_S = \sum_i (-1)^i \boldsymbol{I}_i$ does not commute with the Hamiltonian. In such cases, zero point oscillation becomes important, and in extreme cases, because of quantum fluctuations, the order can be destroyed, and a homogeneous quantum liquid may arise.

Now, let us estimate explicitly the fluctuation around the ordered state due to Goldstone modes for the Hamiltonian (3.2.6) or the action in (3.2.7). First, we calculate the decrease in the order parameter

$$\begin{aligned}\langle\cos\theta_i\rangle_0 &= \left\langle \frac{\mathrm{e}^{\mathrm{i}(\langle\theta\rangle+\delta\theta_i)} + \mathrm{e}^{-\mathrm{i}(\langle\theta\rangle+\delta\theta_i)}}{2}\right\rangle_0 \\ &= \frac{1}{2}\left[\mathrm{e}^{\mathrm{i}\langle\theta\rangle}\left\langle \mathrm{e}^{\mathrm{i}\delta\theta_i}\right\rangle_0 + \mathrm{e}^{-\mathrm{i}\langle\theta\rangle}\left\langle \mathrm{e}^{-\mathrm{i}\delta\theta_i}\right\rangle_0\right] , \end{aligned} \tag{3.2.12}$$

and similarly for $\langle\sin\theta_i\rangle_0$. Here, $\langle\ \rangle_0$ is given in terms of the action S_0 (3.2.10) by

$$\langle A\rangle_0 = \int \mathcal{D}\delta\theta\left[\mathrm{e}^{-S_0}A\right] \Big/ \int \mathcal{D}\delta\theta\, \mathrm{e}^{-S_0} \ . \tag{3.2.13}$$

Because, finally, this leads to averaging in a multiple Gauss distribution, we obtain

$$\left\langle \mathrm{e}^{\mathrm{i}\delta\theta_i}\right\rangle_0 = \left\langle \mathrm{e}^{-\mathrm{i}\delta\theta_i}\right\rangle_0 = \mathrm{e}^{(-1/2)\langle(\delta\theta_i)^2\rangle_0} , \tag{3.2.14}$$

and from (3.2.12)

$$\langle\cos\theta_i\rangle = \exp[(-1/2)\langle(\delta\theta_i)^2\rangle_0]\cos\langle\theta\rangle , \tag{3.2.15}$$

$$\langle\sin\theta_i\rangle = \exp[(-1/2)\langle(\delta\theta_i)^2\rangle_0]\sin\langle\theta\rangle . \tag{3.2.16}$$

With (3.2.9) and (3.2.10), we obtain

$$\begin{aligned}\left\langle(\delta\theta_i)^2\right\rangle_0 &= \frac{1}{\beta}\frac{1}{N}\sum_{\mathrm{i}\omega_n}\sum_{\boldsymbol{k}}\langle\theta(\mathrm{i}\omega_n,\boldsymbol{k})\theta(-\mathrm{i}\omega_n,-\boldsymbol{k})\rangle_0 \\ &= \frac{1}{\beta}\sum_{\mathrm{i}\omega_n}\int\frac{\mathrm{d}^2\boldsymbol{k}}{(2\pi)^d}\frac{1}{C(\omega_n^2+\omega_{\boldsymbol{k}}^2)} . \end{aligned} \tag{3.2.17}$$

Here, we make a somewhat tedious remark. Because, originally, $\delta\theta_i(\tau)$ was real, $\theta(\mathrm{i}\omega_n,\boldsymbol{k})$ and $\theta(-\mathrm{i}\omega_n,-\boldsymbol{k})$ are not independent. Writing $\theta'(\mathrm{i}\omega_n,\boldsymbol{k})$ and $\theta''(\mathrm{i}\omega_n,\boldsymbol{k})$ for the real and complex part of $\theta(\mathrm{i}\omega_n,\boldsymbol{k})$, respectively, we obtain the relations

$$\begin{aligned}\theta'(\mathrm{i}\omega_n,\boldsymbol{k}) &= \theta'(-\mathrm{i}\omega_n,-\boldsymbol{k}) , \\ \theta''(\mathrm{i}\omega_n,\boldsymbol{k}) &= -\theta''(-\mathrm{i}\omega_n,-\boldsymbol{k}) , \end{aligned} \tag{3.2.18}$$

and (3.2.10) becomes

$$S_0 = \sum_{\mathrm{i}\omega_n}\sum_{\boldsymbol{k}} \frac{C}{2}\left(\omega_n^2+\omega_{\boldsymbol{k}}^2\right)\left\{\left(\theta'(\mathrm{i}\omega_n,\boldsymbol{k})\right)^2+\left(\theta''(\mathrm{i}\omega_n,\boldsymbol{k})\right)^2\right\}$$
$$= \sum_{\mathrm{i}\omega_n}\sum_{\boldsymbol{k},k_x>0} C\left(\omega_n^2+\omega_{\boldsymbol{k}}^2\right)\left\{\left(\theta'(\mathrm{i}\omega_n,\boldsymbol{k})\right)^2+\left(\theta''(\mathrm{i}\omega_n,\boldsymbol{k})\right)^2\right\} \; . \quad (3.2.19)$$

Here, by restricting the summation to $k_x > 0$, $\theta'(\mathrm{i}\omega_n, \boldsymbol{k})$ and $\theta''(\mathrm{i}\omega_n, \boldsymbol{k})$ on the right-hand side of (3.2.19) are independent and real integrands. Therefore, writing (3.2.17) in a little more detail, we obtain

$$\left\langle(\delta\theta_i)^2\right\rangle_0 = \frac{2}{\beta}\frac{1}{N}\sum_{\mathrm{i}\omega_n}\sum_{\boldsymbol{k},k_x>0}\left\langle(\theta'(\mathrm{i}\omega_n,\boldsymbol{k}))^2+(\theta''(\mathrm{i}\omega_n,\boldsymbol{k}))^2\right\rangle_0$$
$$= \frac{2}{\beta}\frac{1}{N}\sum_{\mathrm{i}\omega_n}\sum_{\boldsymbol{k},k_x>0}\frac{1}{2C(\omega_n^2+\omega_{\boldsymbol{k}}^2)}\times 2$$
$$= \frac{1}{\beta}\frac{1}{N}\sum_{\mathrm{i}\omega_n}\sum_{\boldsymbol{k},k_x>0}\frac{2}{C(\omega_n^2+\omega_{\boldsymbol{k}}^2)}$$
$$= \frac{1}{\beta}\frac{1}{N}\sum_{\mathrm{i}\omega_n}\sum_{\boldsymbol{k}}\frac{1}{C(\omega_n^2+\omega_{\boldsymbol{k}}^2)}$$
$$= \frac{a^d}{\beta}\sum_{\mathrm{i}\omega_n}\int\frac{\mathrm{d}^d\boldsymbol{k}}{(2\pi)^d}\frac{1}{C(\omega_n^2+\omega_{\boldsymbol{k}}^2)} \; . \quad (3.2.20)$$

Notice that $\langle x^2\rangle = 1/2\alpha$ holds under the $\mathrm{e}^{-\alpha x^2}$ distribution. Furthermore, in the limit $N \to \infty$ we assumed that

$$\frac{1}{N}\sum_{\boldsymbol{k}} \xrightarrow[N\to\infty]{} \int\frac{\mathrm{d}^d\boldsymbol{k}}{(2\pi)^d}$$

holds.

Next, we have to perform the summation of the $\mathrm{i}\omega_n$ in (3.2.17). Here, the ω_n are discrete frequencies with distance $2\pi T$ to each other, given by (3.2.3). This is due to the fact that in the direction of the imaginary time, the sample is bounded by $1/T = \beta$ for the case of finite temperature, and only in the limit of zero temperature does the sum in $\mathrm{i}\omega_n$ transform into an integral.

Fortunately, for the case of finite temperature, the exists a trick to rewrite the sum in an equivalent integral. The trick is to use the residual theorem of complex analysis. With $g(z)$ being a complex function and C_0 a closed path enclosing the complex plane anti-clockwise, the following equation holds:

$$\oint_{C_0} g(z)\,\mathrm{d}z = 2\pi\mathrm{i}\sum_i \mathrm{Res}\, g(z_i) \; . \quad (3.2.21)$$

Here, $\mathrm{Res}\, g(z_i)$ is the so-called residue at the pole z_i. For the case when the pole at z_i is of order n [that is, in the vicinity of z_i the function $g(z)$ behaves like $g(z) \propto 1/(z-z_i)^n$], it is given by

$$\operatorname{Res} g(z_i) = \lim_{z \to z_i} \frac{1}{(n-1)!} \left(\frac{\mathrm{d}}{\mathrm{d}z} \right)^{n-1} [(z - z_i)^n g(z)] \ . \tag{3.2.22}$$

The sum over the poles z_i contains only poles inside the closed path C_0.

The idea is to regard $\sum_{\mathrm{i}\omega_n}$ as the sum over poles, and to write down on the left-hand side an appropriate complex function. In order to do so, we still have to construct a complex function with poles at $z_n = \mathrm{i}\omega_n = \mathrm{i}(2\pi/\beta)n$. The function $(\mathrm{e}^{\beta z} - 1)^{-1}$ has poles of order one in z_n, and the residual is independent of n and given by β^{-1}. Using this fact, we can write the relation

$$\frac{1}{\beta} \sum_{\mathrm{i}\omega_n} f(\mathrm{i}\omega_n) = \frac{1}{2\pi\mathrm{i}} \oint_{C_0} \frac{f(z)}{\mathrm{e}^{\beta z} - 1} \mathrm{d}z \ . \tag{3.2.23}$$

However, in the above equation, attention must be paid to several points. First, $f(z)$ must be finite at $z = z_n = \mathrm{i}\omega_n$. Then, $(\mathrm{e}^{\beta z} - 1)^{-1}$ has a first-order pole at z_n, and using (3.2.22), we obtain

$$\begin{aligned} \operatorname{Res} \left[\frac{f(z)}{\mathrm{e}^{\beta z} - 1} \right]_{z = z_n} &= \lim_{z \to z_n} \left[\frac{z - z_n}{\mathrm{e}^{\beta z} - 1} f(z) \right] \\ &= \lim_{z \to z_n} \frac{z - z_n}{\beta (z - z_n)} f(z) = \frac{1}{\beta} f(z_n) \ . \end{aligned} \tag{3.2.24}$$

Next, the path C_0 must contain all z_n, but no pole of the function $f(z)$. Corresponding to (3.2.17), we choose

$$f(z) = \frac{1}{C(-z^2 + \omega_{\boldsymbol{k}}^2)} \ . \tag{3.2.25}$$

However, at $z = \pm\omega_{\boldsymbol{k}}$ there are poles of first order. Choosing the path C_0 as shown in the Fig. 3.7, this condition is fulfilled. Next, we use Cauchy's integral theorem and change the integration path from C_0 to C_1. Now, at $z = \pm\omega_{\boldsymbol{k}}$ the function $(\mathrm{e}^{\beta z} - 1)^{-1}$ is regular; however, $f(z)$ has first-order poles, and finally, we obtain

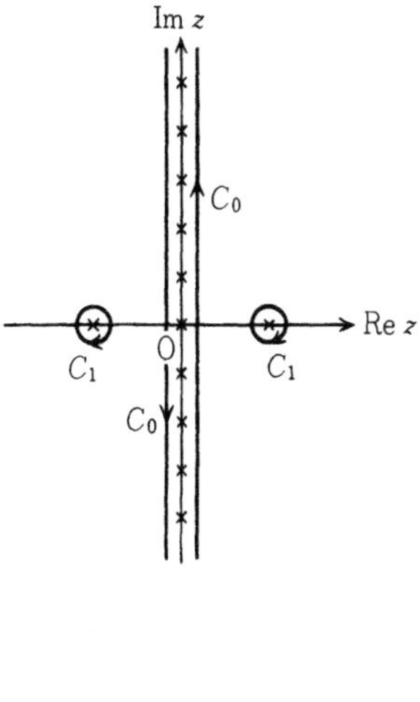

Fig. 3.7. Two different paths in the complex plane for the integration of (3.2.23)

$$\frac{1}{\beta}\sum_{\mathrm{i}\omega_n}\frac{1}{C(\omega_n^2+\omega_{\boldsymbol{k}}^2)}=\frac{1}{2C\omega_{\boldsymbol{k}}}\left[\frac{1}{\mathrm{e}^{\beta\omega_{\boldsymbol{k}}}-1}-\frac{1}{\mathrm{e}^{-\beta\omega_{\boldsymbol{k}}}-1}\right]$$
$$=\frac{1}{C\omega_{\boldsymbol{k}}}\left[n_{\mathrm{B}}(\omega_{\boldsymbol{k}})+\frac{1}{2}\right] . \tag{3.2.26}$$

$n_{\mathrm{B}}(E)$ is the Bose distribution $(\mathrm{e}^{\beta E}-1)^{-1}$, and for $E \ll T$, the approximation $n_{\mathrm{B}}(E) \sim T/E$ holds.

From the above considerations, we conclude

$$\langle(\delta\theta_i)^2\rangle=\int\frac{\mathrm{d}^d\boldsymbol{k}}{(2\pi)^d}\cdot\frac{a^d}{C\omega_{\boldsymbol{k}}}\left[n_{\mathrm{B}}(\omega_{\boldsymbol{k}})+\frac{1}{2}\right] . \tag{3.2.27}$$

There is the possibility that the contribution of the above integral for small $\boldsymbol{k}$ diverges. To examine this, we set

$$\int\mathrm{d}^d\boldsymbol{k}\to\int_0^{k_c}|\boldsymbol{k}|^{d-1}\mathrm{d}|\boldsymbol{k}| , \qquad \omega_{\boldsymbol{k}}\to|\boldsymbol{k}|$$
$$n_{\mathrm{B}}(\omega_{\boldsymbol{k}})\to T/|\boldsymbol{k}| .$$

where k_c is a cut-off of the order of a^{-1}.

Using this behaviour, we estimate from (3.2.27)

$$\langle(\delta\theta_i)^2\rangle\sim\int|\boldsymbol{k}|^{d-1}\mathrm{d}|\boldsymbol{k}|\frac{1}{|\boldsymbol{k}|}\frac{T}{|\boldsymbol{k}|} \quad \text{(finite temperature)} , \tag{3.2.28}$$
$$\sim\int|\boldsymbol{k}|^{d-1}\mathrm{d}|\boldsymbol{k}|\frac{1}{|\boldsymbol{k}|} \quad \text{(zero temperature)} . \tag{3.2.29}$$

We conclude that for $d \leq 2$ at finite temperature, and for $d \leq 1$ at zero temperature, the expressions diverge, respectively. Especially for $d = 2$ ($T > 0$) and $d = 1$ ($T = 0$), logarithmic divergence occurs. In this manner, when the dimension is lower than a critical value d_{l}, the order disappears. This d_{l} is called the lower critical dimension. The gentlemen who proved exactly the inexistence of order in a low-dimensional system gave their name to this theorem; it is called Mermin–Wagner theorem.

Now, we understood that the order parameter is zero for $d = 2$ ($T > 0$) or $d = 1$ ($T = 0$), so what about the correlation function in this case? Will it decrease exponentially as in (3.1.8) when no order is present? We will examine this issue next. In the same manner as in (3.2.12), we can write

$$C(\boldsymbol{R}_i-\boldsymbol{R}_j)=I^2\langle\cos(\theta_i-\theta_j)\rangle_0$$
$$=\frac{I^2}{2}\left[\left\langle\mathrm{e}^{\mathrm{i}(\delta\theta_i-\delta\theta_j)}\right\rangle_0+\left\langle\mathrm{e}^{-\mathrm{i}(\delta\theta_i-\delta\theta_j)}\right\rangle_0\right]$$
$$=I^2\,\mathrm{e}^{(-1/2)\langle(\delta\theta_i-\delta\theta_j)^2\rangle_0} , \tag{3.2.30}$$

and instead of (3.2.17) and (3.2.27), we obtain

$$\begin{aligned}
&\langle(\delta\theta_i - \delta\theta_j)^2\rangle \\
&= \frac{1}{\beta}\sum_n \frac{1}{N}\sum_{\boldsymbol{k}} 2\left[-1\cos(\boldsymbol{k}\cdot(\boldsymbol{R}_i - \boldsymbol{R}_j))\right]\langle\theta(\mathrm{i}\omega_n, \boldsymbol{k})\theta(-\mathrm{i}\omega_n, -\boldsymbol{k})\rangle_0 \\
&= \int \frac{d^d\boldsymbol{k}}{(2\pi)^d}\frac{2a^d}{C\omega_{\boldsymbol{k}}}\left[n_{\mathrm{B}}(\omega_{\boldsymbol{k}}) + \frac{1}{2}\right](1 - \cos[\boldsymbol{k}\cdot(\boldsymbol{R}_i - \boldsymbol{R}_j)]) \ . \qquad (3.2.31)
\end{aligned}$$

For large distances ($|\boldsymbol{R}_i - \boldsymbol{R}_j| \gg a$) the above integral is dominated by the contribution of the region $|\boldsymbol{k}| \ll a^{-1}$.

Different from (3.2.27), $1 - \cos[\boldsymbol{k}(\boldsymbol{R}_i - \boldsymbol{R}_j)]$ behaves like $|\boldsymbol{k}^2|$ for small $|\boldsymbol{k}|$, and therefore the $|\boldsymbol{k}|$-integral that was divergent without this factor now receives an infrared cut-off ($\propto |\boldsymbol{R}_i - \boldsymbol{R}_j|^{-1}$) due to this factor. We obtain the estimation

$$\langle(\delta\theta_i - \delta\theta_j)^2\rangle_0 \sim \int_{|\boldsymbol{R}_i - \boldsymbol{R}_j|^{-1}}^{k_c} |\boldsymbol{k}|^{d-1}\mathrm{d}|\boldsymbol{k}|\frac{T}{|\boldsymbol{k}|^2} \qquad (T > 0) \qquad (3.2.32)$$

$$\sim \int_{|\boldsymbol{R}_i - \boldsymbol{R}_j|^{-1}}^{k_c} |\boldsymbol{k}|^{d-1}\mathrm{d}|\boldsymbol{k}|\frac{1}{|\boldsymbol{k}|} \qquad (T = 0) \ . \qquad (3.2.33)$$

Even more explicitly, for $T > 0$ (3.2.32) and (3.2.33) become

$$\langle(\delta\theta_1 - \delta\theta_j)^2\rangle_0 \sim \int_{|\boldsymbol{R}_i - \boldsymbol{R}_j|^{-1}}^{k_c} \frac{T\mathrm{d}|\boldsymbol{k}|}{|\boldsymbol{k}|^2} \sim T|\boldsymbol{R}_i - \boldsymbol{R}_j| \qquad (d = 1) \qquad (3.2.34)$$

$$\sim \int_{|\boldsymbol{R}_i - \boldsymbol{R}_j|^{-1}}^{k_c} \frac{T\mathrm{d}|\boldsymbol{k}|}{|\boldsymbol{k}|} \sim T\ln|\boldsymbol{R}_i - \boldsymbol{R}_j| \qquad (d = 2) \quad (3.2.35)$$

and for $T = 0$

$$\langle(\delta\theta_i - \delta\theta_j)^2\rangle_0 \sim \int_{|\boldsymbol{R}_i - \boldsymbol{R}_j|^{-1}}^{k_c} \frac{\mathrm{d}|\boldsymbol{k}|}{|\boldsymbol{k}|} \sim \ln|\boldsymbol{R}_i - \boldsymbol{R}_j| \qquad (d = 1) \ . \qquad (3.2.36)$$

Inserting these results into (3.2.30), we obtain for the case $T > 0$ and $d = 1$ the behaviour $C(\boldsymbol{R}_i - \boldsymbol{R}_j) \propto \mathrm{e}^{-\gamma T|\boldsymbol{R}_i - \boldsymbol{R}_j|}$, for the case $d = 2$ we obtain $C(\boldsymbol{R}_i - \boldsymbol{R}_j) \propto |\boldsymbol{R}_i - \boldsymbol{R}_j|^{-\eta T}$ and for the case $T = 0$ and $d = 1$ we obtain $C(\boldsymbol{R}_i - \boldsymbol{R}_j) \propto |\boldsymbol{R}_i - \boldsymbol{R}_j|^{-\delta}$. Here γ, η and δ are some constants.

Therefore, at the lower critical dimension, the correlation function shows a power-law behaviour, and even if it decays, this decay is extremely slow compared with the exponential decay. The correlation function therefore reaches long distances. Indeed, in terms of ξ as defined in (3.1.8), we obtain correspondence to $\xi = \infty$. Comparing with the high-temperature expansion of the classical model discussed in the first part of the chapter, consider a $d = 2$ classical XY model at finite temperature. Although the order parameters $\langle\cos\theta_i\rangle$ and $\langle\sin\theta_i\rangle$ are always zero, when studying the correlation function, we obtain as result a power-law decay at low temperature and exponential decay at high temperature, suggesting that between both, a phase transition (of a new type, because the definition of on order parameter is difficult) occurs. This phase transition is called the Kosterlitz–Thouless transition, being an extremely important phase transition, where the crucial point are topological defects. Because this idea strongly underlies gauge theory on a lattice, we will discuss it in detail in the next sections.

3.3 Kosterlitz–Thouless Transition

We now understand the properties of symmetry breaking at low temperature and of the order parameter characterizing the phase transition discussed in Sect. 3.1. In the present and the following sections, we examine a phase transition where it is difficult to define such an order parameter.

We consider a classical XY model on a two-dimensional orthogonal lattice, described by the following Hamiltonian:

$$\mathcal{H} = -J \sum_{\langle ij \rangle} \cos(\theta_i - \theta_j) \ , \tag{3.3.1}$$

where $\langle ij \rangle$ indicates nearest-neighbour sites. We already demonstrated in Sects. 3.1 and 3.2 that the correlation length of the correlation function is characterized by a power behaviour at low temperature

$$\left\langle e^{i(\theta_i - \theta_j)} \right\rangle \sim |\boldsymbol{R}_i - \boldsymbol{R}_j|^{-T/2\pi J} \ , \tag{3.3.2}$$

and exponential decay at high temperature

$$\left\langle e^{i(\theta_i - \theta_j)} \right\rangle \sim e^{-|\boldsymbol{R}_i - \boldsymbol{R}_j|/\xi} \ . \tag{3.3.3}$$

For the derivation of (3.3.2), the assumption has been made that in

$$\begin{aligned} \cos(\theta_i - \theta_j) &\approx 1 - \frac{1}{2}(\theta_i - \theta_j)^2 \\ &\approx 1 - \frac{1}{2}\left[(\boldsymbol{R}_i - \boldsymbol{R}_j) \cdot \nabla\theta\left(\frac{\boldsymbol{R}_i + \boldsymbol{R}_j}{2}\right)\right]^2 \end{aligned} \tag{3.3.4}$$

the difference in θ between neighbouring sites is small compared with π (the spin wave approximation).

In this approximation, the Hamiltonian becomes quadratic and we always obtain (3.3.2), and no phase transition would occur. Therefore, we conclude that the high-temperature phase transition described by (3.3.3) occurs, because configurations start to be excited where spins at neighbouring sites do make a difference in θ of magnitude π, and can no longer be described by a continuous function $\theta(\boldsymbol{R})$. Indeed, for the derivation of (3.3.3) in Sect. 3.1 we used the high-temperature expansion where (i) the angle θ is defined on the lattice and (ii) the range of θ is limited to the interval from 0 to 2π (reflecting the 2π periodicity in θ).

So, what kind of distribution will it be that cannot be described by a continuous function $\theta(\boldsymbol{R})$? We have to handle three points to answer this question. First, when θ_i is expressed by $\theta(\boldsymbol{R})$ in the continuum limit, owing to (i), $\theta(\boldsymbol{R})$ may also have singularities $\boldsymbol{R} = \boldsymbol{R}_0$. That is, when the singularity is placed inside the plaquette, where originally no spin has been defined, we do not run into difficulties. Second, owing to (ii), $\theta(\boldsymbol{R})$ may also be multi-valued. Of course, this multi-valueness is limited at one point $\boldsymbol{R}$ to be an integer multiplied by 2π.

The last point is that in the continuum limit, besides the singularity at $\boldsymbol{R} = \boldsymbol{R}_0$, the function $\theta(\boldsymbol{R})$ is determined by the Hamiltonian

$$\mathcal{H} = \frac{J}{2} \int (\nabla\theta(\boldsymbol{R}))^2 d^2\boldsymbol{R} \tag{3.3.5}$$

as a solution of the variational equation $\delta\mathcal{H} = 0$. The trivial solution which for all sides i, $\theta_i = \theta_0$, obeys $\delta\mathcal{H} = 0$. However, we are looking for different, non-homogeneous solutions. It is evident that $\delta\mathcal{H}$ is given explicitly by the Laplace equation

$$\nabla^2\theta(\boldsymbol{R}) = 0 \ . \tag{3.3.6}$$

In two dimensions, the Laplace equation can be derived by the Cauchy–Riemann equation. Recall that both the real part and the imaginary part of a regular function of $z = x + \mathrm{i}y$ ($\boldsymbol{R} = (x, y)$) obey (3.3.6).

We conclude that the solution can be given by

$$\theta(\boldsymbol{R}) = \theta(z) = \pm \operatorname{Im} \ln(z - z_0) \ . \tag{3.3.7}$$

Here, $z_0 = x_0 + \mathrm{i}y_0$ is the position of the singularity, and by moving once anticlockwise around the singularity, the phase of $z - z_0$ gains 2π, and $\theta(\boldsymbol{R})$ just changes by $\pm 2\pi$. In such a way, a whirl emerges around the point $z = z_0$, being the so-called vortex.

Next, we derive the excitation energy corresponding to the solution (3.3.7). It is sufficient to insert (3.3.7) into (3.3.5), with the result

$$\begin{aligned} E_{\text{vortex}} &= \frac{J}{2} \int_a^{R_\mathrm{c}} 2\pi R \, \mathrm{d}R \frac{1}{R^2} \\ &= \pi J \ln \frac{R_\mathrm{c}}{a} \ . \end{aligned} \tag{3.3.8}$$

Here, a is the smallest size where the continuum limit (3.3.5) is valid, that is, the lattice spacing. R_c is the size of the sample. Then, the vortex energy is logarithmically diverging with respect to the sample size!

Therefore, can we forget about the configuration (3.3.5) in the limit $R_\mathrm{c} \to \infty$? In this context, recall the discussion that followed equation (3.1.5). That is, not the energy E, but the free energy $F = E - TS$ has to be considered. Therefore, we have to calculate the entropy S_{vortex} of one vortex. With W being the number of all possible microstates, the entropy is given by $S =$

$\ln W$. From the number $W \sim R_c^2/a^2$ of possibilities of placing the centre of the vortex, we again obtain a logarithmic dependence:

$$S_{\text{vortex}} = \ln\left(\frac{R_c}{a}\right)^2 = 2\ln\frac{R_c}{a} \ . \tag{3.3.9}$$

Using (3.3.8) and (3.3.9) we obtain for the free energy of the vortex

$$\begin{aligned} F_{\text{vortex}} &= E_{\text{vortex}} - TS_{\text{vortex}} \\ &= (\pi J - 2T)\ln\frac{R_c}{a} \ . \end{aligned} \tag{3.3.10}$$

From the sign of the coefficient of $\ln(R_c/a)$, we conclude that at $T_c = \pi J/2$, a phase transition occurs between the phases where a vortex does or does not occur due to thermal excitation. This transition is the so-called Kosterlitz–Thouless transition (KT transition), the unique phase transition where a vortex, i.e. a topological defect, plays the main role.

However, the above discussion is incomplete because only one vortex has been considered, and also the discussion whether thermal excitation arises or not is incomplete. In order to clarify this point, we have to consider a system with many vortices, and so for a more detailed mathematical investigation, we will introduce the so-called duality mapping.

We return to (3.3.1) and discuss again the partition function

$$Z = \int \mathrm{d}\theta_1 \cdots \int \mathrm{d}\theta_N \exp\left[\beta J \sum_{\langle ij\rangle} \cos(\theta_i - \theta_j)\right] \ . \tag{3.3.11}$$

Consider one pair $\langle ij\rangle$ corresponding to one link. $\exp[\beta J\cos(\theta_i - \theta_j)]$ is 2π-periodic in $\theta_i - \theta_j$, and for every $\theta_i - \theta_j = 2\pi m + \varepsilon$ ($\varepsilon \ll \pi$), the integral becomes Gaussian, $\mathrm{e}^{\beta J}\mathrm{e}^{-(\beta J/2)\varepsilon^2}$.

We replace $\exp[\beta J\cos(\theta_i - \theta_j)]$ by a function that fulfils these two properties and that can be handled more easily:

$$\mathrm{e}^{\beta J\cos(\theta_i-\theta_j)} \to \sum_{m=-\infty}^{\infty} \mathrm{e}^{\beta J} \exp\left[-\left(\frac{\beta J}{2}\right)(\theta_i - \theta_j - 2\pi m)^2\right] \ . \tag{3.3.12}$$

Near every minimum $2\pi m$, the right-hand side of (3.3.12) equals the above approximation. Furthermore, using Poisson's equation

$$\sum_{m=-\infty}^{\infty} h(m) = \sum_{l=-\infty}^{\infty} \int_{-\infty}^{\infty} \mathrm{d}\phi\, h(\phi)\, \mathrm{e}^{2\pi i l\phi} \ , \tag{3.3.13}$$

(3.3.12) becomes

$$
\begin{aligned}
&\mathrm{e}^{\beta J\cos(\theta_i-\theta_j)}\\
&\quad\to \sum_{l_{ij}=-\infty}^{\infty}\int \mathrm{d}\phi\,\mathrm{e}^{\beta J}\exp\left[-\left(\frac{\beta J}{2}\right)(\theta_i-\theta_j-2\pi\phi)^2+2\pi\mathrm{i}l_{ij}\phi\right]\\
&\quad=\frac{1}{\sqrt{2\pi\beta J}}\sum_{l_{ij}=-\infty}^{\infty}\mathrm{e}^{\beta J}\exp[\mathrm{i}l_{ij}(\theta_i-\theta_j)]\exp[-l_{ij}^2/2\beta J]\ . \qquad (3.3.12')
\end{aligned}
$$

Inserting (3.3.12′) into (3.3.11), up to some constant factor, we obtain

$$
Z=\int \mathrm{d}\theta_1\cdots\int \mathrm{d}\theta_N\sum_{\{l_{ij}\}}\exp\left(-\sum_{\langle ij\rangle}\left[\frac{l_{ij}^2}{2\beta J}-\mathrm{i}l_{ij}\cdot(\theta_i-\theta_j)\right]\right)\ . \qquad (3.3.14)
$$

Here, l_{ij} is defined on every link, and we interpret it as a vector field $l_\mu(r)$ ($\mu=x,y$) that is directed from the starting point r, the left-hand side or the lower side of the link between i and j, to the other side of the link. Then, the argument of the exponent in (3.3.14) becomes

$$
-\sum_{r,\mu}\left[\frac{l_\mu(r)^2}{2\beta J}-\mathrm{i}l_\mu(r)\cdot(\theta(r)-\theta(r+\mu))\right]\ . \qquad (3.3.15)
$$

Here, r runs over all lattice points, and μ is the sum in x and y. We can rewrite the second term in (3.3.15) as

$$
-\mathrm{i}\sum_{r,\mu}l_\mu(r)\cdot(\theta(r)-\theta(r+\mu))=-\mathrm{i}\sum_{r,\mu}(l_\mu(r)-l_\mu(r-\mu))\theta(r) \qquad (3.3.16)
$$

and can therefore perform the $\theta(r)$-integration from 0 to 2π. In addition to numerical factors, we obtain from (3.3.14)

$$
Z=\sum_{\{l_\mu(r)\}}\exp\left(-\sum_{r,\mu}\frac{l_\mu(r)^2}{2\beta J}\right)\prod_r\delta_{\sum_\mu(l_\mu(r)-l_\mu(r-\mu)),0}\ . \qquad (3.3.17)
$$

In (3.3.17), the constraint given by the delta function

$$
\sum_\mu(l_\mu(r)-l_\mu(r-\mu))=0 \qquad (3.3.18)
$$

is the discrete version of $\operatorname{div}\boldsymbol{l}=0$ of the vector $\boldsymbol{l}(r)=(l_x(r),l_y(r))$.

Then, by analogy, a vector field $\boldsymbol{n}(r)$ satisfying $\boldsymbol{l}(r)=\operatorname{rot}\boldsymbol{n}(r)$ should exist, and because $\boldsymbol{l}(r)$ is two-dimensional in this case, using only the z-component of $\boldsymbol{n}$, we write $l_x(r)=\partial_y n(r)$, $l_y(r)=-\partial_x n(r)$. Indeed, the discrete version of these equations

$$
\begin{aligned}
l_x(r)&=n(r)-n(r-y)\ ,\\
l_y(r)&=-n(r)+n(r-x)\ ,
\end{aligned} \qquad (3.3.19)
$$

inserted into the left-hand side of (3.3.18) leads to

$$\begin{aligned} &l_x(r) - l_x(r-x) + l_y(r) - l_y(r-y) \\ &\quad = [n(r) - n(r-y)] - [n(r-x) - n(r-x-y)] \\ &\qquad + [-n(r) + n(r-x)] - [-n(r-y) + n(r-x-y)] \\ &\quad = 0 \ . \end{aligned} \tag{3.3.20}$$

Next, we consider the number of degrees of freedom. With N being the number of lattice points, the number of different $l_\mu(r)$ is $2N$. The number of conditions (3.3.18) equals the number of points r, that is, N, and therefore only $2N - N = N$ vectors are independent. Indeed, because the number of $n(r)$ is also given by N, the result is consistent. Inserting (3.3.19) into (3.3.17), we obtain

$$Z = \sum_{\{n(r)\}} \exp\left(-\sum_{r,\mu} \frac{1}{2\beta J} (n(r) - n(r-\mu))^2 \right) \ . \tag{3.3.21}$$

Interpreting $n(r)$ as the height of the atom layer at position r, then (3.3.21) describes a model where with increasing height difference at neighbouring positions, the energy becomes larger. The model describes the roughening transition of the surface. Notice that in (3.3.21), β appears in the denominator, and therefore the high-temperature (low-temperature) phase of the original XY model corresponds to the low-temperature (high-temperature) phase of this model.

Let us again rewrite the sum (3.3.21) running over integers $n(r)$ using the Poisson equation (3.3.13):

$$Z = \int_{-\infty}^{\infty} \prod_r \mathrm{d}\phi(r) \sum_{m(r)=-\infty}^{\infty} \exp\left[-\frac{1}{2\beta J} \sum_{r,\mu} (\Delta_\mu \phi(r))^2 + 2\pi \mathrm{i} \sum_r m(r)\phi(r) \right] \ . \tag{3.3.22}$$

Here, we defined $\Delta_\mu \phi(r) = \phi(r) - \phi(r-\mu)$. The integral (3.3.22) can be performed when $\phi(r)$ is Fourier transformed, with the result

$$Z = Z_{\mathrm{SW}} \sum_{m(r)=-\infty}^{\infty} \exp\left\{ -2\pi^2 \beta J \sum_{r,r'} m(r) G(r-r') m(r') \right\} \ . \tag{3.3.23}$$

Here, Z_{SW} is the sum of states of the spin waves, and $m(r)$ indicates the presence of $m(r)$ vortices ($= 0, \pm 1, \pm 2, \ldots$) at position r. Furthermore, $G(r-r')$ is given by

$$G(r-r') = \int_{-\pi}^{\pi} \frac{\mathrm{d}k_x}{2\pi} \int_{-\pi}^{\pi} \frac{\mathrm{d}k_y}{2\pi} \frac{\mathrm{e}^{\mathrm{i}k(r-r')}}{(4 - 2\cos k_x - 2\cos k_y)} \ . \tag{3.3.24}$$

For large $|r - r'|$, it behaves like

$$G(r-r') \approx -\frac{1}{2\pi}\ln\left(\frac{|r-r'|}{a}\right) - \frac{1}{4} + G(0) \tag{3.3.25}$$

and $G(0)$ is estimated by the logarithmic dependence $\ln R_c$, where the lower range of $k = \sqrt{k_x + k_y}$ is given by π/R_c.

Splitting $G(r-r')$ into two parts,

$$G(r-r') = G(0) + G'(r-r') \ , \tag{3.3.26}$$

where by definition only the first term contains divergent terms, we can write (3.3.23) as

$$\begin{aligned} Z = Z_{\mathrm{SW}} \sum_{m(r)=-\infty}^{\infty} & \exp\left\{-2\pi^2\beta J G(0)\left[\sum_r m(r)\right]^2\right\} \\ & \times \exp\left\{ -2\pi^2\beta J \sum_{r,r'} m(r)G'(r-r')m(r')\right\} \end{aligned} \tag{3.3.27}$$

and conclude that there is only a contribution to the sum of states in the case when the term $[\sum_r m(r)]^2$ that is multiplied by $G(0)$ vanishes. As mentioned, the absolute value of $m(r)$ indicates the vortex number, and the sign its direction. Interpreting $m(r)$ as an electric charge at position r, and identifying the logarithmic potential (3.3.25) with the Coulomb potential in two dimensions, then $\sum_r m(r) = 0$ can be interpreted as the neutrality condition of the whole system. That is, (3.3.27) signifies that the XY model can be split into a degree of freedom of spin waves and a degree of freedom of vortices; the latter is equivalent to a two-dimensional Coulomb gas.

From this fact, we can deduce the following physical picture. At low temperature, even when vortices are excited, they must emerge as a $+/-$ pair, forming a dipole, but no free charge, and therefore the system is in the insulator phase (dielectric substance). The lowering of the Coulomb force due to the dielectric constant ε_0 does not affect it being a long-distance force, and therefore the fact that the charge must be bound as a plus or minus charge pair is a self-consistent description. However, when the temperature becomes higher, the number of charges becomes larger and larger, and therefore the screening effect gains importance. Therefore, conversely, because of the existence of free charges, the Coulomb force becomes a short-range force due to screening, and therefore free charges can exist, and the metallic state emerges self-consistently.

The KT transition is the phase transition between the metallic state and the insulator state. We need to obtain a more quantitative picture:

$$\begin{aligned} &\sum_{r,r'} m(r)G'(r-r')m(r') \\ &\qquad = \sum_{r\neq r'} m(r)G'(r-r')m(r') \end{aligned}$$

$$\approx -\frac{1}{4}\sum_{r\neq r'} m(r)m(r') - \frac{1}{2\pi}\sum_{r\neq r'} m(r)\ln\left(\frac{|r-r'|}{a}\right) m(r')$$
$$= \frac{1}{4}\sum_{r} m(r)^2 - \frac{1}{2\pi}\sum_{r\neq r'} m(r)\ln\left(\frac{|r-r'|}{a}\right) m(r') \ . \qquad (3.3.28)$$

Here, we have used $\sum_r m(r) = 0$, $G'(0) = 0$ and the approximation (3.3.25) of $G'(r-r')$. Therefore, from (3.3.27) we obtain

$$Z = Z_{\text{SW}} \sum_{m(r)=-\infty}^{\infty} \exp\left[\ln y\cdot\sum_{r} m(r)^2 - \pi\beta J\sum_{r\neq r'} m(r)\ln\left(\frac{|r-r'|}{a}\right) m(r')\right] . \qquad (3.3.29)$$

Here, y is the so-called fugacity, given by $e^{\beta\mu}$, with μ being the chemical potential. In the present case, μ is given by $\mu = -\pi^2 J/2$.

When y is small enough, the absolute value of $m(r)$ cannot become very large. We now discuss the properties of the system in this dilute state limit. In order to do so, we return to the step (3.3.22) and add by hand the term $\ln(y)\sum_r m(r)^2$:

$$Z = \int \prod_r \mathrm{d}\phi(r) \sum_{m(r)=-\infty}^{\infty} \exp\left[-\frac{1}{2\beta J}\sum_{r,\mu}(\Delta_\mu\phi(r))^2 + \ln y\cdot\sum_r m(r)^2 + 2\pi\mathrm{i}\sum_r m(r)\phi(r)\right] . \qquad (3.3.30)$$

Here, because y is small, we only sum over $0, \pm 1$ in the sum of $m(r)$. We obtain approximately

$$\sum_{m(r)=0,\pm1} \exp\left[\ln y\cdot m(r)^2 + 2\pi\mathrm{i}m(r)\phi(r)\right] = 1 + 2y\cos(2\pi\phi(r))$$
$$\cong \exp\left[2y\cos(2\pi\phi(r))\right] \ . \qquad (3.3.31)$$

Then, (3.3.30) becomes

$$Z \cong \int \prod_r \mathrm{d}\phi(r)\exp\left[-\frac{1}{2\beta J}\sum_{r,\mu}(\Delta_\mu\phi(r))^2 + 2y\sum_r \cos(2\pi\phi(r))\right] . \qquad (3.3.32)$$

Performing the continuum limit at this stage, we obtain the so-called Sine–Gordon model, given by

$$Z \cong \int \mathcal{D}\phi(\boldsymbol{r})\exp\left[-\int \mathrm{d}^2\boldsymbol{r}\left(\frac{1}{2\beta J}|\nabla\phi(\boldsymbol{r})|^2 - 2y\cos(2\pi\phi(\boldsymbol{r}))\right)\right]$$
$$= \int \mathcal{D}\phi(\boldsymbol{r})\exp\left[-\int \mathrm{d}^2\boldsymbol{r}\left(\frac{1}{2}|\nabla\phi(\boldsymbol{r})|^2 - 2y\cos(2\pi\sqrt{\beta J}\,\phi(\boldsymbol{r}))\right)\right] . \qquad (3.3.33)$$

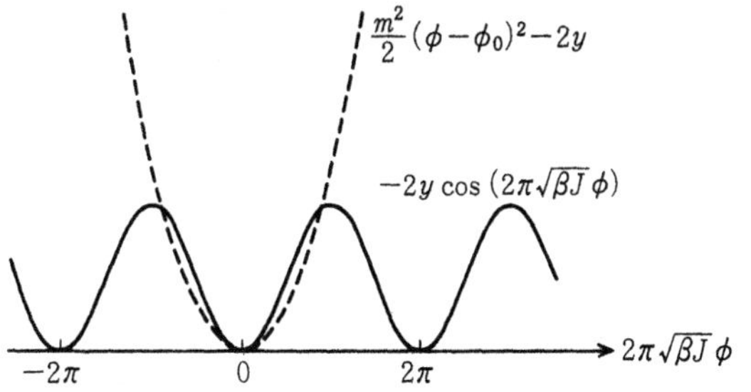

Fig. 3.8. The potential of the phase ϕ

Here, we redefined the integration variable $\phi(\boldsymbol{r})$ in a suitable manner. With $\phi(\boldsymbol{r}) = \phi_0$ (constant) being the minimum of the action, we obtain an infinite number of possibilities $2\pi\sqrt{\beta J}\phi_0 = 2\pi m$ (m constant), as indicated in Fig. 3.8. The question is: Will $\phi(\boldsymbol{r})$ rest in the vicinity of one valley or not?

We apply the variation method discussed in (2.2.20) and (2.2.21) to discuss this question. We choose as the trial action the action S_0 with the dotted line in Fig. 3.8 as the potential, obeying the quadratic equation

$$S_0 = \int \mathrm{d}^2\boldsymbol{r} \left\{ \frac{1}{2}|\nabla\phi(\boldsymbol{r})|^2 + \frac{m^2}{2}(\phi(\boldsymbol{r}) - \phi_0)^2 \right\} . \tag{3.3.34}$$

Using the curvature m^2 of the potential as the variation parameter, it can be determined whether $\phi(\boldsymbol{r})$ is bounded in the vicinity of ϕ_0 or not, depending on whether m is zero or finite. Let us compute $f(m) = -T\ln Z_0 + T\langle S - S_0\rangle$, the right-hand side of (2.2.21). In order to do so, we perform a Fourier transformation of S_0 in (3.3.34). With $\phi_{\boldsymbol{k}}$ being the Fourier component of $\phi(\boldsymbol{r}) - \phi_0$, we can write

$$\begin{aligned} S_0 &= \sum_{\boldsymbol{k}} \frac{1}{2}(\boldsymbol{k}^2 + m^2)\phi_{\boldsymbol{k}}\phi_{-\boldsymbol{k}} \\ &= \sum_{\boldsymbol{k}} \frac{1}{2}(\boldsymbol{k}^2 + m^2)\left\{(\mathrm{Re}\,\phi_{\boldsymbol{k}})^2 + (\mathrm{Im}\,\phi_{\boldsymbol{k}})^2\right\} \\ &= \sum_{\boldsymbol{k}:\mathrm{half}} (\boldsymbol{k}^2 + m^2)\left\{(\mathrm{Re}\,\phi_{\boldsymbol{k}})^2 + (\mathrm{Im}\,\phi_{\boldsymbol{k}})^2\right\} . \end{aligned} \tag{3.3.35}$$

Notice that because $\phi(\boldsymbol{r})$ is real, the relation $\phi_{\boldsymbol{k}}^* = \phi_{-\boldsymbol{k}}$ holds, and that when the degrees of freedom are assigned both to the real and imaginary parts for every $\boldsymbol{k}$, then it is sufficient to sum only over one half of the $\boldsymbol{k}$'s.

With this action S_0 in the exponential, performing the Gauss integral with the integration measure

$$\prod_{\boldsymbol{k}:\mathrm{half}} \int \mathrm{d}(\mathrm{Re}\,\phi_{\boldsymbol{k}})\,\mathrm{d}(\mathrm{Im}\,\phi_{\boldsymbol{k}})$$

gives the right expression for $\langle\ \rangle_0$, leading to

$$\begin{aligned}\langle\phi_{\boldsymbol{k}}\phi_{-\boldsymbol{k}}\rangle_0 &= \left\langle(\mathrm{Re}\,\phi_{\boldsymbol{k}})^2\right\rangle_0 + \left\langle(\mathrm{Im}\,\phi_{\boldsymbol{k}})^2\right\rangle_0 \\ &= \frac{1}{2(\boldsymbol{k}^2+m^2)} + \frac{1}{2(\boldsymbol{k}^2+m^2)} \\ &= \frac{1}{\boldsymbol{k}^2+m^2} \ . \end{aligned} \tag{3.3.36}$$

Using the following equation derived from the Gauss integral

$$\begin{aligned}\langle\exp[\mathrm{i}a(\phi(\boldsymbol{r})-\phi_0)]\rangle_0 &= \exp\left[-\frac{1}{2}a^2\left\langle(\phi(\boldsymbol{r})-\phi_0)^2\right\rangle_0\right] \\ &= \exp\left[-\frac{1}{2}a^2\sum_{\boldsymbol{k}}\langle\phi_{\boldsymbol{k}}\phi_{-\boldsymbol{k}}\rangle_0\right] \\ &= \exp\left[-\frac{1}{2}a^2\sum_{\boldsymbol{k}}\frac{1}{\boldsymbol{k}^2+m^2}\right]\end{aligned} \tag{3.3.37}$$

we can compute every term in $f(m)$, and the result is (when performing the k integration, a cut-off k_c of magnitude of the inverse lattice constant is introduced, and terms with higher powers in m^2/k_c^2 as well as terms independent of m^2 are ignored)

$$f(m) \cong T\left[m^2 - 2y\left(\frac{m^2}{k_c^2}\right)^{(\pi/2)\beta J}\right] \ . \tag{3.3.38}$$

Notice that the term $\sum_{\boldsymbol{k}} 1/(\boldsymbol{k}^2+m^2)$ in (3.3.37) causes the $\ln(k_c^2/m^2)$ dependence, leading finally to the particular exponent in the second term in (3.3.38).

The behaviour of $f(m)$ is determined by the relation between the powers of the first and second terms. That is, as shown in Fig. 3.9, for the case $(\pi/2)\beta J > 1$, the minimum is at $m^2 = 0$, whereas for $(\pi/2)\beta J < 1$, a minimum at a finite value of m^2 emerges. Therefore, the temperature obeying the equation $(\pi/2)\beta J = 1$ corresponds to the phase transition point $T = T_c = \pi J/2$. At temperatures lower than T_c, the cos term in (3.3.33) can effectively be ignored and the system can be described by the spin waves only (in the

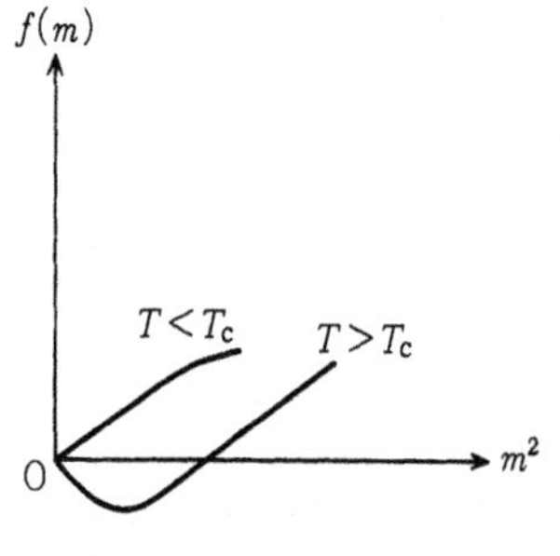

Fig. 3.9. The free energy $f(m)$ as obtained by the variation method. T_c is given by $T_c = \pi J/2$

picture of the Coulomb gas this corresponds to the insulator phase). On the other hand, at higher temperatures the potential arising due to the cos term leads to confining of phase ϕ, and as a result, the Coulomb force becomes a short-range force with the range $1/m$. This corresponds to the metallic phase, where the Coulomb force is screened.

In this manner, using a simple variational method, we have discussed whether the cos term effectively plays a role at low energy. This approach is quite similar to the discussion of relevant/irrelevant terms in the theory of renormalization. Indeed, an excellent paper [G.10] has been written about the application of Wilson's style renormalization group to the model (3.3.33), which is recommended to the reader.

We end this section by discussing briefly the three-dimensional XY model. In three dimensions, it is also possible to perform the duality transformation starting from the reformulation (3.3.12). However, $l_\mu(r)$ introduced below (3.3.14) is now a three-component vector $\boldsymbol{l}(r)$, and therefore the vector field $\boldsymbol{n}$ that describes $\boldsymbol{l}$ through $\boldsymbol{l} = \operatorname{rot} \boldsymbol{n}$ also has three components. Therefore, the integers vector $\boldsymbol{m} = m_\mu(r)$ appearing when rewriting the sum for every component in an integral using the Poisson equation also has three components. For details, the literature [19,20] is recommanded. As can be found there, the sum of states finally reads

$$Z = \prod_{r,\mu} \sum_{m_\mu(r)=-\infty}^{\infty} \prod_{r} \delta_{\Delta_\mu m_\mu(r),0} \exp\left[-4\pi^2 \beta J \sum_{r,r',\mu} m_\mu(r) v(r-r') m_\mu(r') \right] . \tag{3.3.39}$$

Here, $v(r-r')$ is the three-dimensional Coulomb potential. From the property of the delta function, it follows that at every site the number of ingoing and outgoing vector fields $\boldsymbol{m}(r)$ are equal (that is, there is no source present).

Therefore, when $m_\mu(r)$ at one link is given by 1, on both neighbouring sites it must be joined with links with $m_{\mu'} = \pm 1$. In this manner, a string of joined links grows, and this string cannot have a starting point; therefore the only possibility is the construction of a closed loop. Comparing the above consideration with the two-dimensional case, where topological defects (vortices) have been points, in the three-dimensional space, topological defect lines (vortex lines) arise that are creating loops themselves. Equation (3.3.39) describes the statistical mechanics of loops created by segments interacting via the Coulomb force. It is important to notice how the form of the topological defects changes depending on the dimension. In the next section, when the gauge field is discussed, we will meet a similar case.

However, it is a mistake to think that a far-reaching force acts between the loops just because the Coulomb interaction acts on the vortex segments. When m_μ in the direction from site i to site $i+\mu$ is given by $+1$, then in the direction from $i+\mu$ to i we have $m_\mu(i) = -1$. Therefore, from a wider point of view, the $+$ and $-$ contributions of m_μ almost cancel. Therefore, we expect an effective short-distance potential for $v(r-r')$.

Having this in mind, we choose one vortex line and try to sketch roughly the phase transition in the three-dimensional XY model. Consider the free energy F of a vortex line of length L. With (3.3.39), the energy is given by

$$E \cong 4\pi^2 J v(0) L \ . \tag{3.3.40}$$

Here, $v(0)$ is a representative finite short-range value of $v(r - r')$ (≈ 0.253). On the other hand, because the entropy S is given by the logarithm of the number of possibilities of constructing a string of length L, it can roughly be estimated as follows. We ignore the condition that the string must be closed, then because at every site the string has the possibility of proceeding further in five directions, which are all directions different from the one it came from, we conclude $W \propto 5^L$. Therefore we obtain

$$F \cong (4\pi^2(0.253)J - T\ln 5)L \tag{3.3.41}$$

and has critical temperature $T_\mathrm{c} = 4\pi^2(0.253)J/\ln 5$. Above this temperature, vortex rings with infinite radius exist, leading to a short-range spin correlation function; on the other hand, at low temperature, all vortex rings have a finite radius, having no influence at a scale above this radius. Above, we described the phase transition of the three-dimensional XY model from the point of view of condensation of topological defects.

3.4 Lattice Gauge Theory and the Confinement Problem

In solid state physics, the existence of a crystal lattice is fundamental. In the tight-binding approximation, at every lattice point i, an atomic wave function φ_i is defined, and the wave function of the electrons in the crystal is determined by linear superposition of the atomic wave functions. This approximation is useful because the originally infinitely many degrees of freedom in the continuous space around the lattice point i are represented only by φ_i, and when the number of lattice points is N, in this manner the problem is reduced to a system with N degrees of freedom.

For simplicity, we consider a two-dimensional square lattice as shown in Fig. 3.10. On the lattice points are orbits, and between neighbouring orbits there is the hopping integral t. The transition process of an electron from i to j is described by the Hamiltonian $-t_{ji}a_j^\dagger a_i$. $a_i^\dagger$ and a_i are the creation and annihilation operators of the orbit φ_i. We require that the theory is invariant when an arbitrary phase $\mathrm{e}^{\mathrm{i}\theta_i}$ is multiplied by φ_i at every point i. This corresponds to the transformation $a_i \to a_i\,\mathrm{e}^{-\mathrm{i}\theta_i}$, $a_i^\dagger \to a_i^\dagger\,\mathrm{e}^{\mathrm{i}\theta_i}$. The Hamiltonian is invariant if t_{ij} transforms as $t_{ij} \to t_{ij}\,\mathrm{e}^{\mathrm{i}(\theta_i - \theta_j)}$. For the case that t_{ij} is a simple number, no such transformation can emerge. It is necessary to think of t_{ij} as a field causing such a phase transformation. This field is nothing but the electromagnetic field.

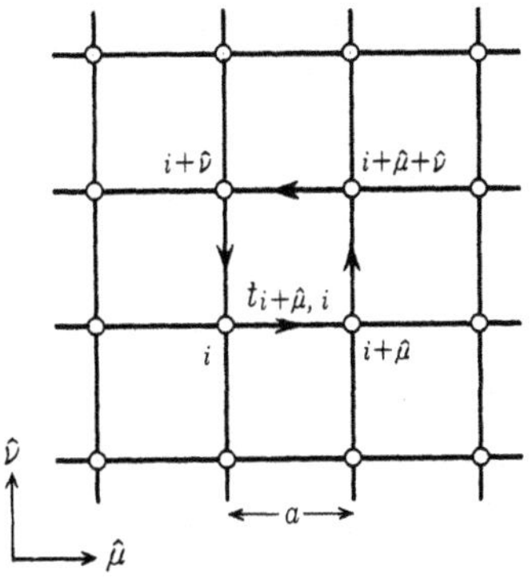

Fig. 3.10. Two-dimensional cubic lattice. It is important to distinguish the lattice points and the links between the lattice points from each other. The matter field is defined on the lattice points, and the gauge field is defined on the links

In order to see this, we discuss how is t_{ji} altered under the presence of the vector potential $\boldsymbol{A}(\boldsymbol{r})$. When the variation in space of $\boldsymbol{A}(\boldsymbol{r})$ is very slow compared with the lattice spacing a, then by replacing t_{ij} by the expression $t_{ij}\exp[\mathrm{i}e(\boldsymbol{r}_i - \boldsymbol{r}_j)\cdot\boldsymbol{A}((\boldsymbol{r}_i + \boldsymbol{r}_j)/2)]$, the effect of the vector potential $\boldsymbol{A}$ can be introduced. This phase factor is called the Peierls phase. We choose as a representative value of $\boldsymbol{A}$ the value at the midpoint between i and j. In this way, the $\boldsymbol{A}(\boldsymbol{r})$ originally defined on a continuous space can be considered as defined on the links between lattice points only.

Then, instead of $\boldsymbol{A}(\boldsymbol{r})$, it is convenient to discuss the transformation of the expression $U_{ji} = \exp[\mathrm{i}e(\boldsymbol{r}_i - \boldsymbol{r}_j)\boldsymbol{A}((\boldsymbol{r}_i + \boldsymbol{r}_j)/2)] \equiv \exp[\mathrm{i}eaA_{ji}] = \exp[-\mathrm{i}eaA_{ij}]$. Using this A_{ij}, the transformation $t_{ji} \to t_{ji}\,\mathrm{e}^{\mathrm{i}(\theta_i-\theta_j)}$ reads $eaA_{ji} \to eaA_{ji} + \theta_i - \theta_j$, and in terms of U_{ji}, we obtain $U_{ji} \to \mathrm{e}^{-\mathrm{i}\theta_j}U_{ji}\,\mathrm{e}^{\mathrm{i}\theta_i}$. This corresponds to $\boldsymbol{A} \to \boldsymbol{A} + (1/e)\nabla\theta$. Because U_{ji} is the crucial variable, being identical for A_{ij} and $A_{ij} + 2\pi/ea$, the system is periodic in A_{ij}. Therefore, when integrating over A_{ij}, it is sufficient to integrate in a finite interval, say between 0 and $2\pi/ea$. When elements are limited to a finite interval, the region is called compact. The continuous group with elements $\exp[-\mathrm{i}eaA_{ij}]$ is called $U(1)$.

Above, we discussed the space components $\boldsymbol{A}$ of the vector potential A_μ. Concerning the time component $A_{\mu=0}$, because the time t or the imaginary time τ are still continuous parameters, at first the above considerations cannot be applied. As an approximation, we introduce a lattice structure on the (imaginary) time axis and consider a $(d+1)$-dimensional lattice, where A_{ij} is defined on the link.

Above, the interaction of the gauge field with the matter field has been discussed; however, the gauge field itself also has dynamics. The dynamics was given by (2.4.1), and now we discuss how to express it using A_{ij}. Rewriting the differentials in $F_{\mu\nu} = \partial_\mu A_\nu - \partial_\nu A_\mu$ in terms of finite differences using A_{ij}, we obtain

$$F_{\mu\nu}(i) \cong (A_{i+\hat{\mu},i+\hat{\mu}+\hat{\nu}} - A_{i,i+\hat{\nu}} + A_{i,i+\hat{\mu}} - A_{i+\hat{\nu},i+\hat{\mu}+\hat{\nu}})/a \ . \tag{3.4.1}$$

Owing to $A_{ij} = -A_{ji}$ (that is, $U^*_{ij} = U_{ji}$), we can rewrite (3.4.1) as

$$(A_{i,i+\hat{\mu}} + A_{i+\hat{\mu},i+\hat{\mu}+\hat{\nu}} + A_{i+\hat{\mu}+\hat{\nu},i+\hat{\nu}} + A_{i+\hat{\nu},i})/a \ . \tag{3.4.2}$$

However, (3.4.1) and (3.4.2) are not invariant under the transformation $A_{ij} \to A_{ij} + 2\pi/ea$. Writing (3.4.2) in the exponential

$$\begin{aligned}
&\exp[iea(A_{i,i+\hat{\mu}} + A_{i+\hat{\mu},i+\hat{\mu}+\hat{\nu}} + A_{i+\hat{\mu}+\hat{\nu},i+\hat{\nu}} + A_{i+\hat{\nu},i})] \\
&\quad = U_{i,i+\hat{\mu}}U_{i+\hat{\mu},i+\hat{\mu}+\hat{\nu}}U_{i+\hat{\mu}+\hat{\nu},i+\hat{\nu}}U_{i+\hat{\nu},i} \\
&\quad \equiv U_{\square}(i;\mu\nu) \ ,
\end{aligned} \tag{3.4.3}$$

we obtain the periodicity in A_{ij}. In order to regain $(1/16\pi)F_{\mu\nu}^2$ in the continuum limit, once more we have to be careful in obtaining the final expression:

$$S = \sum_i \sum_{\mu\nu} \frac{1}{8\pi e^2 a^{3-d}} [1 - \mathrm{Re}\, U_{\square}(i;\mu\nu)] \ , \tag{3.4.4}$$

where d is the space dimension. As can be seen easily, $U_{\square}(i)$ is invariant under the gauge transformation on the lattice points, because it "turns round once" in the square (following the arrows in Fig. 3.10). Therefore, S also is invariant. We call the smallest unit square around links a plaquette.

Above, we introduced the compact $U(1)$-lattice gauge field theory. As described, one element of the group $U(1)$ is defined on every link between lattice points. It is possible to consider gauge fields transforming under a group other than $U(1)$. For example, the group Z_2 with the two elements $\{+1, -1\}$ leads to the so-called Ising gauge theory with the action given by ($\sigma_{ij} = \pm 1$)

$$S = J \sum_i \sum_{\mu,\nu} \sigma_{i,i+\hat{\mu}}\sigma_{i+\hat{\mu},i+\hat{\mu}+\hat{\nu}}\sigma_{i+\hat{\mu}+\hat{\nu},i+\hat{\nu}}\sigma_{i+\hat{\nu},i} \ . \tag{3.4.5}$$

Furthermore, groups like $SU(2)$ with non-commuting element [the so-called non-Abelian groups] can be considered, leading to a non-commuting gauge field [the Yang–Mills field, QCD belongs to this category]. Owing to the non-commutativity of the gauge fields, very many distinguishing properties have been discovered. However, we will restrict our discussion to commuting gauge fields.

Now, our main point of interest will be the question of phase transitions occurring in lattice gauge theories. As described in Sect. 3.1, in a common phase transition the symmetry is broken in the ordered phase, and to describe it, the order parameter is introduced. In the case of the XY model, the symmetry was the invariance of the Hamiltonian $\mathcal{H}$ under the spin rotation $\theta_i \to \theta_i + \theta_0$. In this case, at all sites i a common rotation about θ_0 has to be performed. In this sense, this symmetry is called a global symmetry. We explained that by applying an external magnetic field h, the symmetry is broken, and that by taking first the limit $N \to \infty$ of the spin number N and then the limit $h \to 0$, a finite value $\langle \boldsymbol{S}_i \rangle$ remains.

Let us couple the XY model to a gauge field and generalize it to a locally gauge invariant system. By setting both the lattice spacing a and the electrical charge e to 1, we obtain

$$S = -K\sum_{i,\mu}\cos(\theta_i - \theta_{i+\mu} - A_{i,\mu})$$
$$-\frac{1}{g^2}\sum_{i,\mu,\nu}\cos(A_{i,\mu} + A_{i+\mu,\nu} - A_{i+\nu,\mu} - A_{i,\nu}) \ . \tag{3.4.6}$$

Notice that the index notation of $A_{i,\mu}$ has altered a little. Up to now, we have defined A_{ij} to be the field on the link connecting i and j. In the above equation, we expressed the field A that connects the link i with $i+\hat{\mu}$ as $A_{i,\mu}$. In what follows, we will use this notation.

In the first line of (3.4.6), the phase of the XY model is coupled to the gauge field; therefore the action is invariant under the transformations

$$\theta_i \to \theta_i + f_i \ , \tag{3.4.7}$$

$$A_{i,\mu} \to A_{i,\mu} + f_i - f_{i+\mu} \ . \tag{3.4.8}$$

At every different lattice point i, f_i can have a different value. This is the sense of "local".

Next, the partition function Z of the system is given by

$$Z(J,K,g) = \int \mathcal{D}\theta\mathcal{D}A \exp\left[-S + J\sum_{i,\mu}\cos A_{i,\mu}\right] \ . \tag{3.4.9}$$

Here, we have introduced an external field J to break the symmetry, as we learned to do in the case of the XY model. Now, we explain the theorem of Elitzur. The theorem states that "local gauge symmetry cannot be spontaneously broken". In the case of (3.4.9), this means

$$\lim_{J\to 0}\lim_{N\to\infty}\langle\cos A_{i,\mu}\rangle = 0 \ .$$

For the details of the theorem, the reader is recommended to read once the original work [17], because it is written in a very clear manner. Here, we only mention the crucial points.

The reason why the "proof" of (3.1.4) $\langle \boldsymbol{I}_i\rangle = \boldsymbol{0}$ is no longer valid in the limit $N \to \infty$ is the following. When shifting θ_i to $\theta_i + \pi$, also all other sides are shifted simultaneously θ_j to $\theta_j + \pi$. Therefore, it is not sufficient to restrict consideration to the site i (or its vicinity). That is, because the number of degrees of freedom of the invariance transformation of the Hamiltonian is small (in our case only one), the whole system is involved. This is the reason why a singularity at $N \to \infty$ emerges. However, for the gauge model (3.4.6), things are different.

In this case, at every point the transformation has one degree of freedom, and the transformation at one site i preserves the action invariant not affecting some separated point at all. Using this fact, we consider the transformations (3.4.7) and (3.4.8) where only at one site i is f_i different from zero. Then, in $J\sum_{i,\mu}\cos A_{i,\mu}$ in the exponent of (3.4.9), only the coefficient i

changes, all others are independent of f_i, that is, are not transformed. Therefore, also in the limit $N \to \infty$, independent of N, the expression $\langle \cos A_{i,\mu} \rangle$ becomes arbitrarily small for small J. This point differs totally from the case of the XY model in Sect. 3.1, where in $h \sum_i \cos\theta_i$ at all sites the transformation $h \sum_i \cos(\theta_i + \theta_0)$ is performed, and in the limit $N \to \infty$ the effect of h is enlarged.

With the same considerations, we conclude that in the gauge model (3.4.6), $\langle \cos\theta_i \rangle$ will not reach a finite value, and it is not possible to describe the phase transition of the gauge model with an order parameter that is not gauge invariant. Now, the next questions are: Does a phase transition occur, and what could characterize this phase transition?

In this book, for simplicity we consider the case where the "matter field" θ_i is not present. That is, we think of a model where the action is given by the second term of (3.4.6):

$$S = -\frac{1}{g^2} \sum_{i,\mu,\nu} \cos(A_{i,\mu} + A_{i+\mu,\nu} - A_{i+\nu,\mu} - A_{i,\nu}) \ . \tag{3.4.10}$$

Now, from the above considerations, we conclude that we have to choose some gauge invariant physical quantity, and if it undergoes a qualitative change, then a phase transition has occurred. As mentioned above, in order to obtain a gauge invariant expression, because when "turning round once", the phase factors f_i occurring in (3.4.8) cancel out, it is sufficient to construct the sum $\sum_C A_{i,\mu}$ of a closed loop C (Fig. 3.11).

On the other hand, because physical quantities must be 2π-periodic in $A_{i,\mu}$, finally

$$\hat{\Gamma}(C) = \exp\left[\mathrm{i} \sum_C A_{i,\mu}\right] \tag{3.4.11}$$

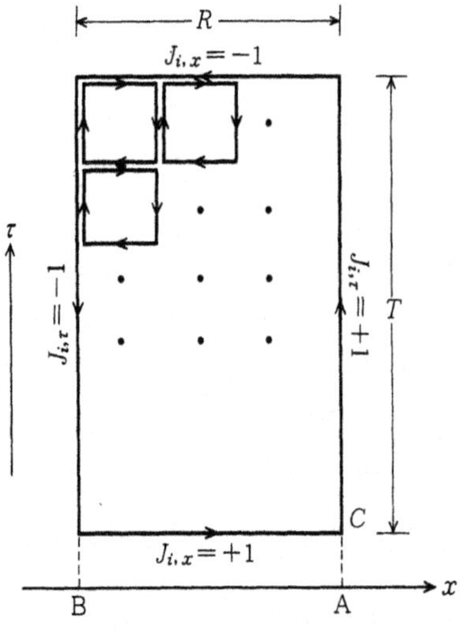

Fig. 3.11. Wilson loop

seems to be a good candidate. The expectation value $\Gamma(C) = \langle \hat{\Gamma}(C) \rangle$ of $\hat{\Gamma}(C)$ is called the Wilson loop. Indeed, the Wilson loop indicates the phase transition between the confining and non-confining phase, but before explaining this, we consider the physical significance of $\Gamma(C)$. Therefore, we consider the partition function defined with the "electrical current" $J_{i,\mu}$:

$$Z[J] = \int \mathcal{D}\theta \mathcal{D}A \exp\left[- S + \mathrm{i} \sum_{i,\mu} J_{i,\mu} A_{i,\mu} \right] . \qquad (3.4.12)$$

Here, $J_{i,\mu}$ is the μ-component of the current flowing from i to $i + \mu$; in particular, for $\mu = \tau$ it corresponds to the charge. The conservation law for the current in the continuum $\partial\rho/\partial t + \operatorname{div} \boldsymbol{J} = 0$ on the lattice reads

$$\sum_\mu \Delta_\mu J_{i,\mu} \equiv \sum_\mu (J_{i,\mu} - J_{i-\mu,\mu}) = 0 . \qquad (3.4.13)$$

Especially for the case of a loop C with surface $T \times R$ around a rectangle in the complex plane spanned by imaginary time and some space direction, as shown in Fig. 3.11, (3.4.13) holds, when corresponding to the direction of the arrows the values $+1$ and -1 are assigned. In the limit $T \to \infty$ this corresponds to a system with one static charge $+1$ at point A and one static charge -1 at point B. Therefore, the potential $V(R)$ between the two charges at a distance R is given by

$$\begin{aligned} V(R) &= - \lim_{T\to\infty} \frac{1}{T} \ln\left[\frac{Z(J)}{Z(0)}\right] \\ &= - \lim_{T\to\infty} \frac{1}{T} \ln\left\langle \exp\left[\mathrm{i}\sum J_{i,\mu} A_{i,\mu}\right]\right\rangle \\ &= - \lim_{T\to\infty} \frac{1}{T} \ln\left\langle \exp\left[\mathrm{i}\sum_C A_{i,\mu}\right]\right\rangle \\ &= - \lim_{T\to\infty} \frac{1}{T} \ln \Gamma(C) . \end{aligned} \qquad (3.4.14)$$

That is, the Wilson loop determines the force law between the two particles with opposite gauge charges.

There exists a limit where the behaviour of $\Gamma(C)$ can be examined easily. This is the case where the "coupling constant" g in (3.4.10) is extremely large ($g \gg 1$). In this case, in the same manner as in the high-temperature expansion of the exponent in (3.1.10), it is useful to expand the exponential. This expansion is called the strong coupling expansion. Totally analogous to the discussion of the high-temperature expansion, it is also possible to use graphs to calculate the expansion of $\Gamma(C)$. However, qualitatively different is the point that in Sect. 3.1 the phase on the lattice points has been integrated, whereas in the present case, $A_{i,\mu}$ on the links is integrated. Furthermore, every term of the action S is represented by one loop around a plaquette.

For a non-vanishing $A_{i,\mu}$ integral, the number of arrows from i to $i+\mu$ must equal the number of arrows from $i+\mu$ to i. Applying this rule to Fig. 3.11, contributing diagrams with lowest power in $1/g^2$ are given by paths around one plaquette, and for covering the interior of the path C, we need only $T \times R$ of such plaquettes, therefore

$$\Gamma(C) \propto \left(\frac{1}{g^2}\right)^{TR} = \exp[-TR\ln g^2] \ .$$

This behaviour of the Wilson loop is called the area law. Inserting this expression into (3.4.14), we obtain

$$V(R) = R\ln g^2 \ . \tag{3.4.15}$$

An attractive potential proportional to R acts on opposite charges, therefore at large scales it will be absolutely impossible to observe independent charges. This is called confinement, a theme that is extremely important for the explanation of the fact that it is absolutely impossible to observe independent quarks.

On the other hand, what is expected to happen in the other limit, the limit of small coupling ($g \ll 1$)? We start with a very simple consideration. $g \ll 1$ corresponds to the low-energy phase of the XY model. Now, we search for the approximation corresponding to the spin wave approximation of the XY model. Because $A_{i,\mu}$ in (3.4.10) changes very slowly compared with the lattice spacing, the continuum approximation

$$\cos(A_{i,\mu} + A_{i+\mu,\nu} - A_{i+\nu,\mu} - A_{i,\nu}) \to 1 - \tfrac{1}{2}(\partial_\mu A_\nu - \partial_\nu A_\mu)^2 \tag{3.4.16}$$

is the corresponding approximation.

Expression (3.4.16) leads of course to the action of the normal electromagnetic field, because we simply reverse the steps performed when introducing the lattice gauge theory. Therefore, the potential $V(R)$ between two static charges (in $3+1$ dimensions) is given by

$$V(R) = -\frac{g^2}{2R} + \text{const.} \qquad (\text{const.} \neq 0) \ . \tag{3.4.17}$$

Here, on the other hand, for T and R large enough, and $T \gg R$, $\ln\Gamma(C)$ becomes proportional to T and independent of R. Considering both T and R symmetrically, we expect that the length $2(T+R)$ of the path C is proportional to $\ln\Gamma(C)$. In contrast to the area law mentioned earlier, this behaviour of the Wilson loop is called the perimeter law.

In such a way, when the qualitative behaviour of $\Gamma(C)$ in the limit $g \gg 1$ and $g \ll 1$ is different, there should exist a boundary value g_c where a phase transition occurs, but will this really be the case? Considering the KT transition of Sect. 3.3 as an example, we demonstrated that at $T < T_c$ vortices arose only in pairs, and no free vortices exist. In the present case, will the topological defects, reflected by the discrete lattice and the periodicity of A, really become ignorable when the coupling g is smaller than some finite g_c? Definitely, this is not an obvious problem, depending on the dimension of the lattice. In this sense, the strong coupling expansion of lattice gauge theory is a natural and gentle expansion, and it is evident that confining occurs for $g \gg 1$; in contrast, care must be taken with regard to the weak coupling expansion.

First, we consider a $1+1$ dimensional space with one space dimension and one time dimension. In this case, the problem can be solved exactly. When choosing an appropriate gauge, the action (3.4.10) can be written as the sum of actions of non-interacting one-dimensional systems. Explicitly, in order to make, say, $A_{i,x}$ vanish in the x direction, it is sufficient to set for the f_i appearing in the gauge transformation (3.4.8) for every τ in the order from the left to the right $f_{i+x} = f_i - A_{i,\mu}$. Then, the action becomes

$$S = - \sum_{i=(i_x, i_\tau)} \frac{1}{g^2} \cos\left(A_{(i_x,i_\tau)T} - A_{(i_x+1,i_\tau)T}\right) \ . \qquad (3.4.18)$$

Interpreting $A_{(i_x,i_\tau)}$ as the phase of the XY model, we see that the action describes decoupled one-dimensional classical XY models at every i_τ. Then, the Wilson loop is given by

$$\Gamma(C) = \prod_{i_\tau=1}^{T} \exp\left[\mathrm{i}\{A_{(i_x,i_\tau)T} - A_{(i_x+R,i_\tau)T}\}\right] \ . \qquad (3.4.19)$$

Following the Mermin–Wagner theorem mentioned in Sect. 3.2, in the one-dimensional XY model, no phase transition at finite temperature (in our case at finite g^2) occurs, and the system is in the high-temperature phase (strong coupling phase); therefore we deduce for $\Gamma(C)$ in (3.4.19) the area law

$$\Gamma(C) \cong \prod_{i_\tau=1}^{T} \mathrm{e}^{-R/\xi} = \mathrm{e}^{-TR/\xi} \ . \qquad (3.4.20)$$

That is, for all g^2, confining occurs.

Next, what about $2+1$ dimensions? Concerning this question, there exists a famous work by A. M. Polyakov. We begin by noting the conclusion: much the same as in $1+1$ dimensions, confining occurs for all g^2. We explain this fact by using the duality transformation that has been introduced during the discussion of the KT transition. The principal idea is to deduce an effective action for topological defects (which will be clarified later in the chapter), and to discuss its effect on $\Gamma(C)$.

Now, as was done in (3.3.12′), we perform the following replacement:

$$\exp\left[\left(\frac{1}{g^2}\right)\cos(\Delta_\mu A_\nu(i) - \Delta_\nu A_\mu(i))\right]$$
$$\to \left(\frac{g^2}{2\pi}\right)^{1/2} \sum_{l_{\mu\nu}(i)=-\infty}^{\infty} \exp\left[\frac{1}{g^2}\right] \exp\left[\mathrm{i} l_{\mu\nu}(i)(\Delta_\mu A_\nu(i) - \Delta_\nu A_\mu(i)) - \frac{g^2}{2} l_{\mu\nu}(i)^2\right] . \tag{3.4.21}$$

Here, $l_{\mu\nu}(i)$ is an integer number defined on the plaquette bounded by the links starting from the origin i in the μ direction, and in the ν direction. $\Delta_\mu A_\nu(i) \equiv A_{i+\hat{\mu},\nu} - A_{i,\nu}$, and $\Delta_\mu A_\nu(i) - \Delta_\nu A_\mu(i)$ is the sum turning around one plaquette. Corresponding to (3.3.17), we obtain

$$Z[J] = \int \prod_{i,\mu} \mathrm{d}A_{i,\mu} \sum_{\{l_{\mu\nu}(i)\}} \exp\left[\sum_{i,\mu,\nu}\left\{ -\frac{g^2}{2} l_{\mu\nu}(i)^2 \right.\right.$$
$$\left.\left. + \mathrm{i} l_{\mu\nu}(i)(\Delta_\mu A_\nu(i) - \Delta_\nu A_\mu(i))\right\} + \mathrm{i}\sum_{i,\mu} J_{i,\mu} A_{i,\mu}\right]$$
$$= \sum_{\{l_{\mu\nu}(i)\}} \exp\left(-\frac{g^2}{2}\sum_{i,\mu,\nu} l_{\mu\nu}(i)^2\right) \prod_{i,\mu} \delta_{\Delta_\nu l_{\mu\nu}(i) + J_{i,\mu},0} . \tag{3.4.22}$$

Here, we defined $\Delta_\nu l_{\mu\nu} \equiv \sum_\nu [l_{\mu\nu}(i) - l_{\mu\nu}(i-\hat{\nu})]$.

In order to make the meaning of the delta function appearing in the above equation more explicit, we integrate the variable $A_{i,z}$ on the link $\mu = z$, as can be seen in Fig. 3.12. Four plaquettes have to be considered, two in the x direction, and two in the y direction, respectively. The arrow on the surface of every plaquette indicates the direction of A_{ij}. When this direction agrees with the direction of $(i, i+z)$, $A_{i,z}$ is multiplied by $+l_{\mu,\nu}$, and when the direction is opposite, it is multiplied by $-l_{\mu,\nu}$. Explicitly, the parts of the exponential of (3.4.22) related to $A_{i,z}$ are given by

$$\mathrm{i}A_{i,z}\left[J_{i,z} + l_{zx}(i) - l_{zx}(i-x) - l_{yz}(i) + l_{yz}(i-y)\right] . \tag{3.4.23}$$

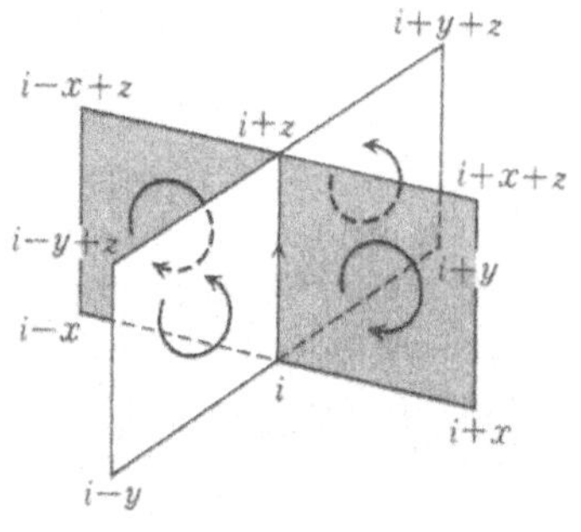

Fig. 3.12. The four plaquettes that are connected with A_{iz}

Using the relation $l_{\mu\nu}(i) = -l_{\nu\mu}(i)$, the part written in [ ] of (3.4.23) can by expressed as $\Delta_\nu l_{z\nu}(i) + J_z(i)$. Now, defining the vector $\boldsymbol{b}(i)$ by $l_{xy}(i) = b_z(i)$, $l_{yz}(i) = b_y(i)$ and $l_{zx}(i) = b_y(i)$, the part in [ ] of (3.4.23) reads $J_z + (\mathrm{rot}\, \boldsymbol{b})_z$. Therefore, the delta function (3.4.22) equals the condition

$$\boldsymbol{J} + \mathrm{rot}\, \boldsymbol{b} = \boldsymbol{0} \ . \qquad (3.4.24)$$

This equation is just (up to some constants) one of the Maxwell equations describing the "magnetic field" occurring due to the presence of a constant current $\boldsymbol{J}$.

Now, let us first consider the case $\boldsymbol{J} = \boldsymbol{0}$. Then, due to (3.4.24), $\boldsymbol{b}$ can be expressed using a scalar potential $\phi(i)$ as

$$\boldsymbol{b} = \mathrm{grad}\, \phi \ , \qquad (3.4.25)$$

and (3.4.22) can be written as

$$Z[\boldsymbol{J} = \boldsymbol{0}] = \prod_i \sum_{\phi(i)} \exp\left[-\frac{g^2}{2} \sum_i (\mathrm{grad}\, \phi(i))^2\right] \ . \qquad (3.4.26)$$

Up to now we have developed the equations in the continuum limit; however, $\phi(i)$ is a field defined on the lattice points reaching only integer numbers. We express the sum over integer values $\phi(i)$ in (3.4.26) using the Poisson formula to obtain an integral expression, as was done in (3.3.22):

$$Z[\boldsymbol{J} = \boldsymbol{0}] = \int \mathcal{D}\psi(r) \sum_{\{m(r)\}} \exp\left[-\frac{g^2}{2} \sum_{r,\mu} (\Delta_\mu \psi(r))^2 + 2\pi \mathrm{i} \sum_r m(r)\psi(r)\right] \ . \qquad (3.4.27)$$

This equation has exactly the same form as (3.3.22), the only difference is that we now discuss the $2+1$ dimensional case. Corresponding to (3.3.22), where $m(r)$ represented vortices, in (3.4.27), $m(r)$ are magnetic monopoles. This is due to the fact that in two dimensions, $\phi(r)$ was the static electric potential; on the other hand, $\psi(r)$ in (3.4.27) is just the static magnetic potential. Therefore, $m(r)$ corresponds to the "magnetic charge".

There are two approaches for the description of magnetic monopoles. From one point of view the magnetic monopole is one end of a dipole, as shown in Fig. 3.13a. On the other hand, recalling that a magnetic dipole is equivalent to a circular micro current, the monopole can be regarded as the end point of a solenoid with infinitesimal small diameter, as shown in Fig. 3.13b. In both cases, the other end is placed in infinity. In the solenoid, the magnetic flux flows to the end of the magnetic monopol, where it is emitted in a spherically symmetric manner.

The amount of magnetic flux must be 2π multiplied by an integer. This is due to the fact that the system is in principle described by cos[magnetic flux going through the plaquette], and is therefore 2π-periodic in the magnetic flux. And only in such a case does the action remain finite for an infinitely

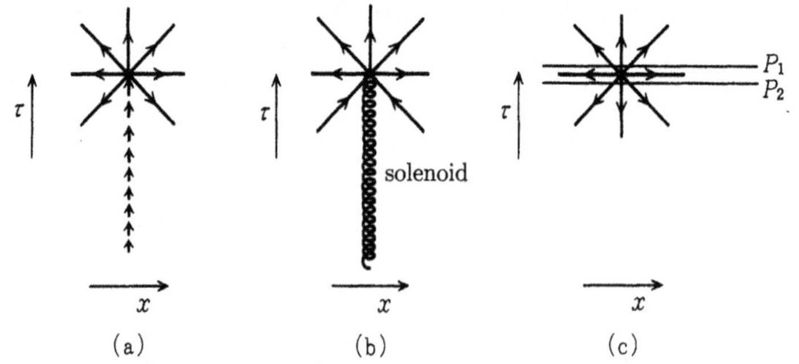

Fig. 3.13a–c. Different viewpoints for magnetic monopoles (instantons)

long solenoid. That is, beyond the vicinity of the magnetic monopole, owing to the 2π-periodicity the system is not at all disturbed. Now, we let the solenoid disappear and consider Fig. 3.13c. From the magnetic monopole, the magnetic flux 2π is emitted. Remember that the z axis in Fig. 3.13 is the imaginary time. We call Φ_1 the magnetic flux through the surface P_1 that is placed in the "future" of the magnetic monopole, and Φ_2 the flux through P_2 placed in its "past". Then we obtain

$$\Phi_1 - \Phi_2 = 2\pi \ . \tag{3.4.28}$$

Therefore, the magnetic monopole is the instanton solution [Sect. 2.1] that represents the tunnelling between two topological sectors with different magnetic flux.

Now, in the same way as the electric charge can screen the electric field, also the magnetic monopole (the instanton) can screen the magnetic field. In the three-dimensional Coulomb gas model there exist of course free charges, and screening does emerge, and it is known that the Coulomb force becomes a short-range force. In terms of the present problem, when free instantons and anti-instantons exist, the magnetic field will become a short-range force. For the case that an external current J is given, the partition function $Z[J]$ behaves like

$$Z[J] \propto \mathrm{e}^{-\,\mathrm{const}\cdot TR} \ . \tag{3.4.29}$$

That is, the system will in general be in the confining phase.

In this manner, Polyakov explained how confinement occurs due to the disturbance of the gauge field due to instantons. We now look briefly at the theory. Up to the dimension, equation (3.4.27) is the same as (3.3.22), therefore we can re-express it in the same manner as the Sine–Gordon model:

$$Z \cong \int \prod_r \mathrm{d}\psi(r) \exp\left[-\frac{g^2}{2} \sum_{r,\mu} [\Delta_\mu \psi(r)]^2 + 2y \sum_r \cos[2\pi\psi(r)] \right] \ . \tag{3.4.30}$$

Here, we have $y = \mathrm{e}^{-\,\mathrm{const}/2g^2}$. We try to apply the variation method to (3.4.30) as was done before. The result is that the integral leading to a singularity $\ln(k_c^2/m^2)$ in (3.3.37) is in three dimensions not infrared divergent for $m \to 0$. Therefore, $\langle \cos[2\pi\psi(r)] \rangle_0$ has a finite value, and in its turn m is also finite. As a result, screening emerges all the time in the system, and the screening length is given by $1/m$.

Now, we consider an "electric current" J around the surface TR of Fig. 3.11 in the $2+1$ dimensional space. Because this current can be considered as the total sum of micro currents inside the surface TR, we can assume equivalently the existence of magnetic dipoles inside the surface. Therefore, in a region of about $1/m$ thickness above and below the surface TR, the "magnetic field" $\boldsymbol{b}$ will be influenced, and the energy will be proportional to TR. Expressing this in the sum of states $Z[J]$, we obtain (3.4.29), signifying that the system is in the confining phase.

Above, we understood that in the $1+1$ and $2+1$ dimensional compact $U(1)$-lattice gauge theory, for all values of the coupling constant, the system is in the confining phase. So, what will occur in $3+1$ dimensions? In this case, the same as in the three-dimensional XY model discussed earlier will occur. That is, the instanton that has been a point in $2+1$ dimensions will gain one dimension in $3+1$ dimensions and will become a loop. The question whether loops of the magnetic field with infinite diameter arise or not corresponds to the question whether confinement occurs or not, and the boundary, i.e. the phase transition point, is given by a finite value $g = g_{\mathrm{c}}$ of the coupling constant. That is to say, in the weak coupling regime $g < g_{\mathrm{c}}$ in $3+1$ dimensions, the description with the usual continuous electromagnetic field is sufficient, and we regain the long-distance Coulomb force. In this case, the gauge field is in the non-confining phase.

4. Simple Examples for the Application of Field Theory

Starting from this chapter, we will solve many different problems that occur in practice in condensed matter physics. First, to warm up, we will choose one problem of a fermionic system and one of a bosonic system. We will examine the RPA theory of the Coulomb gas and the Bogoliubov theory of superfluidity using path integral methods.

4.1 The RPA Theory of a Coulomb Gas

Starting from this chapter, the techniques of quantum field theory that have been introduced in the foregoing chapters will be applied to several problems occurring in condensed matter physics.

First, we will introduce the Gell-Mann–Brückner theory, or RPA theory, describing a Coulomb gas. In the Coulomb gas model, electrons interact by the long-range Coulomb force, moving in a homogeneous, positively charged background. The Hamiltonian is given by

$$\hat{H} = \sum_{i=1}^{N} \frac{\hat{\boldsymbol{p}}_i^2}{2m} + \sum_{i<j} \frac{e^2}{|\hat{\boldsymbol{r}}_i - \hat{\boldsymbol{r}}_j|} \quad . \tag{4.1.1}$$

Defining the density operator as

$$\hat{\rho}(\boldsymbol{r}) = \sum_{i=1}^{N} \delta(\boldsymbol{r} - \hat{\boldsymbol{r}}_i) \quad , \tag{4.1.2}$$

then its Fourier transformation reads

$$\hat{\rho}(\boldsymbol{k}) = \int \mathrm{d}\boldsymbol{r}\, \mathrm{e}^{-\mathrm{i}\boldsymbol{k}\cdot\boldsymbol{r}} \hat{\rho}(\boldsymbol{r}) = \sum_{i=1}^{N} \mathrm{e}^{-\mathrm{i}\boldsymbol{k}\cdot\hat{\boldsymbol{r}}_i} \quad . \tag{4.1.3}$$

Using this $\hat{\rho}(\boldsymbol{k})$, we can reformulate the Coulomb interaction. In order to do so, we apply the Fourier transformation formula:

$$\frac{e^2}{|\boldsymbol{r}|} = \frac{1}{V} \sum_{\boldsymbol{q}} \frac{4\pi e^2}{|\boldsymbol{q}|^2} \mathrm{e}^{\mathrm{i}\boldsymbol{q}\cdot\boldsymbol{r}} \quad . \tag{4.1.4}$$

This equation is formulated for a cube with side length L (volume $V = L^3$) with periodic boundary conditions. The wave vector $\boldsymbol{q}$ is quantized as

$$\boldsymbol{q} = \frac{2\pi}{L}(n_x, n_y, n_z) \ , \tag{4.1.5}$$

with n_x, n_y, n_z being integers. Because in the volume $(2\pi/L)^3$ of the $\boldsymbol{q}$-space, one $\boldsymbol{q}$ value is allowed, we obtain the relation

$$\sum_{\boldsymbol{q}} = \left(\frac{L}{2\pi}\right)^3 \int \mathrm{d}^3\boldsymbol{q} = V \int \frac{\mathrm{d}^3\boldsymbol{q}}{(2\pi)^3} \ . \tag{4.1.6}$$

Now, because the term $\boldsymbol{q} = 0$ in the Fourier expansion (4.1.4) corresponds to an integral over the whole space, it diverges. However, the integral over the Coulomb interaction in the whole space should vanish because the negatively charged electrons and the positively charged background cancel each other. Having this consent in mind, we ignore the term $\boldsymbol{q} = 0$ in the $\boldsymbol{q}$-sum in (4.1.4). Then, we obtain

$$\begin{aligned}
\sum_{i<j} \frac{e^2}{|\hat{\boldsymbol{r}}_i - \hat{\boldsymbol{r}}_j|} &= \sum_{i<j} \frac{1}{V} \sum_{\boldsymbol{q}\neq 0} \frac{4\pi e^2}{|\boldsymbol{q}|^2} \mathrm{e}^{\mathrm{i}\boldsymbol{q}\cdot(\hat{\boldsymbol{r}}_i - \hat{\boldsymbol{r}}_j)} \\
&= \frac{1}{V} \sum_{\boldsymbol{q}\neq 0} \frac{4\pi e^2}{|\boldsymbol{q}|^2} \left\{ \frac{1}{2} \sum_j \mathrm{e}^{-\mathrm{i}\boldsymbol{q}\cdot\hat{\boldsymbol{r}}_j} \sum_i \mathrm{e}^{\mathrm{i}\boldsymbol{q}\cdot\hat{\boldsymbol{r}}_i} - \frac{N}{2} \right\} \\
&= \frac{1}{V} \sum_{\boldsymbol{q}\neq 0} \frac{2\pi e^2}{|\boldsymbol{q}|^2} \{\hat{\rho}(\boldsymbol{q})\hat{\rho}(-\boldsymbol{q}) - N\} \ .
\end{aligned} \tag{4.1.7}$$

Now, we apply second quantization and obtain

$$\begin{aligned}
\hat{H} = &\int \sum_\sigma \hat{\psi}_\sigma^\dagger(\boldsymbol{r}) \left[-\frac{\hbar^2}{2m}\nabla^2 - \mu \right] \hat{\psi}_\sigma(\boldsymbol{r})\, \mathrm{d}\boldsymbol{r} \\
&+ \frac{1}{V} \sum_{\boldsymbol{q}\neq 0} \frac{2\pi e^2}{|\boldsymbol{q}|^2} \{\hat{\rho}(\boldsymbol{q})\hat{\rho}(-\boldsymbol{q}) - N\} \ .
\end{aligned} \tag{4.1.8}$$

The density operator $\hat{\rho}(\boldsymbol{r})$ is expressed as the sum over the spins σ

$$\hat{\rho}(\boldsymbol{r}) = \sum_\sigma \hat{\psi}_\sigma^\dagger(\boldsymbol{r}) \hat{\psi}_\sigma(\boldsymbol{r}) \ . \tag{4.1.9}$$

We decompose the field operators $\hat{\psi}_\sigma$ and $\hat{\psi}_\sigma^\dagger$ into their Fourier components:

$$\hat{\psi}_\sigma(\boldsymbol{r}) = \sum_{\boldsymbol{k}} \frac{\mathrm{e}^{\mathrm{i}\boldsymbol{k}\cdot\boldsymbol{r}}}{\sqrt{V}} \hat{C}_{\boldsymbol{k}\sigma} \ , \tag{4.1.10}$$

$$\hat{\psi}_\sigma^\dagger(\boldsymbol{r}) = \sum_{\boldsymbol{k}} \frac{\mathrm{e}^{-\mathrm{i}\boldsymbol{k}\cdot\boldsymbol{r}}}{\sqrt{V}} \hat{C}_{\boldsymbol{k}\sigma}^\dagger \ . \tag{4.1.11}$$

This step corresponds to choosing an orthonormal basis of plane waves

$$\phi_n(\boldsymbol{r}) \to \frac{1}{\sqrt{V}} \mathrm{e}^{\mathrm{i}\boldsymbol{k}\cdot\boldsymbol{r}} \tag{4.1.12}$$

in (1.2.16). Using this, the Fourier expansion $\hat{\rho}(\boldsymbol{q})$ of $\hat{\rho}(\boldsymbol{r})$ can be written as

$$\begin{aligned}\hat{\rho}(\boldsymbol{q}) &= \int \mathrm{e}^{-\mathrm{i}\boldsymbol{q}\cdot\boldsymbol{r}} \hat{\rho}(\boldsymbol{r})\,\mathrm{d}^3\boldsymbol{r} \\ &= \sum_{\sigma,\boldsymbol{k},\boldsymbol{k}'} \frac{1}{V} \int \mathrm{e}^{-\mathrm{i}\boldsymbol{q}\cdot\boldsymbol{r}-\mathrm{i}\boldsymbol{k}\cdot\boldsymbol{r}+\mathrm{i}\boldsymbol{k}'\boldsymbol{r}}\,\mathrm{d}^3\boldsymbol{r} \cdot \hat{C}^\dagger_{\boldsymbol{k}\sigma}\hat{C}_{\boldsymbol{k}'\sigma} \\ &= \sum_\sigma \sum_{\boldsymbol{k}} \hat{C}^\dagger_{\boldsymbol{k}\sigma}\hat{C}_{\boldsymbol{k}+\boldsymbol{k}\sigma} \;. \end{aligned} \tag{4.1.13}$$

Finally, the Hamiltonian becomes

$$\begin{aligned}\hat{\mathcal{H}} &= \sum_{\boldsymbol{k},\sigma}\left(\frac{\hbar^2\boldsymbol{k}^2}{2m}-\mu\right)\hat{C}^\dagger_{\boldsymbol{k}\sigma}\hat{C}_{\boldsymbol{k}\sigma} + \frac{1}{V}\sum_{\boldsymbol{q}\neq\boldsymbol{0}}\frac{2\pi e^2}{|\boldsymbol{q}|^2}\sum_{\substack{\boldsymbol{k},\boldsymbol{k}'\\ \sigma,\sigma'}}\hat{C}^\dagger_{\boldsymbol{k}+\boldsymbol{q}\sigma}\hat{C}^\dagger_{\boldsymbol{k}'-\boldsymbol{q}\sigma'}\hat{C}_{\boldsymbol{k}'\sigma'}\hat{C}_{\boldsymbol{k}\sigma} \\ &= \sum_{\boldsymbol{k},\sigma}\left(\frac{\hbar^2\boldsymbol{k}^2}{2m}-\mu\right)\hat{C}^\dagger_{\boldsymbol{k}\sigma}\hat{C}_{\boldsymbol{k}\sigma} + \frac{1}{V}\sum_{\boldsymbol{q}\neq\boldsymbol{0}}\frac{2\pi e^2}{|\boldsymbol{q}|^2}\{\hat{\rho}(\boldsymbol{q})\hat{\rho}(-\boldsymbol{q})-N\}\;. \end{aligned}\tag{4.1.14}$$

Now, before we investigate (4.1.14) using path integral methods, we need to make some preliminary considerations. We return to the Hamiltonian (4.1.1) and estimate the magnitude of the kinetic energy in the first term and the potential energy in the second term. The method is similar to the considerations made in (1.1.54) and thereafter, where in the present case r should be considered as the average distance between the particles. That is, writing

$$\frac{4\pi}{3}r^3 N = V \tag{4.1.15}$$

the kinetic energy E_k of one particle is approximately

$$E_\mathrm{k} \sim \frac{\hbar^2}{2mr^2} \;. \tag{4.1.16}$$

On the other hand, the Coulomb energy E_C of one particle is approximately

$$E_\mathrm{C} \sim \frac{e^2}{r} \;. \tag{4.1.17}$$

With r_B being the Bohr radius, the ratio of both is given by

$$E_\mathrm{C}/E_\mathrm{k} \sim \frac{me^2}{\hbar^2}r \sim \frac{r}{r_\mathrm{B}} = r_\mathrm{s} \;. \tag{4.1.18}$$

Here, we have introduced the dimensionless particle distance parameter r_s.

In the limit of high density $r_\mathrm{s} \ll 1$, the kinetic energy is larger than the Coulomb energy, and we expect that a perturbative expansion in e^2 or in r_s

works well. At first sight, the reader might expect that because the Coulomb force becomes stronger at high density, it will be dominant. However, when r becomes small, owing to the uncertainty principle the increase in the kinetic energy is even stronger. On the other hand, at low density the Coulomb force is dominant. The electrons build an ordered crystal lattice (the Wigner crystal), and the kinetic energy gives rise to the zero point excitation of the crystal lattice. The theory that will be developed in the following corresponds to the high-density limit $r_s \ll 1$. As is clear from (4.1.14), because the Coulomb interaction has the form $2\pi e^2 / |\boldsymbol{q}^2|$, the contribution of small $\boldsymbol{q}$ also is important, because e^2 is small. Because the integral diverges, simple perturbation theory cannot be applied, therefore the problem is definitely non-trivial. Normally, the RPA theory is developed by introducing the diagram method, where for the divergent parts a corresponding infinitesimal series of diagrams is added. In the language of path integrals, this method can be deduced in a very compact manner.

Following Sect. 2.3, the partition function of the system can be written as

$$Z = \int \mathcal{D}\bar{\psi}\mathcal{D}\psi e^{-S} \ . \tag{4.1.19}$$

Here, the action S is given by ($\hbar = 1$)

$$\begin{aligned} S = \int_0^\beta \mathrm{d}\tau \bigg\{ \int \mathrm{d}\boldsymbol{r} \sum_\sigma \bar{\psi}_\sigma(\boldsymbol{r},\tau) \left[\partial_\tau - \frac{1}{2m}\nabla^2 - \mu \right] \psi_\sigma(\boldsymbol{r},\tau) \\ + \frac{1}{V} \sum_{\boldsymbol{q}\neq 0} \frac{2\pi e^2}{|\boldsymbol{q}|^2} \{ \rho(\boldsymbol{q},\tau)\rho(-\boldsymbol{q},\tau) - N \} \bigg\} \ , \end{aligned} \tag{4.1.20}$$

where $\bar{\psi}_\sigma$ and ψ_σ are Grassmann integration variables, and corresponding to (4.1.13), the density $\rho(\boldsymbol{q})$ is given by

$$\rho(\boldsymbol{q}) = \sum_\sigma \sum_{\boldsymbol{k}} \bar{C}_{\boldsymbol{k}\sigma} C_{\boldsymbol{k}+\boldsymbol{q}\sigma} \ . \tag{4.1.21}$$

The problem in the action (4.1.20) is that $\rho(\boldsymbol{q})\rho(-\boldsymbol{q})$ is a non-linear term of fourth power when expressed in $\bar{C}$ and C, and that for this reason the path integral cannot be performed. Using the Stratonovich–Hubbard transformation, we rewrite the non-linear term in another way. The identity

$$\begin{aligned} &\exp\left[- \int_0^\beta \mathrm{d}\tau \frac{1}{V} \sum_{\boldsymbol{q}\neq 0} \frac{2\pi e^2}{|\boldsymbol{q}|^2} \rho(\boldsymbol{q},\tau)\rho(-\boldsymbol{q},\tau) \right] \\ &= \int \mathcal{D}\varphi(\boldsymbol{q},\tau) \exp\left[- \frac{1}{8\pi} \int_0^\beta \mathrm{d}\tau \sum_{\boldsymbol{q}\neq 0} |\boldsymbol{q}|^2 \varphi(\boldsymbol{q},\tau)\varphi(-\boldsymbol{q},\tau) \right. \\ &\quad \left. - \int_0^\beta \mathrm{d}\tau \frac{\mathrm{i}e}{2\sqrt{V}} \sum_{\boldsymbol{q}\neq 0} \{ \varphi(\boldsymbol{q},\tau)\rho(-\boldsymbol{q},\tau) + \rho(\boldsymbol{q},\tau)\varphi(-\boldsymbol{q},\tau) \} \right] \end{aligned} \tag{4.1.22}$$

is called the Stratonovich–Hubbard transformation. Because the $\varphi(\boldsymbol{q},\tau)$ integral is a Gauss integral, (4.1.22) can be proved by completing the square, disregarding the occurring pre-factor. Here, because $\rho(\boldsymbol{r})$ is real, we obtain $[\rho(\boldsymbol{q})]^* = \rho(-\boldsymbol{q})$, and similarly $[\varphi(\boldsymbol{q})]^* = \varphi(-\boldsymbol{q})$. Writing (4.1.22) in real variables, we finally obtain for the partition function

$$Z = \int \mathcal{D}\bar{\psi}\mathcal{D}\psi\mathcal{D}\varphi\, \mathrm{e}^{-S(\bar{\psi},\psi,\varphi)} \ , \tag{4.1.23}$$

$$\begin{aligned} S(\bar{\psi},\psi,\varphi) = \int_0^\beta \mathrm{d}\tau \int \mathrm{d}\boldsymbol{r} \Big\{ &\frac{1}{8\pi}[\nabla\varphi(\boldsymbol{r},\tau)]^2 \\ &+ \sum_\sigma \bar{\psi}(\boldsymbol{r},\tau)\left[\partial_\tau - \frac{1}{2m}\nabla^2 - \mu + \mathrm{i}e\varphi(\boldsymbol{r},\tau)\right]\psi_\sigma(\boldsymbol{r},\tau)\Big\} + \text{const.} \end{aligned} \tag{4.1.24}$$

Here, const. stands for the term $-\frac{1}{V}\sum_{\boldsymbol{q}\neq 0}(2\pi e^2)/(|\boldsymbol{q}^2|)$ in (4.1.20). For a while, we will omit this term. The most striking advantage of this form of the equation is that it is quadratic in $\bar{\psi}$ and ψ, and that the path integral can (at least formally) be performed. From (2.3.3), we obtain

$$\begin{aligned} Z = \int \mathcal{D}\varphi \exp\left[-\int_0^\beta \mathrm{d}\tau \int \mathrm{d}\boldsymbol{r} \frac{(\nabla\varphi)^2}{8\pi}\right] \\ \times \left(\det\left[\partial_\tau - \frac{1}{2m}\nabla^2 - \mu + \mathrm{i}e\varphi(\boldsymbol{r},\tau)\right]\right)^2 \ . \end{aligned} \tag{4.1.25}$$

We now explain the meaning of the determinant. Regarding the second term of (4.1.24) as the quadratic form of the vector $\psi_\sigma(\boldsymbol{r},\tau)$ with components $(\boldsymbol{r},\tau)$, then the determinant runs over its coefficient matrix. Because the spin σ can be up $\uparrow$ or down $\downarrow$, the determinant has a power of two.

Because the determinant is not affected by a change of basis, the matrix element can be expressed in $(\boldsymbol{r},\tau)$ components or in its Fourier components $(\boldsymbol{k},\omega_n)$:

$$\begin{aligned} &\left[\partial_\tau - \frac{1}{2m}\nabla^2 - \mu + \mathrm{i}e\varphi(\boldsymbol{r},\tau)\right]_{(\boldsymbol{k},\omega_n)(\boldsymbol{k}',\omega_m)} \\ &= \left[-\mathrm{i}\omega_n + \frac{1}{2m}\boldsymbol{k}^2 - \mu\right]\delta_{\boldsymbol{k},\boldsymbol{k}'}\delta_{\omega_n,\omega_m} + \frac{\mathrm{i}e}{(\beta V)^{1/2}}\varphi(\boldsymbol{k}-\boldsymbol{k}',\omega_n-\omega_m) \ . \end{aligned} \tag{4.1.26}$$

Here, $\omega_n = (2n+1)\pi/\beta$ are the Matsubara frequencies of the fermions. We have introduced the Fourier transformation

$$\varphi(\boldsymbol{r},\tau) = \frac{1}{(\beta V)^{1/2}}\sum_{\boldsymbol{q}}\sum_{\omega_l} \mathrm{e}^{-\mathrm{i}\omega_l\tau + \mathrm{i}\boldsymbol{q}\cdot\boldsymbol{r}}\varphi(\boldsymbol{q},\omega_l) \ , \tag{4.1.27}$$

where φ is a c-number field fulfilling the periodic boundary conditions $\varphi(\boldsymbol{r}, \tau+\beta) = \varphi(\boldsymbol{r}, \tau)$. Therefore, the Matsubara frequencies are quantized as

$$\omega_l = \frac{2\pi l}{\beta} \qquad (l : \text{integer}) \ . \tag{4.1.28}$$

Using the notation $k = (\boldsymbol{k}, \omega_n)$, we write for (4.1.26)

$$\begin{aligned} M_{k,k'} &= (M_0)_{k,k'} + (M_1)_{k,k'} \\ &= -G_0^{-1}(k)\delta_{\boldsymbol{k},\boldsymbol{k}'} + \frac{\mathrm{i}e}{(\beta V)^{1/2}}\varphi(k-k') \ . \end{aligned} \tag{4.1.29}$$

Here, $G_0(k)$ is the Green function of the free fermions

$$G_0(k) = G_0(\boldsymbol{k}, \omega_n) = \frac{1}{\mathrm{i}\omega_n - \xi_{\boldsymbol{k}}} \ , \tag{4.1.30}$$

where $\xi_{\boldsymbol{k}}$ is the energy measured relative to the chemical potential, $\xi_{\boldsymbol{k}} = (\boldsymbol{k}^2)/(2m) - \mu$.

Now, writing the $(\det)^2$ in (4.1.25) in the exponential, the partition function becomes

$$Z = \int \mathcal{D}\varphi \, \mathrm{e}^{-S_{\text{eff}}(\varphi)} \ , \tag{4.1.31}$$

$$S_{\text{eff}}(\varphi) = \int_0^\beta \mathrm{d}\tau \int \mathrm{d}\boldsymbol{r} \frac{(\nabla\varphi(\boldsymbol{r},\tau))^2}{8\pi} - 2\ln\det M \ . \tag{4.1.32}$$

On the other hand, for the matrix M

$$\ln\det M = \operatorname{Tr}\ln M \tag{4.1.33}$$

holds. In the present case of a Hermitian matrix, this equation can be proved by applying an appropriate unitary transformation U to diagonalize M

$$M = U \operatorname{diag}(\dots m_i \dots) U^\dagger \ ,$$

with a diagonal matrix $\operatorname{diag}(\dots m_i \dots)$ of the eigenvalues $\{m_i\}$ of the matrix M. Then, it is easy to see that both sides equal $\sum_i \ln m_i$.

Using (4.1.33), we write

$$\begin{aligned} \ln\det M &= \operatorname{Tr}\ln M = \operatorname{Tr}\ln(M_0 + M_1) \\ &= \operatorname{Tr}\ln(-G_0^{-1} + M_1) = \operatorname{Tr}\ln[(-G_0^{-1})(1 - G_0 M_1)] \\ &= \operatorname{Tr}\ln(-G_0^{-1}) + \operatorname{Tr}\ln(1 - G_0 M_1) \ . \end{aligned} \tag{4.1.34}$$

$\ln(AB) = \ln A + \ln B$ under the trace Tr can be proved similarly to the proof of (4.1.33). Because the first term in (4.1.34) is independent of φ, it is expressed by the sum of states Z_0 at the vanishing interaction $e^2 = 0$. Therefore, we obtain

$$\frac{Z}{Z_0} = \int \mathcal{D}\varphi \exp\left\{ - \int_0^\beta \mathrm{d}\tau \left[\frac{1}{8\pi} (\nabla\varphi)^2 + 2 \operatorname{Tr} \ln(1 - G_0^{-1} M_1) \right] + \frac{\beta}{V} \sum_{q \neq 0} \frac{2\pi e^2}{|\boldsymbol{q}|^2} N \right\} \Big/ \int \mathcal{D}\varphi \exp\left\{ - \int_0^\beta \mathrm{d}\tau \frac{1}{8\pi} (\nabla\varphi)^2 \right\}. \quad (4.1.35)$$

We redefine $-S_{\text{eff}}$ to be the exponent of the numerator. φ is the field that mediates the Coulomb interaction, and is nothing but the time component A_0 of the electromagnetic field introduced in Chap. 1.4, namely the scalar potential in the Coulomb gauge. (4.1.35) is a functional of the scalar potential describing the system by the action S_{eff}.

Here, we might consider expanding S_{eff} in terms of e. Because M_1 in (4.1.29) is proportional to e, it would be sufficient to expand $\operatorname{Tr}\ln(1 - G_0 M_1)$ in M_1. Using the Taylor expansion

$$\ln(1 - x) = - \sum_{n=1}^{\infty} \frac{1}{n} x^n \ , \quad (4.1.36)$$

we obtain

$$- \operatorname{Tr} \ln(1 - G_0 M_1) = \sum_{n=1}^{\infty} \frac{1}{n} \operatorname{Tr} (G_0 M_1)^n \ . \quad (4.1.37)$$

First, we discuss the lowest order term $n = 1$:

$$\begin{aligned} \operatorname{Tr} G_0 M_1 &= \sum_{k,k'} (G_0)_{kk'} (M_1)_{k'k} = \sum_{k,k'} G_0(k) \delta_{k,k'} (M_1)_{k'k} \\ &= \sum_k G_0(k) (M_1)_{kk} = \sum_k G_0(k) \left(\frac{\mathrm{i}e}{(\beta V)^{1/2}} \varphi(0) \right) \ . \end{aligned} \quad (4.1.38)$$

Owing to the neutrality condition with the positively charged background, we omit the interaction at $\boldsymbol{q} = \boldsymbol{0}$. As is clear from (4.1.22), we should therefore set $\varphi(0) = 0$. Then, the right-hand side of (4.1.38) vanishes. Therefore, the lowest order term in e is the $n = 2$ contribution:

$$\begin{aligned} \frac{1}{2} \operatorname{Tr} (G_0 M_1)^2 &= \frac{1}{2} \sum_{k,k'} G_0(k) (M_1)_{k,k'} G_0(k') (M_1)_{k'k} \\ &= \frac{1}{2} \sum_{q,k} G_0(k) (M_1)_{k,k+q} G_0(k+q) (M_1)_{k+q,k} \\ &= -\frac{1}{2} \sum_q \frac{e^2}{\beta V} \left(\sum_k G_0(k) G_0(k+q) \right) \varphi(q) \varphi(-q) \ . \end{aligned} \quad (4.1.39)$$

Defining the polarization function $\pi(q)$ as

$$\pi(q) = +\frac{2}{\beta V} \sum_k G_0(k) G_0(k+q) \ , \quad (4.1.40)$$

then, with regard to (4.1.39), S_{eff} can be written as (4.1.41) + terms of fourth or higher order in $e\varphi$:

$$S_{\text{eff}} = \sum_q \left[\frac{|\boldsymbol{q}|^2}{8\pi} - \frac{1}{2} e^2 \pi(q) \right] \varphi(q)\varphi(-q) + o(e^4\varphi^4) \ . \tag{4.1.41}$$

Notice that because the interaction was at first proportional to e^2, terms with odd power of e vanish.

Ignoring terms in fourth or higher order in $e\varphi$, and writing S_{eff} quadratic in φ corresponds to the so-called RPA approximation. In the limit $r_s \ll 1$, this approximation is justified. Because the equation is quadratic, every single q-component decouples, and every integral becomes a simple Gauss integral. Then, as was discussed in Sect. 2.2, the coefficient of $\varphi(q)\varphi(-q)$ is (half of) the inverse of the Green function of the potential φ:

$$D^{-1}(q) = \frac{|\boldsymbol{q}|^2}{4\pi} - e^2\pi(q) \ . \tag{4.1.42}$$

Using the Green function $D_0(q) = 4\pi/\left|\boldsymbol{q}^2\right|$ of the non-interacting case, and using (4.1.42) we obtain

$$D(q) = \left(D_0(q)^{-1} - e^2\pi(q)\right)^{-1} = D_0(q)\left(1 - e^2 D_0(q)\pi(q)\right)^{-1} \ . \tag{4.1.43}$$

From (4.1.43), we deduce the Dyson equation

$$D(q) = D_0(q) + D_0(q)e^2\pi(q)D(q) \ . \tag{4.1.44}$$

For readers who are familiar with the framework of diagrams, the correspondence between (4.1.43), (4.1.44) and Fig. 4.1 should be evident.

One advantage of (4.1.42) is that contrary to $D(q)$, $D^{-1}(q)$ can be calculated directly, and the trouble with an infinite series occurring when using the Dyson equation can be circumvented. Saying it the other way round, not $D(q)$, but rather $D^{-1}(q)$ has a natural interpretation as the coefficient matrix of a quadratic form.

We now come to the calculation of the polarization function $\pi(q)$. Writing (4.1.40) explicitly, we obtain

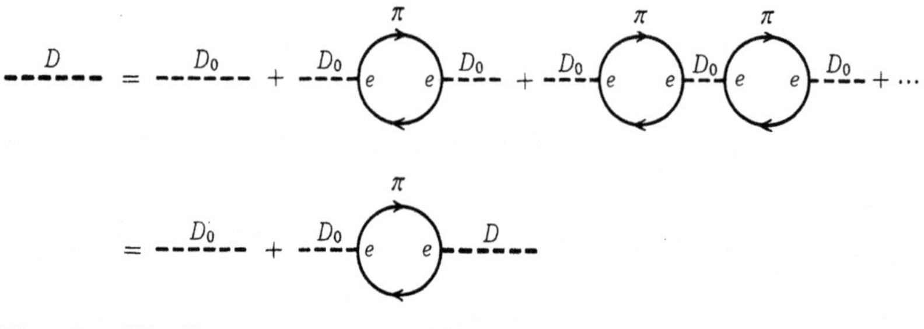

Fig. 4.1. The diagrams represented by the Dyson equation

$$\pi(\boldsymbol{q},\omega_l) = 2\frac{1}{V}\sum_{\boldsymbol{k}}\frac{1}{\beta}\sum_{\omega_n}\frac{1}{\mathrm{i}(\omega_n+\omega_l)-\xi_{\boldsymbol{k}+\boldsymbol{q}}}\cdot\frac{1}{\mathrm{i}\omega_n-\xi_{\boldsymbol{k}}} \quad . \tag{4.1.45}$$

The sum of Matsubara frequencies of the fermions can be performed by using a variation of the discussion of (3.2.21) and thereafter

$$\begin{aligned}\frac{1}{\beta}\sum_{\omega_n}&\frac{1}{\mathrm{i}(\omega_n+\omega_l)-\xi_{\boldsymbol{k}+\boldsymbol{q}}}\cdot\frac{1}{\mathrm{i}\omega_n-\xi_{\boldsymbol{k}}}\\ &= -\oint_{C_0}\frac{\mathrm{d}z}{2\pi\mathrm{i}}f(z)\frac{1}{z+\mathrm{i}\omega_l-\xi_{\boldsymbol{k}+\boldsymbol{q}}}\cdot\frac{1}{z-\xi_{\boldsymbol{k}}}\\ &= \frac{f(\xi_{\boldsymbol{k}})-f(\xi_{\boldsymbol{k}+\boldsymbol{q}})}{\mathrm{i}\omega_l+\xi_{\boldsymbol{k}}-\xi_{\boldsymbol{k}+\boldsymbol{q}}} \quad .\end{aligned} \tag{4.1.46}$$

Here, $f(z) = (\mathrm{e}^{\beta z}+1)^{-1}$ is the Fermi distribution. We are left with the $\boldsymbol{k}$-integration; writing again the formula for $\pi(q)$, and shifting $\boldsymbol{k} \to \boldsymbol{k}-\boldsymbol{q}/2$, we obtain

$$\pi(\boldsymbol{q},\omega_l) = 2\frac{1}{V}\sum_{\boldsymbol{k}}\frac{f(\xi_{\boldsymbol{k}-\boldsymbol{q}/2})-f(\xi_{\boldsymbol{k}+\boldsymbol{q}/2})}{\mathrm{i}\omega_l+\xi_{\boldsymbol{k}-\boldsymbol{q}/2}-\xi_{\boldsymbol{k}+\boldsymbol{q}/2}} \quad . \tag{4.1.47}$$

Here, we expand $\boldsymbol{q}$ for $|\boldsymbol{q}| \ll k_{\mathrm{F}}$:

$$\begin{aligned}&f(\xi_{\boldsymbol{k}+\boldsymbol{q}/2})-f(\xi_{\boldsymbol{k}-\boldsymbol{q}/2})\\ &\quad= \left[f(\xi_{\boldsymbol{k}})+\frac{\partial f(\xi_{\boldsymbol{k}})}{\partial\xi_{\boldsymbol{k}}}(\xi_{\boldsymbol{k}+\boldsymbol{q}/2}-\xi_{\boldsymbol{k}})+\frac{1}{2}\frac{\partial^2 f(\xi_{\boldsymbol{k}})}{\partial\xi_{\boldsymbol{k}}^2}(\xi_{\boldsymbol{k}+\boldsymbol{q}/2}-\xi_{\boldsymbol{k}})^2+\cdots\right]\\ &\qquad-\left[f(\xi_{\boldsymbol{k}})+\frac{\partial f(\xi_{\boldsymbol{k}})}{\partial\xi_{\boldsymbol{k}}}(\xi_{\boldsymbol{k}-\boldsymbol{q}/2}-\xi_{\boldsymbol{k}})+\frac{1}{2}\frac{\partial^2 f(\xi_{\boldsymbol{k}})}{\partial\xi_{\boldsymbol{k}}^2}(\xi_{\boldsymbol{k}-\boldsymbol{q}/2}-\xi_{\boldsymbol{k}})^2+\cdots\right]\\ &\quad= \frac{\partial f(\xi_{\boldsymbol{k}})}{\partial\xi_{\boldsymbol{k}}}\frac{\boldsymbol{k}\cdot\boldsymbol{q}}{m}+O(|\boldsymbol{q}|^3) \ ,\end{aligned} \tag{4.1.48}$$

where we have used

$$\xi_{\boldsymbol{k}\pm\boldsymbol{q}/2} = \xi_{\boldsymbol{k}} \pm \frac{\boldsymbol{k}\cdot\boldsymbol{q}}{2m}+\frac{|\boldsymbol{q}|^2}{8m} \tag{4.1.49}$$

in order to derive the above equation. For the case of zero temperature, we obtain

$$\frac{\partial f(\xi_{\boldsymbol{k}})}{\partial\xi_{\boldsymbol{k}}} = -\delta(\xi_{\boldsymbol{k}}-\varepsilon_{\mathrm{F}}) = -\delta\left(\frac{|\boldsymbol{k}|^2-k_{\mathrm{F}}^2}{2m}\right) = -\frac{m}{k_{\mathrm{F}}}\delta(|\boldsymbol{k}|-k_{\mathrm{F}}) \ . \tag{4.1.50}$$

Using $\boldsymbol{k}\boldsymbol{q} = |\boldsymbol{k}|\,|\boldsymbol{q}|\cos\theta$, (4.1.47) becomes

$$
\begin{aligned}
&\pi(\boldsymbol{q},\omega_l) \\
&\cong \frac{2}{(2\pi)^2}\int_0^\infty |\boldsymbol{k}|^2\,\mathrm{d}|\boldsymbol{k}|\int_{-1}^{1}\mathrm{d}(\cos\theta)\frac{\frac{m}{k_\mathrm{F}}\delta(|\boldsymbol{k}|-k_\mathrm{F})\frac{|\boldsymbol{k}||\boldsymbol{q}|}{m}\cos\theta}{\mathrm{i}\omega_l-\frac{|\boldsymbol{k}||\boldsymbol{q}|}{m}\cos\theta}+O(|\boldsymbol{q}|^3) \\
&= \frac{2}{(2\pi)^2}mk_\mathrm{F}\int_{-1}^{1}\mathrm{d}(\cos\theta)\frac{\frac{k_\mathrm{F}|\boldsymbol{q}|}{m}\cos\theta}{\mathrm{i}\omega_l-\frac{k_\mathrm{F}|\boldsymbol{q}|}{m}\cos\theta}+O(|\boldsymbol{q}|^3)\ . \qquad (4.1.51)
\end{aligned}
$$

Also the $\cos\theta$ integral can be performed immediately, leading to

$$
\begin{aligned}
\pi(\boldsymbol{q},\omega_l) &\cong -2\rho_0\left(1+\frac{\mathrm{i}\omega_l}{2v_\mathrm{F}|\boldsymbol{q}|}\ln\left[\frac{\mathrm{i}\omega_l-v_\mathrm{F}|\boldsymbol{q}|}{\mathrm{i}\omega_l+v_\mathrm{F}|\boldsymbol{q}|}\right]\right) \\
&= -2\rho_0\left(1-\frac{\omega_l}{v_\mathrm{F}|\boldsymbol{q}|}\tan^{-1}\left[\frac{v_\mathrm{F}|\boldsymbol{q}|}{\omega_l}\right]\right)\ . \qquad (4.1.52)
\end{aligned}
$$

Here, $v_\mathrm{F} = k_\mathrm{F}/m$ is the Fermi velocity, and $\rho_0 = 2mk_\mathrm{F}/(2\pi)^2$ is the unit volume at the Fermi energy ε_F, the state density around spins. We investigate (4.1.52) further for two different cases.

Because $|\omega_l|$ is the energy scale of the present problem, and $v_\mathrm{F}\,|\boldsymbol{q}|$ the energy of the exited state with wave vector $\boldsymbol{q}$, the ratio of both determines two different limits, namely the static limit $|\omega_l| \ll v_\mathrm{F}\,|\boldsymbol{q}|$, and the dynamic limit $|\omega_l| \gg v_\mathrm{F}\,|\boldsymbol{q}|$. For the static limit ($|\omega_l| \ll v_\mathrm{F}\,|\boldsymbol{q}|$), we use the expansion $\tan^{-1}x \cong \frac{\pi}{2}\,\mathrm{sgn}\,x(|x| \gg 1)$ to write

$$
\pi(\boldsymbol{q},\omega_l) \cong -2\rho_0\left[1-\frac{\pi}{2}\frac{|\omega_l|}{v_\mathrm{F}|\boldsymbol{q}|}\right]\ . \qquad (4.1.53)
$$

The Green function $D(q)$ of φ becomes

$$
D(\boldsymbol{q},\omega_l) \cong \frac{4\pi}{|\boldsymbol{q}|^2+8\pi e^2\rho_0\left[1-\frac{\pi}{2}\frac{|\omega_l|}{v_\mathrm{F}|\boldsymbol{q}|}\right]}\ . \qquad (4.1.54)
$$

Writing $\mathrm{i}\omega_l \to \omega+\mathrm{i}\delta$, we obtain the retarded Green function by analytical continuation in the upper half plane:

$$
D^\mathrm{R}(\boldsymbol{q},\omega) \cong \frac{4\pi}{|\boldsymbol{q}|^2+8\pi e^2\rho_0\left[1+\frac{i\pi\omega}{2v_\mathrm{F}|\boldsymbol{q}|}\right]}\ . \qquad (4.1.55)
$$

In this equation, the second term in [ ] expresses damping and corresponds to the imaginary part of the self-energy. It becomes clear that for $|\omega_l| < v_\mathrm{F}\,|\boldsymbol{q}|$, the excitation of electron–hole pairs gives rise to a finite life time of φ.

For the exactly static case $\omega = 0$, writing $\lambda = [8\pi e^2 \rho_0]^{-1/2}$, we obtain

$$D(\boldsymbol{q}, \omega_l = 0) = D^{\mathrm{R}}(\boldsymbol{q}, \omega = 0) = \frac{4\pi}{|\boldsymbol{q}|^2 + \lambda^{-2}} \quad . \tag{4.1.56}$$

Performing a Fourier transformation, we obtain

$$D(\boldsymbol{r}, \omega_l = 0) = D^{\mathrm{R}}(\boldsymbol{r}, \omega = 0) = \frac{\mathrm{e}^{-|\boldsymbol{r}|/\lambda}}{|\boldsymbol{r}|} \quad . \tag{4.1.57}$$

Equation (4.1.57) signifies that the Coulomb force does not reach far, but will be damped at a distance of about the magnitude of λ. The physical interpretation can be described as follows. We focus on one electron. Because it is negatively charged, other negative charges around it will tend to move away. As a result, a positively charged cloud with radius λ emerges which is just balancing the negative charge of the original electron. Looking from a distance larger than λ, the system behaves like a neutral particle, and the interaction becomes a short-range interaction. This phenomenon is called screening, and λ is called the screening length.

On the other hand, in the dynamic limit ($|\omega_l| \gg v_F |\boldsymbol{q}|$), using the approximation $\tan^{-1} x \cong x$ for $|x| \ll 1$ we obtain

$$\pi(\boldsymbol{q}, \omega_l) \cong -\frac{2}{3}\rho_0 \frac{v_{\mathrm{F}}^2 |\boldsymbol{q}|^2}{\omega_l} \quad . \tag{4.1.58}$$

With n being the electron density N/V, we make use of the relation

$$\rho_0 v_{\mathrm{F}}^2 = \frac{2k_{\mathrm{F}}^2}{(2\pi)^2 m} = \frac{3n}{2m} \tag{4.1.59}$$

to obtain with $\omega_{\mathrm{p}}^2 = 4\pi n e^2/m$

$$D(\boldsymbol{q}, \omega_l) = \frac{4\pi}{|\boldsymbol{q}|^2 \left[1 + \omega_{\mathrm{p}}^2/\omega_l^2\right]} \quad . \tag{4.1.60}$$

Once again by analytical continuation, we obtain (δ is a positive, infinitesimal constant)

$$D^{\mathrm{R}}(\boldsymbol{q}, \omega) = \frac{4\pi}{|\boldsymbol{q}|^2 \left[1 - \omega_{\mathrm{p}}^2/(\omega + \mathrm{i}\delta)^2\right]} \quad . \tag{4.1.61}$$

At $\omega = \omega_{\mathrm{p}}$, the collective excitement with undamped frequency ω_{p} emerges. This oscillation mode is called plasma oscillation, being an excitation mode where the positively charged background and the negatively charged electrons are moving uniformly ($|\boldsymbol{q}| \to 0$) against each other. ω_{p} is called the plasma frequency.

We conclude that in the two limits $|\omega_l| \ll v_{\mathrm{F}} |\boldsymbol{q}|$ and $|\omega_l| \gg v_{\mathrm{F}} |\boldsymbol{q}|$ the polarization function $\pi(\boldsymbol{q}, \omega_l)$ contains the physical phenomenona of screening and plasma excitation. The ground state energy E_G will be determined by the integral over the whole $q = (\boldsymbol{q}, \omega_l)$ region. We obtain with (4.1.35) and (4.1.41) from

$$\lim_{\beta\to\infty}\left(-\frac{1}{\beta}\ln Z\right) = E_G \tag{4.1.62}$$

the expression

$$E_G - E_{G0} = -\sum_{\boldsymbol{q}} \frac{2\pi e^2}{|\boldsymbol{q}|^2} n + \lim_{\beta\to\infty}\frac{1}{2\beta}\sum_{\boldsymbol{q}}\sum_{\omega_l}\ln\left(\frac{|\boldsymbol{q}|^2/8\pi - \frac{1}{2}e^2\pi(\boldsymbol{q},\omega_l)}{|\boldsymbol{q}|^2/8\pi}\right) . \tag{4.1.63}$$

Here, E_{G0} is the ground state energy of non-interacting ($e = 0$) free electrons. We use the fact that $\frac{1}{\beta}\sum_{\omega_l} \to \int_{-\infty}^{+\infty}(\mathrm{d}\omega)/(2\pi)$ holds for $\beta \to \infty$ to write

$$E_G = E_{G0} + \sum_{\boldsymbol{q}}\left(\int_{-\infty}^{\infty}\frac{\mathrm{d}\omega}{4\pi}\ln\left[1 - \frac{4\pi e^2}{|\boldsymbol{q}|^2}\pi(\boldsymbol{q},\omega)\right] - \frac{2\pi e^2}{|\boldsymbol{q}|^2}n\right) . \tag{4.1.64}$$

This is just the equation that Gell-Mann and Brückner obtained, and the asymptotic expansion in r_s gives rise to

$$E_G = N\left(\frac{2.21}{r_s^2} - \frac{0.916}{r_s} + 0.062\ln r_s - 0.096 + \cdots\right) , \tag{4.1.65}$$

where the Rydberg constant R_H is set to be one. A more detailed discussion can be found in the literature [G.13].

4.2 The Bogoliubov Theory of Superfluidity

Bogoliubov was the first to study a bosonic system with repulsive short-range interaction as a model of superfluidity. Here, we will discuss his theory from the point of view of path integrals. The sum of states of the system can be expressed with the c-number fields $\bar{\psi}(\boldsymbol{r},\tau)$, $\psi(\boldsymbol{r},\tau)$ obeying the periodic boundary conditions $\bar{\psi}(\boldsymbol{r},\tau+\beta) = \bar{\psi}(\boldsymbol{r},\tau)$ and $\psi(\boldsymbol{r},\tau+\beta) = \psi(\boldsymbol{r},\tau)$:

$$Z = \int \mathcal{D}\bar{\psi}\mathcal{D}\psi e^{-s} , \tag{4.2.1}$$

$$\begin{aligned} S = \int_0^\beta \mathrm{d}\tau \int \mathrm{d}\boldsymbol{r}\Big\{&\bar{\psi}(\boldsymbol{r},\tau)\left[\partial_\tau - \frac{1}{2m}\nabla^2 - \mu\right]\psi(\boldsymbol{r},\tau) \\ &+ \frac{1}{2}g[\bar{\psi}(\boldsymbol{r},\tau)\psi(\boldsymbol{r},\tau)]^2\Big\} , \end{aligned} \tag{4.2.2}$$

where g is the interaction constant between the bosons; the repulsive interaction case corresponds to $g > 0$. Equation (4.2.2) equals (2.2.4) for the choice $v(\boldsymbol{r} - \boldsymbol{r}') = g\delta(r - r')$.

As is well known, a non-interacting bosonic system ($g = 0$) undergoes a phase transition at a given temperature T_0 leading to Bose condensation. Because (4.2.2) is a quadratic problem for $g = 0$, it can be solved exactly. In this case, the chemical potential is determined by the condition that the total number of particles is given by N. Using the Green function (2.2.14), the particle number at wave vector $\boldsymbol{k}$ is determined by

$$\begin{aligned} n_{\boldsymbol{k}} = \langle \hat{a}^\dagger_{\boldsymbol{k}} \hat{a}_{\boldsymbol{k}} \rangle &= + \lim_{\tau \to 0-} \langle T \hat{a}_{\boldsymbol{k}}(\tau) \hat{a}^\dagger_{\boldsymbol{k}}(0) \rangle \\ &= - \lim_{\tau \to 0-} G(\boldsymbol{k}, \tau) \\ &= - \lim_{\tau \to 0-} \frac{1}{\beta} \sum_{\omega_n} \frac{\mathrm{e}^{-\mathrm{i}\omega_n \tau}}{\mathrm{i}\omega_n - \xi_{\boldsymbol{k}}} \ . \end{aligned} \tag{4.2.3}$$

Here, we make some remarks about the sign of τ. When rewriting the sum that runs over ω_n in (4.2.3) as a complex integral, there is the subtlety whether the integral running over the half circle at $|z| \to \infty$ vanishes or not. Explicitly, the function $g(z)$ having a first-order pole at $z = \mathrm{i}\omega_n$ must fulfil the condition

$$\lim_{|z| \to \infty} g(z) \, \mathrm{e}^{-z\tau} = 0 \ .$$

In the present case, with $\omega_n = 2\pi n T$ for $\tau < 0$, when using the Bose distribution $n(z) = (\mathrm{e}^{\beta z} - 1)^{-1}$ for $g(z)$, the above equation is fulfilled. However, for $\tau > 0$, it is necessary to use $g(z) = n(z) + 1 = (1 - \mathrm{e}^{\beta z})^{-1}$. The same remarks apply to the fermionic case as well.

Having this remark in mind, (4.2.3) leads to

$$n_{\boldsymbol{k}} = n(\xi_{\boldsymbol{k}}) \tag{4.2.4}$$

in the same manner as in Sect. 3.2.

$$\sum_{\boldsymbol{k}} n_{\boldsymbol{k}} = \sum_{\boldsymbol{k}} n(\xi_{\boldsymbol{k}}) \equiv N(\mu) = N \ . \tag{4.2.5}$$

This equation determines μ. Because $N(\mu)$ has the form of the Bose distribution, it is an increasing function in μ and a decreasing function in $\beta = T^{-1}$. Therefore, at lower temperature (for larger β), μ also increases. However, there exists an upper limit μ_c for the value of μ. As can be understood directly from (4.2.2), this upper limit is given by $\mu_\mathrm{c} = 0$. For $g = 0$, (4.2.2) becomes

$$S = \sum_{\omega_n} \sum_{\boldsymbol{k}} \left[-\mathrm{i}\omega_n + \frac{|\boldsymbol{k}|^2}{2m} - \mu \right] \bar{\psi}(\boldsymbol{k}, \omega_n) \psi(\boldsymbol{k}, \omega_n) \ . \tag{4.2.6}$$

In particular the $\omega_n = 0$, $\boldsymbol{k} = \boldsymbol{0}$ component is given by $-\mu\bar{\psi}(0)\psi(0)$. If μ were positive, then the action (energy) would become small for infinitely large $\bar{\psi}(0)\psi(0)$, and the system unstable. Therefore, $\mu = \mu(T)$ is negative and approaches zero when the temperature decreases, and finally becomes $\mu(T_0) = 0$ at some temperature T_0.

When lowering the temperature below T_0, μ is still zero; however, $\bar{\psi}(\boldsymbol{0}, 0)$ and $\psi(\boldsymbol{0}, 0)$ reach a finite value. That is, splitting $\bar{\psi}(\boldsymbol{r}, 0)$ and $\psi(\boldsymbol{r}, 0)$ into a condensed part $\bar{\psi}_0, \psi_0$ and a non-condensed part $\bar{\psi}_1, \psi_1$

$$\begin{aligned} \bar{\psi}(\boldsymbol{r}, \tau) &= \psi_0^* + \bar{\psi}_1(\boldsymbol{r}, \tau) \ , \\ \psi(\boldsymbol{r}, \tau) &= \psi_0 + \psi_1(\boldsymbol{r}, \tau) \ , \end{aligned} \tag{4.2.7}$$

(4.2.6) becomes

$$S = \sum_{\omega_n} \sum_{\boldsymbol{k}} \left[-\mathrm{i}\omega_n + \frac{|\boldsymbol{k}|^2}{2m} \right] \bar{\psi}_1(\boldsymbol{k}, \omega_n) \psi_1(\boldsymbol{k}, \omega_n) \tag{4.2.8}$$

and (4.2.5) becomes an equation that determines instead of μ now the value of $|\psi_0|$:

$$|\psi_0|^2 + \sum_{\boldsymbol{k} \neq \boldsymbol{0}} n(\xi_{\boldsymbol{k}})|_{\mu=0} = N \ . \tag{4.2.9}$$

Above, we gave a short review of the Bose condensation and, as will become clear in the second half of this section, for $g = 0$, superfluidity cannot occur. Here, we only want to stress that following (4.2.8), the excited states in the Bose condensation phase obey the dispersion relation $\omega_{\boldsymbol{k}} = |\boldsymbol{k}|^2 / 2m$.

For the case when the interaction g is finite, for large $\bar{\psi}\psi$, the non-linear term in (4.2.2) surely dominates the quadratic term in $\bar{\psi}, \psi$ and guarantees the stability of the system. Therefore, μ also is expected to reach positive values. In this case, the part S_0 of the action that depends only on ψ_0^*, ψ_0 reads in the notation of (4.2.7)

$$S_0 = \beta V \left[-\mu|\psi_0|^2 + \tfrac{1}{2} g |\psi_0|^4 \right] \ . \tag{4.2.10}$$

As shown in Fig. 4.2, the potential is shaped like the bottom of a wine bottle, and reaches a degenerate minimum at the finite value $|\psi_0|^2 = \mu/g$.

In the present case, the meaning of spontaneous symmetry breaking, as discussed in Sect. 3.1, is the fixing of the phase of ψ_0. The Bogoliubov theory consists of a Gaussian approximation up to second order in the small fluctuations $\bar{\psi}_1$ and ψ_1 around this minimum. Here, for simplicity, we fix the phase ψ_0 by requiring $\psi_0 =$ real, without loss of generality. Expanding the action up to second order in $\bar{\psi}_1$ and ψ_1 leads to

$$\begin{aligned} S \cong S_0 + \int_0^\beta \mathrm{d}\tau \int \mathrm{d}\boldsymbol{r} \Bigg\{ & \bar{\psi}_1 \left[\partial_\tau - \frac{1}{2m}\nabla^2 - \mu \right] \psi_1 \\ & + \frac{1}{2} g \psi_0^2 (\bar{\psi}_1^2 + \psi_1^2 + 4\bar{\psi}_1\psi_1) \Bigg\} \ . \end{aligned} \tag{4.2.11}$$

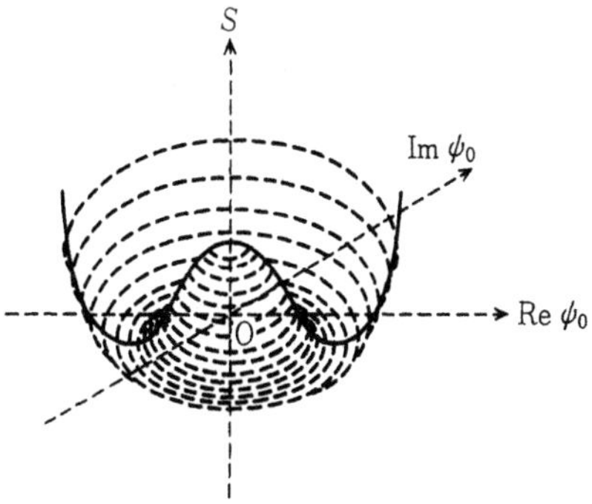

Fig. 4.2. The 'potential' for the homogeneous part ψ_0

Here, μ and ψ_0 must be determined by requiring that the particle number is N; however, we will postpone this calculation and assume μ and ψ_0 as given. Here, only the relation $\mu = g\psi_0^2$ between both is necessary. Then, from (4.2.11) we obtain

$$S \cong S_0 + \int_0^\beta d\tau \int d\boldsymbol{r} \left\{ \bar{\psi}_1 \left[\partial_\tau - \frac{1}{2m}\nabla^2 \right] \psi_1 + \frac{1}{2} g\psi_0^2 (\bar{\psi}_1 + \psi_1)^2 \right\} . \quad (4.2.12)$$

Introducing the real fields A and P for $\bar{\psi}_1$ and ψ_1 by writing

$$\psi_1(\boldsymbol{r}, \tau) = A(\boldsymbol{r}, \tau) + \mathrm{i}P(\boldsymbol{r}, \tau) \quad (4.2.13)$$

and

$$\bar{\psi}_1(\boldsymbol{r}, \tau) = A(\boldsymbol{r}, \tau) - \mathrm{i}P(\boldsymbol{r}, \tau) \ , \quad (4.2.14)$$

we obtain

$$S \cong S_0 + \int_0^\beta d\tau \int d\boldsymbol{r} \left\{ A \left[-\frac{1}{2m}\nabla^2 + 2g\psi_0^2 \right] A + P \left[-\frac{1}{2m}\nabla^2 \right] P + 2\mathrm{i}A\partial_\tau P \right\} . \quad (4.2.15)$$

In order to derive (4.2.15), we used Gauss theorem

$$\int d\boldsymbol{r} \, (A\nabla^2 P - P\nabla^2 A) = \int d\boldsymbol{r} \, \nabla \cdot (A\nabla P - P\nabla A) = 0$$

and

$$\int_0^\beta d\tau \, A\partial_\tau A = \frac{1}{2}[A^2]_0^\beta = 0 \ .$$

As is clear from Fig. 4.3, the meaning of the fields A and B introduced in (4.2.13) and (4.2.14) is the following. A corresponds to the degree of freedom of increasing and decreasing the amplitude of ψ_0. P is orthogonal to it, corresponding to the degree of freedom that alters the phase a little. The

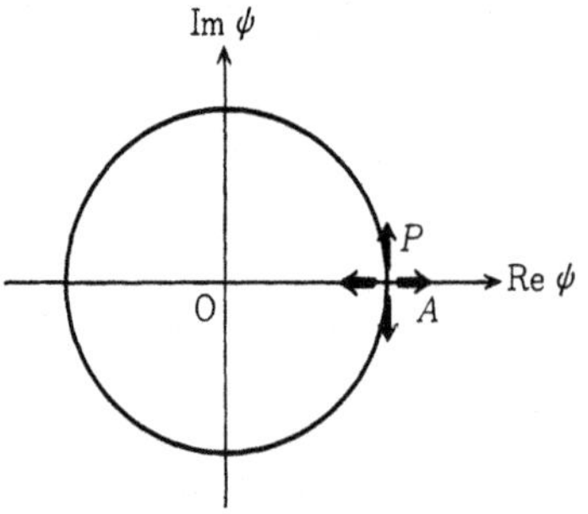

Fig. 4.3. The excitation mode A and the phase mode B

third term in { } of (4.2.15) shows that A and P are canonical conjugate to each other, and both build up a harmonic oscillator together. This mode corresponds to the movement in the continuous minimum of the potential of Fig. 4.2, and is nothing but the Goldstone mode introduced in Sect. 3.2. Performing a Fourier transformation in (4.2.15), we obtain

$$S \cong S_0 + \sum_{\boldsymbol{k},\omega_n} {}^{\mathrm{t}}\begin{bmatrix} A(-\boldsymbol{k},-\omega_n) \\ P(-\boldsymbol{k},-\omega_n) \end{bmatrix} \begin{bmatrix} \dfrac{|\boldsymbol{k}|^2}{2m} + 2g\psi_0^2 & \omega_n \\ -\omega_n & \dfrac{|\boldsymbol{k}|^2}{2m} \end{bmatrix} \begin{bmatrix} A(\boldsymbol{k},\omega_n) \\ P(\boldsymbol{k},\omega_n) \end{bmatrix} . \tag{4.2.16}$$

We can determine the Green function

$$\begin{aligned} -\left\langle \begin{bmatrix} A(\boldsymbol{k},\omega_n) \\ P(\boldsymbol{k},\omega_n) \end{bmatrix} {}^{\mathrm{t}}\begin{bmatrix} A(-\boldsymbol{k},-\omega_n) \\ P(-\boldsymbol{k},-\omega_n) \end{bmatrix} \right\rangle &= -\frac{1}{2} \begin{bmatrix} \dfrac{|\boldsymbol{k}|^2}{2m} + 2g\psi_0^2 & \omega_n \\ -\omega_n & \dfrac{|\boldsymbol{k}|^2}{2m} \end{bmatrix}^{-1} \\ &= -\frac{1}{2} \frac{1}{\dfrac{|\boldsymbol{k}|^2}{2m}\left(\dfrac{|\boldsymbol{k}|^2}{2m} + 2g\psi_0^2\right) + \omega_n^2} \begin{bmatrix} \dfrac{|\boldsymbol{k}|^2}{2m} & -\omega_n \\ \omega_n & \dfrac{|\boldsymbol{k}|^2}{2m} + 2g\psi_0^2 \end{bmatrix} . \end{aligned} \tag{4.2.17}$$

The Green function has a first-order pole at

$$\mathrm{i}\omega_n = \pm\sqrt{\frac{|\boldsymbol{k}|^2}{2m}\left(\frac{|\boldsymbol{k}|^2}{2m} + 2g\psi_0^2\right)} = \pm\frac{|\boldsymbol{k}|\sqrt{|\boldsymbol{k}|^2 + k_0^2}}{2m} \equiv \pm\omega_{\boldsymbol{k}} . \tag{4.2.18}$$

This is the dispersion relation of the excited states. For small $|\boldsymbol{k}|$ ($|\boldsymbol{k}| \ll k_0 \equiv 2\sqrt{mg}\psi_0$), it approaches the dispersion relation of acoustic waves,

$$\omega_{\boldsymbol{k}} \cong \sqrt{\frac{g}{m}}\psi_0|\boldsymbol{k}| = c|\boldsymbol{k}| , \tag{4.2.19}$$

and for large $|\boldsymbol{k}|$ ($|\boldsymbol{k}| \gg k_0 \equiv 2\sqrt{mg}\psi_0$), it approaches the dispersion relation of a free particle,

$$\omega_{\boldsymbol{k}} \cong \frac{|\boldsymbol{k}|^2}{2m} \ . \tag{4.2.20}$$

Bogoliubov was the first to derive the dispersion relation (4.2.18), showing that owing to the interaction g, the dispersion $\propto |\boldsymbol{k}|^2$ alters to the dispersion of acoustic waves $\propto |\boldsymbol{k}|$. This fact leads to superfluidity, which will be explained from a slightly different point of view at the end of this section.

To conclude the Bogoliubov theory, we determine μ and ψ_0. From (4.2.13) and (4.2.14), we obtain

$$\begin{aligned}\langle\psi_1^\dagger(\boldsymbol{k})\psi_1(\boldsymbol{k})\rangle &= \langle A(-\boldsymbol{k})A(\boldsymbol{k})\rangle + \langle P(-\boldsymbol{k})P(\boldsymbol{k})\rangle \\ &\quad + \mathrm{i}\langle A(-\boldsymbol{k})P(\boldsymbol{k})\rangle - \mathrm{i}\langle P(-\boldsymbol{k})A(\boldsymbol{k})\rangle \ .\end{aligned} \tag{4.2.21}$$

Using (4.2.17), with δ being an infinitesimal small positive number, we conclude

$$\begin{aligned}&\left\langle \begin{bmatrix} A(\boldsymbol{k}) \\ P(\boldsymbol{k}) \end{bmatrix}^{\mathrm{t}} \begin{bmatrix} A(-\boldsymbol{k}) \\ P(-\boldsymbol{k}) \end{bmatrix} \right\rangle \\ &= \frac{1}{\beta}\sum_{\omega_n} \mathrm{e}^{\mathrm{i}\omega_n\delta} \left\langle \begin{bmatrix} A(\boldsymbol{k},\omega_n) \\ P(\boldsymbol{k},\omega_n) \end{bmatrix}^{\mathrm{t}} \begin{bmatrix} A(-\boldsymbol{k},-\omega_n) \\ P(-\boldsymbol{k},-\omega_n) \end{bmatrix} \right\rangle \\ &= \frac{1}{\beta}\sum_{\omega_n} \frac{\mathrm{e}^{\mathrm{i}\omega_n\delta}}{2(\omega_{\boldsymbol{k}}^2+\omega_n^2)} \begin{bmatrix} \dfrac{|\boldsymbol{k}|^2}{2m} & -\omega_n \\ \omega_n & \dfrac{|\boldsymbol{k}|^2}{2m} + 2g\psi_0^2 \end{bmatrix} \ .\end{aligned} \tag{4.2.22}$$

We conclude from (4.2.21) that

$$n_{\boldsymbol{k}} = \langle\hat\psi_1^\dagger(\boldsymbol{k})\hat\psi_1(\boldsymbol{k})\rangle = \frac{1}{\beta}\sum_{\omega_n} \frac{\mathrm{e}^{\mathrm{i}\omega_n\delta}}{\omega_{\boldsymbol{k}}^2+\omega_n^2}\left\{\frac{|\boldsymbol{k}|^2}{2m} + g\psi_0^2 + \mathrm{i}\omega_n\right\} \ . \tag{4.2.23}$$

Setting $g = 0$, we obtain $\omega_{\boldsymbol{k}} = \left|\boldsymbol{k}^2\right|/2m$, therefore (4.2.23) leads back to (4.2.3) in the case $\mu = 0$.

For simplicity, we restrict our discussion to zero temperature, where the sum in (4.2.23) becomes an integral, which can be evaluated as the complex line integral, leading to

$$n_{\boldsymbol{k}} = \langle\hat\psi_1^\dagger(\boldsymbol{k})\hat\psi_1(\boldsymbol{k})\rangle = \frac{|\boldsymbol{k}|^2/2m + g\psi_0^2 - \omega_{\boldsymbol{k}}}{2\omega_{\boldsymbol{k}}} = \frac{1}{2}\left(\frac{|\boldsymbol{k}|^2 + k_0^2/2}{|\boldsymbol{k}|\sqrt{|\boldsymbol{k}|^2+k_0^2}} - 1\right) \ .$$

Integrating over $\boldsymbol{k}$, we obtain

$$\sum_{\boldsymbol{k}} n_{\boldsymbol{k}} = \frac{V}{(2\pi)^3}\cdot 4\pi\int_0^\infty |\boldsymbol{k}|^2\,\mathrm{d}|\boldsymbol{k}|\,n_{\boldsymbol{k}} = \gamma V k_0^3 \ . \tag{4.2.24}$$

Here, γ is a constant that can be determined by the following converging integral:

$$\gamma = \frac{1}{4\pi^2}\int_0^\infty \mathrm{d}x\, x\left(\frac{x^2+1/2}{\sqrt{x^2+1}} - x\right) = \frac{1}{24\pi^2} \ .$$

Therefore, the condensed part ψ_0^2 can be determined by the equation

$$N = V\psi_0^2 + \sum_{\boldsymbol{k}} n_{\boldsymbol{k}} = V\left[\psi_0^2 + \frac{1}{3\pi^2}(mg)^{3/2}\psi_0^3\right] \ . \tag{4.2.25}$$

The first term on the right-hand side of the above equation is the classical solution, and the second term corresponds to the Gaussian fluctuation around it. We performed an expansion where the fluctuations are assumed to be small, therefore the second term must be smaller than the first term. This leads to the condition

$$\psi_0 \sim \left(\frac{N}{V}\right)^{1/2} \ll (mg)^{-3/2} \ . \tag{4.2.26}$$

Under this condition we can write approximately

$$\psi_0^2 \approx \frac{N}{V} - \frac{1}{3\pi^2}(mg)^{3/2}\left(\frac{N}{V}\right)^{3/2} \ . \tag{4.2.27}$$

The Bogoliubov theory as described above is appropriate in the dilute limit or when the coupling is small. The fluctuations around the classical, symmetry breaking solution (the saddle-point) $\psi = \psi_0$ and $\bar{\psi} = \psi_0^*$ are assumed to be small, and can therefore be expanded up to second order and treated as Gaussian fluctuations. This path integral evaluation method using the classical solution plus the Gaussian fluctuation is called the saddle-point method in mathematics, and can in physical terms be described as the "mean field + RPA" method.

However, as discussed in Sect. 3.2, and mentioned in Sect. 3.4 as Elitzur's theorem in relation to lattice gauge theory, in low-dimensional systems, or for the case when the system is locally gauge invariant, no spontaneous symmetry breaking occurs, and the mean field approximation becomes even qualitatively meaningless. Of course, also the assumption breaks down that the fluctuations around the mean field are small. Therefore, it is necessary to build up another theoretical description. In order to do so, we refer back to Fig. 4.2. The potential of $|\psi_0|^2$ has a minimum at $|\psi_0|^2 = \mu/g$, and when $|\psi_0|^2$ is altered, a retrospective force is acting. However, there is no energy loss for a global phase transformation ($\boldsymbol{k} = 0$), and long waves have only a small excitation energy (the Goldstone mode of Sect. 3.2).

Explicitly, using the phase and amplitude of ψ and $\bar{\psi}$, we write

$$\begin{aligned}\psi(\boldsymbol{r},\tau) &= [\rho(\boldsymbol{r},\tau)]^{1/2}\,\mathrm{e}^{\mathrm{i}\theta(\boldsymbol{r},\tau)}\\ \bar{\psi}(\boldsymbol{r},\tau) &= [\rho(\boldsymbol{r},\tau)]^{1/2}\,\mathrm{e}^{-\mathrm{i}\theta(\boldsymbol{r},\tau)}\end{aligned} \tag{4.2.28}$$

and insert these expressions into the action (4.2.2). We obtain

$$S = \int_0^\beta \mathrm{d}\tau \int \mathrm{d}\boldsymbol{r} \left\{ \mathrm{i}\rho\,\partial_\tau\theta + \frac{1}{2m}\left[\frac{1}{4\rho}(\nabla\rho)^2 + \rho(\nabla\theta)^2\right] - (\mu+\delta\mu)\rho + \frac{1}{2}g\rho^2 \right\} . \tag{4.2.29}$$

Here, $i\rho\partial_\tau\theta$ is the Berry phase that was discussed in Chap. 2, signifying that the particle number (density) ρ and the phase θ are canonical conjugate.

In order to extract the information about ρ in (4.2.29), we write the chemical potential as the sum of a constant value μ and a test field $\delta\mu(\boldsymbol{r},\tau)$. Now, we consider the case when μ is positive (low temperature). Then, ρ can be expressed as

$$\rho = \psi_0^2 + \delta\rho = \frac{\mu}{g} + \delta\rho \ , \tag{4.2.30}$$

and although the fluctuation $\delta\rho$ is assumed to be small, qualitatively this can be a good estimation. On the other hand, we are interested in the case of low energy, so we assume ω_n and $\boldsymbol{k}$, or ∂_τ and ∇ to be small. Writing approximately for these terms up to second order $\int \mathrm{d}\tau \int \mathrm{d}\boldsymbol{r}\,\delta\rho = \int \mathrm{d}\tau \int \mathrm{d}\boldsymbol{r}\,\delta\mu = 0$ and ignoring the term $(\nabla\delta\rho)^2$, up to constants, the action is given by

$$S \cong \int_0^\beta \mathrm{d}\tau \int \mathrm{d}\boldsymbol{r} \left\{ \mathrm{i}\delta\rho\,\partial_\tau\theta + \frac{\psi_0^2}{2m}(\nabla\theta)^2 + \frac{1}{2}g(\delta\rho)^2 - \delta\mu\,\delta\rho \right\} . \tag{4.2.31}$$

Notice that it has been assumed that the derivative of θ is small; however, θ itself may reach large values. Therefore, (4.2.31) can also describe the scenario described in Sect. 3.2, where the fluctuation of the phase becomes large and finally the order disappears. Therefore, the state $\langle\psi\rangle = \langle\bar{\psi}\rangle = 0$ can also be described [see (5.3.9)].

In fact, (4.2.31) describes thoroughly the low-energy physics of a superfluid. The first term $\mathrm{i}\delta\rho\,\partial_\tau\theta$ has already been discussed. The second term arises due to the effect of $\psi_0 \neq 0$, representing the rigidity of θ [see Sect. 3.1]. That is, an elastic energy proportional to the square of the phase difference $\nabla\theta$ emerges. The phase becomes solid as if it were a rigid body. The third term signifies that the fluctuation of ρ feels a finite retrospective force, and the fourth term is the test field with respect to $\delta\rho$.

Before we discuss this action, we determine the particle current density $\boldsymbol{j}(\boldsymbol{r},\tau)$. From (1.3.20), we obtain

$$\begin{aligned} \boldsymbol{j}(\boldsymbol{r},\tau) &= \frac{1}{2mi}\left\{\bar{\psi}(\boldsymbol{r},\tau)\nabla\psi(\boldsymbol{r},\tau) - \nabla\bar{\psi}(\boldsymbol{r},\tau)\psi(\boldsymbol{r},\tau)\right\} \\ &= \frac{\rho(\boldsymbol{r},\tau)}{m}\nabla\theta(\boldsymbol{r},\tau) \cong \frac{\psi_0^2}{m}\nabla\theta(\boldsymbol{r},\tau) \ . \end{aligned} \tag{4.2.32}$$

Here, we have applied the same approximation as to the action and considered only the lowest-order term.

Now, because (4.2.31) is a kind of harmonic oscillator action, the Heisenberg equations of motion of the field operators that can be deduced from this action following from $\delta S = 0$ read

$$\frac{\partial\hat{\theta}(\boldsymbol{r},t)}{\partial t} = \delta\mu(\boldsymbol{r},t) - g\delta\hat{\rho}(\boldsymbol{r},t) \equiv \delta\hat{\mu}_{\text{eff}}(\boldsymbol{r},t) \ , \tag{4.2.33}$$

$$\frac{\partial\hat{\rho}(\boldsymbol{r},t)}{\partial t} = -\frac{\psi_0^2}{m}\nabla^2\hat{\theta}(\boldsymbol{r},t) = -\nabla\cdot\hat{\boldsymbol{j}}(\boldsymbol{r},t) \ . \tag{4.2.34}$$

Here, we returned to the real-time formalism, $\tau \to it$. In $\delta\hat{\mu}_{\text{eff}} \equiv \delta\mu(\boldsymbol{r},t) - g\delta\hat{\rho}(\boldsymbol{r},t)$, the part coming from the interaction is implemented in the effective chemical potential that is subtracted from the average value μ.

Equations (4.2.33) and (4.2.34) are the Josephson equations describing superfluidity. For example, even when the chemical potential is totally homogeneous ($\delta\mu = 0$), $\langle\hat{\theta}(\boldsymbol{r},t)\rangle = \theta(\boldsymbol{r})$ is time independent, but space dependent. Therefore, a constant particle current $\langle\hat{\jmath}(\boldsymbol{r},t)\rangle = j(\boldsymbol{r})$ is flowing.

Performing the Fourier transformation of (4.2.31), writing $k = (\boldsymbol{k},\omega_n)$ and setting $\delta\mu = 0$, we obtain

$$S \cong \sum_{k} {}^{\mathrm{t}}\begin{bmatrix}\delta\rho(-k)\\ \theta(-k)\end{bmatrix}\begin{bmatrix}\frac{1}{2}g & \frac{\omega_n}{2}\\ -\frac{\omega_n}{2} & \frac{\psi_0^2}{2m}|\boldsymbol{k}|^2\end{bmatrix}\begin{bmatrix}\delta\rho(k)\\ \theta(k)\end{bmatrix} \ . \tag{4.2.35}$$

In the same manner as for (4.2.17), from this expression we obtain

$$\begin{aligned}\left\langle\begin{bmatrix}\delta\rho(k)\\ \theta(k)\end{bmatrix}{}^{\mathrm{t}}\begin{bmatrix}\delta\rho(-k)\\ \theta(-k)\end{bmatrix}\right\rangle &= \frac{1}{2}\frac{1}{\frac{g\psi_0^2}{4m}|\boldsymbol{k}|^2 + \frac{\omega_n^2}{4}}\begin{bmatrix}\frac{\psi_0^2}{2m}|\boldsymbol{k}|^2 & -\frac{\omega_n}{2}\\ \frac{\omega_n}{2} & \frac{1}{2}g\end{bmatrix}\\ &= \frac{1}{\omega_n^2+\omega_{\boldsymbol{k}}^2}\begin{bmatrix}\frac{\psi_0^2}{m}|\boldsymbol{k}|^2 & -\omega_n\\ \omega_n & g\end{bmatrix} \ .\end{aligned} \tag{4.2.36}$$

In this case, $\omega_{\boldsymbol{k}}$ is given by

$$\omega_{\boldsymbol{k}} = \sqrt{\frac{g}{m}}\psi_0|\boldsymbol{k}| = c|\boldsymbol{k}| \ . \tag{4.2.37}$$

This expression equals the long-wavelength limit (4.2.19) of (4.2.18), which is obvious due to the properties of the approximation described above. The dispersion relation of this acoustic wave model and rigidity have a one-to-one correspondence. Therefore, in the case $g = 0$, even when Bose condensation occurs, superfluidity does not arise.

Finally, we calculate the equal time density correlation function $S(\boldsymbol{k})$:

$$\begin{aligned}\frac{N}{V}S(\boldsymbol{k}) = \langle\delta\rho(\boldsymbol{k})\delta\rho(-\boldsymbol{k})\rangle &= \frac{1}{\beta}\sum_{\omega_n}\frac{\frac{\psi_0^2}{m}|\boldsymbol{k}|^2}{\omega_n^2+\omega_{\boldsymbol{k}}^2}\\ &\xrightarrow[\beta\to 0]{} \frac{\frac{\psi_0^2}{2m}|\boldsymbol{k}|^2}{\omega_{\boldsymbol{k}}} \ .\end{aligned} \tag{4.2.38}$$

In the approximation $\psi_0^2 = N/V$, the relation between $S(\boldsymbol{k})$ and $\omega_{\boldsymbol{k}}$ at zero temperature is therefore given by

$$\omega_{\boldsymbol{k}} = \frac{|\boldsymbol{k}|^2}{2mS(\boldsymbol{k})} \ . \tag{4.2.39}$$

This equation signifies that the long-range correlation ($S(\boldsymbol{k}) \propto |\boldsymbol{k}|$) of the boson density leads to the acoustic wave dispersion $\omega_{\boldsymbol{k}} = c\,|\boldsymbol{k}|$. This equation is called the Feynman relation.

Above, we examined the effective action describing the low-energy dynamics of the phase. We conclude that also for $\langle\psi\rangle = \langle\bar{\psi}\rangle = 0$ the two-dimensional bosonic system can become superfluid. However, because in the above discussion the degree of freedom of vortices has not been included, it is necessary to make an even more detailed analysis. This will be done in Sect. 5.3, when two-dimensional superconductors will be discussed.

5. Problems Related to Superconductivity

In solid state physics, superconductivity is an extremely important topic, because it is one of the phenomena with universality. The quantum mechanical phase of the electrons in some sense gains rigidity as if it were a rigid body, and as a result the properties of the quantum wave show up at the macroscopic level. In this chapter, several problems related to this quantal phase are discussed.

5.1 Superconductivity and Path Integrals

In Sect. 4.1 we discussed the problem of the Coulomb interaction of an electron gas. We reduced the problem to the field theory of a bosonic field $\hat{\rho} = \hat{\psi}_\sigma^\dagger \psi_\sigma$, constructed from the fermionic fields $\hat{\psi}_\sigma^\dagger$ and ψ_σ, and the conjugate potential $\hat{\varphi}$ of this bosonic field. This description can be regarded as a close-up of the density in the particle picture. So, which kind of bosonic field can describe the wave picture of the electronic system? We start our discussion with this problem.

As mentioned in Sect. 1.2, $\mathrm{e}^{-\mathrm{i}m\theta/\hbar}$ and $\mathrm{e}^{\mathrm{i}m\theta/\hbar}$ are operators increasing and decreasing, respectively, the particle number by m. We search for a bosonic operator constructed from two fermionic operators having this property. The simplest objects that can be constructed are the product of two creation operators, or two annihilation operators:

$$\hat{\Phi}^\dagger_{\sigma_1\sigma_2}(\boldsymbol{r}_1, \boldsymbol{r}_2) = \hat{\psi}^\dagger_{\sigma_1}(\boldsymbol{r}_1)\hat{\psi}^\dagger_{\sigma_2}(\boldsymbol{r}_2) \ , \tag{5.1.1}$$

$$\hat{\Phi}_{\sigma_1\sigma_2}(\boldsymbol{r}_1, \boldsymbol{r}_2) = \hat{\psi}_{\sigma_2}(\boldsymbol{r}_2)\hat{\psi}_{\sigma_1}(\boldsymbol{r}_1) \ . \tag{5.1.2}$$

These operators act by raising or lowering the electron number by 2. Introducing in (5.1.1) and (5.1.2) instead of $\boldsymbol{r}_1$ and $\boldsymbol{r}_2$ the relative coordinate $\boldsymbol{r}$ and the centre of mass coordinate $\boldsymbol{R}$

$$\begin{aligned} \boldsymbol{r} &= \boldsymbol{r}_1 - \boldsymbol{r}_2 \ , \\ \boldsymbol{R} &= \frac{\boldsymbol{r}_1 + \boldsymbol{r}_2}{2} \ , \end{aligned} \tag{5.1.3}$$

we obtain

$$\begin{aligned}
&\hat{\Phi}^{\dagger}_{\sigma_1\sigma_2}(\boldsymbol{r},\boldsymbol{R}) \\
&= \hat{\psi}^{\dagger}_{\sigma_1}\left(\boldsymbol{R}+\frac{\boldsymbol{r}}{2}\right)\hat{\psi}^{\dagger}_{\sigma_2}\left(\boldsymbol{R}-\frac{\boldsymbol{r}}{2}\right) \\
&= \frac{1}{V}\sum_{\boldsymbol{k}_1,\boldsymbol{k}_2}\exp[-\mathrm{i}(\boldsymbol{k}_1+\boldsymbol{k}_2)\cdot\boldsymbol{R}]\,\exp[-\frac{\mathrm{i}}{2}(\boldsymbol{k}_1-\boldsymbol{k}_2)\cdot\boldsymbol{r}]\,\hat{C}^{\dagger}_{\sigma_1}(\boldsymbol{k}_1)\hat{C}^{\dagger}_{\sigma_2}(\boldsymbol{k}_2)\ .
\end{aligned} \tag{5.1.4}$$

This is the description of the processes of creation of an electron with spin σ_1 and wave number $\boldsymbol{k}_1$ and an electron with spin σ_2 and wave number $\boldsymbol{k}_2$. The product of both is represented by $\hat{\Phi}^{\dagger}$. Performing a Fourier transformation with respect to the centre of mass coordinate, for (5.1.4) we can write

$$\hat{\Phi}^{\dagger}_{\sigma_1\sigma_2}(\boldsymbol{r},\boldsymbol{R}) = \frac{1}{\sqrt{V}}\sum_{q}\mathrm{e}^{-\mathrm{i}\boldsymbol{Q}\cdot\boldsymbol{R}}\,\hat{\Phi}^{\dagger}_{\sigma_1\sigma_2}(\boldsymbol{r},\boldsymbol{Q})\ , \tag{5.1.5}$$

$$\hat{\Phi}^{\dagger}_{\sigma_1\sigma_2}(\boldsymbol{r},\boldsymbol{Q}) = \frac{1}{\sqrt{V}}\sum_{\boldsymbol{k}}\mathrm{e}^{-\mathrm{i}(\boldsymbol{k}-\boldsymbol{Q}/2)\cdot\boldsymbol{r}}\hat{C}^{\dagger}_{\sigma_1}(\boldsymbol{k})\hat{C}^{\dagger}_{\sigma_2}(\boldsymbol{Q}-\boldsymbol{k})\ . \tag{5.1.6}$$

At low temperature $k_{\mathrm{B}}T \ll \varepsilon_{\mathrm{F}}$, Fermi degeneracy occurs in the fermionic system, and all states from wave vector $|\boldsymbol{k}| = 0$ to k_{F} are occupied. A new electron can only be created with a wave vector larger than k_{F} ($|\boldsymbol{k}| \geq k_{\mathrm{F}}, |\boldsymbol{Q}-\boldsymbol{k}| \geq k_{\mathrm{F}}$). As a result, $\boldsymbol{k}-\boldsymbol{Q}/2$ in (5.1.6) is of the order of the wave vector k_{F}, and for the relative coordinate $\boldsymbol{r}$, the region $|\boldsymbol{r}| \simeq k_{\mathrm{F}}^{-1}$ is relevant. For a typical metal, k_{F}^{-1} is about the size of the lattice constant a. Therefore, the typical scale of the relative coordinate is given by $\boldsymbol{r} \simeq a$. The reader might think that the length scale characterizing the relative coordinate can be identified with the correlation length ξ of the superconductor that will be introduced later on. This is also an important scale; when considering the wave number space, the states in the range of $\pm\xi^{-1}$ around the Fermi wave number k_{F} do contribute. In the present discussion, the scale that determines the oscillations with $\sim k_{\mathrm{F}}$ is important.

Now, on the basis of the above discussion, we reconsider the interaction between the electrons. The Hamiltonian can be expressed in terms of the density $\hat{\rho}$, or the fields $\hat{\Phi}^{\dagger}$ and $\hat{\Phi}$:

$$\begin{aligned}
\mathcal{H}_{\mathrm{int}} &= \frac{1}{2}\int \mathrm{d}\boldsymbol{r}_1\,\mathrm{d}\boldsymbol{r}_2\,\hat{\psi}^{\dagger}_{\sigma_1}(\boldsymbol{r}_1)\hat{\psi}^{\dagger}_{\sigma_2}(\boldsymbol{r}_2)v(\boldsymbol{r}_1-\boldsymbol{r}_2)\hat{\psi}_{\sigma_2}(\boldsymbol{r}_2)\hat{\psi}_{\sigma_1}(\boldsymbol{r}_1) \\
&= \frac{1}{2}\int \mathrm{d}\boldsymbol{r}_1\,\mathrm{d}\boldsymbol{r}_2\,v(\boldsymbol{r}_1-\boldsymbol{r}_2)\,\{\hat{\rho}(\boldsymbol{r}_1)\hat{\rho}(\boldsymbol{r}_2)-\delta(\boldsymbol{r}_1-\boldsymbol{r}_2)\hat{\rho}(\boldsymbol{r}_1)\} \quad (=\mathcal{H}_1) \\
&= \frac{1}{2}\int \mathrm{d}\boldsymbol{r}\,\mathrm{d}\boldsymbol{R}\,\hat{\Phi}^{\dagger}_{\sigma_1\sigma_2}(\boldsymbol{r},\boldsymbol{R})v(\boldsymbol{r})\hat{\Phi}_{\sigma_1\sigma_2}(\boldsymbol{r},\boldsymbol{R}) \quad (=\mathcal{H}_2)\ .
\end{aligned} \tag{5.1.7}$$

As can be seen in (5.1.7), there are two different ways to express $\mathcal{H}_{\mathrm{int}}$, signifying that it contains two different types of physics, corresponding to the particle picture and the wave picture, respectively. Because (5.1.7) is exact, when proceeding in the calculation without approximations, the result should be the same regardless of which expression for the Hamiltonian has been used. However, because in Sect. 4.2 the approximation "mean field +RPA" has been applied, the result differs depending on the expression. Here, a physical picture or intuition is necessary, because no general method exists.

Having two descriptions as is the case in (5.1.7), the most physically reasonable approximation is

$$\mathcal{H}_{\text{int}} \to \mathcal{H}_1 + \mathcal{H}_2 \ . \tag{5.1.8}$$

We perform the Fourier transformation for $\rho(\boldsymbol{r})$ with respect to $\boldsymbol{r}$ in $\mathcal{H}_1$, and for $\hat{\Phi}(\boldsymbol{r}, \boldsymbol{R})$ with respect to $\boldsymbol{R}$ in $\mathcal{H}_2$ and call the wave number $\boldsymbol{Q}$, respectively. We restrict the summation over $\boldsymbol{Q}$ in $\mathcal{H}_1$ and $\mathcal{H}_2$ on the right-hand side of (5.1.7) to the region where $|\boldsymbol{Q}|$ is small. Then, when writing $\mathcal{H}_1$ and $\mathcal{H}_2$ in terms of the original wave numbers of the fermions, the summation runs over almost disjunct regions. If the summation over $\boldsymbol{Q}$ were performed over the whole region, then we would obtain $\mathcal{H}_{\text{int}} = \frac{1}{2}(\mathcal{H}_1 + \mathcal{H}_2)$. However, in the approximate discussion that will be applied in what follows, for the correct description of the two kinds of physics that are contained in $\mathcal{H}_{\text{int}}$, the factor 1/2 must be omitted.

$\mathcal{H}_1$ has already been discussed in Sect. 4.1. In this chapter, we will investigate $\mathcal{H}_2$. First, we introduce an approximation to $\mathcal{H}_2$.

As explained above, for the relative coordinate $\boldsymbol{r}$ occurring in $\hat{\Phi}^\dagger$ and $\hat{\Phi}$, the dominant contribution comes from $|\boldsymbol{r}| \sim a$. Therefore, only the region $|\boldsymbol{r}| \simeq a$ of the interaction $v(\boldsymbol{r})$ in $\mathcal{H}_2$ is relevant. Even when the Coulomb interaction $v(\boldsymbol{r}) = e^2/|\boldsymbol{r}|$ is a long-range interaction, this is not important for $\mathcal{H}_2$. When only the properties of wave lengths longer than a are considered, we might replace $v(\boldsymbol{r})$ by an effective delta function potential:

$$v(\boldsymbol{r}) \to U\delta(\boldsymbol{r}) \ . \tag{5.1.9}$$

Then, the $\boldsymbol{r}$ integral in $\mathcal{H}_2$ can be performed, leading to

$$\mathcal{H}_2 = \frac{U}{2} \int \mathrm{d}\boldsymbol{R}\, \hat{\Phi}^\dagger_{\sigma_1\sigma_2}(\boldsymbol{0}, \boldsymbol{R}) \Phi_{\sigma_1\sigma_2}(\boldsymbol{0}, \boldsymbol{R}) \ . \tag{5.1.10}$$

Because of $\boldsymbol{r} = \boldsymbol{0}$, for the case $\sigma_1 = \sigma_2$, $\hat{\Phi}_{\sigma_1\sigma_2}(\boldsymbol{0}, \boldsymbol{R}) = 0$ automatically holds.

$$\mathcal{H}_2 = U \int \mathrm{d}\boldsymbol{R}\, \hat{\Phi}^\dagger_{\uparrow\downarrow}(\boldsymbol{0}, \boldsymbol{R}) \hat{\Phi}_{\uparrow\downarrow}(\boldsymbol{0}, \boldsymbol{R}) = U \sum_{\boldsymbol{q}} \hat{\Phi}^\dagger_{\uparrow\downarrow}(\boldsymbol{0}, \boldsymbol{Q}) \hat{\Phi}_{\uparrow\downarrow}(\boldsymbol{0}, \boldsymbol{Q}) \ . \tag{5.1.11}$$

Since the Coulomb interaction is repulsive, we might think that U is positive as a matter of course. However, in the crystal the lattice oscillation and the electrons are coupled, leading to an attractive force due to phonons for energies smaller than the Debye frequency ω_{D}. In order to describe this feature, we integrate out successively the high-energy excitations of the electronic system and derive an effective action describing the phenomenon at the energy scale below ω_{D}. First, for excitations with energy higher than ω_{D}, because

U is positive, we perform the Stratonovich–Hubbart transformation as was done in Sect. 4.1:

$$\begin{aligned}
&\exp\left[-U\int_0^\beta d\tau\int d\boldsymbol{R}\bar{\Phi}_{\uparrow\downarrow}(\boldsymbol{0},\boldsymbol{R},\tau)\Phi_{\uparrow\downarrow}(\boldsymbol{0},\boldsymbol{R},\tau)\right]\\
&=\int\mathcal{D}\bar{\Delta}\mathcal{D}\Delta\exp\left[-\int_0^\beta d\tau\int d\boldsymbol{R}\left\{\frac{1}{U}\bar{\Delta}(\boldsymbol{R},\tau)\Delta(\boldsymbol{R},\tau)\right.\right.\\
&\quad\left.\left.+\,i\bar{\Delta}(\boldsymbol{R},\tau)\Phi_{\uparrow\downarrow}(\boldsymbol{0},\boldsymbol{R},\tau)+i\Delta(\boldsymbol{R},\tau)\bar{\Phi}_{\uparrow\downarrow}(\boldsymbol{0},\boldsymbol{R},\tau)\right\}\right]\\
&=\int\mathcal{D}\bar{\Delta}\mathcal{D}\Delta\exp\left[-\int_0^\beta d\tau\int d\boldsymbol{R}\left\{\frac{1}{U}\bar{\Delta}(\boldsymbol{R},\tau)\Delta(\boldsymbol{R},\tau)\right.\right.\\
&\quad\left.\left.+\,i\bar{\Delta}(\boldsymbol{R},\tau)\psi_\downarrow(\boldsymbol{R},\tau)\psi_\uparrow(\boldsymbol{R},\tau)+i\Delta(\boldsymbol{R},\tau)\bar{\psi}_\uparrow(\boldsymbol{R},\tau)\bar{\psi}_\downarrow(\boldsymbol{R},\tau)\right\}\right]\ .
\end{aligned} \tag{5.1.12}$$

Here, we introduce the cut-off Λ_0 for the energy $\xi_{\boldsymbol{k}} = |\boldsymbol{k}|^2/2m - \varepsilon_F$:

$$\begin{aligned}
\psi_\sigma(\boldsymbol{R},\tau) &= \frac{1}{\sqrt{\beta V}}\sum_{\omega_n}\sum_{|\xi_{\boldsymbol{k}}|<\Lambda_0} e^{+i\boldsymbol{k}\cdot\boldsymbol{R}-i\omega_n\tau}\psi_\sigma(\boldsymbol{k},\omega_n)\ ,\\
\bar{\psi}_\sigma(\boldsymbol{R},\tau) &= \frac{1}{\sqrt{\beta V}}\sum_{\omega_n}\sum_{|\xi_{\boldsymbol{k}}|<\Lambda_0} e^{-i\boldsymbol{k}\cdot\boldsymbol{R}+i\omega_n\tau}\bar{\psi}_\sigma(\boldsymbol{k},\omega_n)\ .
\end{aligned} \tag{5.1.13}$$

First, we choose Λ_0 to be of the order of the Fermi energy ε_F. The partition function becomes

$$Z = \int\mathcal{D}\bar{\psi}_\sigma\mathcal{D}\psi_\sigma\mathcal{D}\bar{\Delta}\mathcal{D}\Delta\, e^{-S(\bar{\psi}_\sigma,\psi_\sigma,\bar{\Delta},\Delta)}\ , \tag{5.1.14}$$

$$\begin{aligned}
&S(\bar{\psi}_\sigma,\psi_\sigma,\bar{\Delta},\Delta)\\
&=\sum_{\omega_l,\boldsymbol{Q}}\frac{1}{U}\bar{\Delta}(\boldsymbol{Q},\omega_l)\Delta(\boldsymbol{Q},\omega_l)\\
&\quad+\sum_{\substack{\boldsymbol{k},\omega_n\\|\xi_{\boldsymbol{k}}|<\Delta_0}}(-i\omega_n+\xi_{\boldsymbol{k}})\bar{\psi}_\sigma(\boldsymbol{k},\omega_n)\psi_\sigma(\boldsymbol{k},\omega_n)\\
&\quad+i\frac{1}{\sqrt{\beta V}}\sum_{\substack{\boldsymbol{k},\omega_n\\|\xi_{\boldsymbol{k}}|<\Lambda_0}}\sum_{\substack{\boldsymbol{Q},\omega_l\\|\xi_{\boldsymbol{Q}-\boldsymbol{k}}|<\Lambda_0}}\bar{\Delta}(\boldsymbol{Q},\omega_l)\psi_\downarrow(\boldsymbol{Q}-\boldsymbol{k},\omega_l-\omega_n)\psi_\uparrow(\boldsymbol{k},\omega_n)\\
&\quad+i\frac{1}{\sqrt{\beta V}}\sum_{\substack{\boldsymbol{k},\omega_n\\|\xi_{\boldsymbol{k}}|<\Lambda_0}}\sum_{\substack{\boldsymbol{k},\omega_l\\|\xi_{\boldsymbol{Q}-\boldsymbol{k}}|<\Lambda_0}}\Delta(\boldsymbol{Q},\omega_l)\bar{\psi}_\uparrow(\boldsymbol{k},\omega_n)\bar{\psi}_\downarrow(\boldsymbol{Q}-\boldsymbol{k},\omega_l-\omega_n)\ .
\end{aligned} \tag{5.1.15}$$

Now, we need to integrate out the high-energy components $\Lambda_0 - \mathrm{d}\Lambda < |\xi_{\boldsymbol{k}}| < \Lambda_0$. Doing so, the factor $1/U$ in the first term on the right-hand side in (5.1.15) will in general be renormalized to $1/U(\boldsymbol{Q}, \omega_l, \Lambda_0 - d\Lambda)$. Because we are now interested in the low-energy dynamics, we expand in $\boldsymbol{Q}$ and ω_l and consider the zeroth-order term, that is, we set $\boldsymbol{Q} = \boldsymbol{0}$ and $\omega_l = 0$. Then, the parts S_0 of the action that are relevant for the renormalization of $U(\Lambda) = U(\boldsymbol{Q} = \boldsymbol{0}, \omega_l = 0, \Lambda)$ are given by

$$\begin{aligned}
S_0(\Lambda) &= \frac{1}{U(\Lambda)}\bar{\Delta}(0)\Delta(0) + \sum_{\substack{\omega_n, \boldsymbol{k} \\ |\xi_{\boldsymbol{k}}|<\Lambda}} (-\mathrm{i}\omega_n + \xi_{\boldsymbol{k}})\bar{\psi}_\sigma(\boldsymbol{k}, \omega_n)\psi_\sigma(\boldsymbol{k}, \omega_n) \\
&\quad + \mathrm{i}\frac{\bar{\Delta}(0)}{\sqrt{\beta V}} \sum_{\substack{\boldsymbol{k}, \omega_n \\ |\xi_{\boldsymbol{k}}|<\Lambda}} \psi_\downarrow(-\boldsymbol{k}, -\omega_n)\psi_\uparrow(\boldsymbol{k}, \omega_n) \\
&\quad + \mathrm{i}\frac{\Delta(0)}{\sqrt{\beta V}} \sum_{\substack{\boldsymbol{k}, \omega_n \\ |\xi_{\boldsymbol{k}}|<\Lambda}} \bar{\psi}_\uparrow(\boldsymbol{k}, \omega_n)\bar{\psi}_\downarrow(-\boldsymbol{k}, -\omega_n) \\
&= \frac{1}{U(\Lambda)}\bar{\Delta}(0)\Delta(0) \\
&\quad + \sum_{\substack{\boldsymbol{k}, \omega_n \\ |\xi_{\boldsymbol{k}}|<\Lambda}} {}^{\mathrm{t}}\begin{bmatrix} \bar{\psi}_\uparrow(\boldsymbol{k}, \omega_n) \\ \psi_\downarrow(-\boldsymbol{k}, -\omega_n) \end{bmatrix} \begin{bmatrix} -\mathrm{i}\omega_n + \xi_{\boldsymbol{k}} & \mathrm{i}\frac{1}{\sqrt{\beta V}}\Delta(0) \\ \mathrm{i}\frac{1}{\sqrt{\beta V}}\bar{\Delta}(0) & -\mathrm{i}\omega_n - \xi_{\boldsymbol{k}} \end{bmatrix} \begin{bmatrix} \psi_\uparrow(\boldsymbol{k}, \omega_n) \\ \bar{\psi}_\downarrow(-\boldsymbol{k}, -\omega_n) \end{bmatrix}.
\end{aligned} \tag{5.1.16}$$

Writing $U(\Lambda = \Lambda_0) = U$ and performing the integral over $\Lambda_0 - \mathrm{d}\Lambda < |\xi_{\boldsymbol{k}}| < \Lambda_0$, from

$$\int_{\Delta - \mathrm{d}\Lambda < |\xi_{\boldsymbol{k}}| < \Lambda} \mathcal{D}\bar{\psi}\mathcal{D}\psi\, \mathrm{e}^{-S_0(\Lambda)} = \mathrm{e}^{-S_0(\Lambda - \mathrm{d}\Lambda)} \tag{5.1.17}$$

we obtain the expression

$$\begin{aligned}
&\frac{1}{U(\Lambda - \mathrm{d}\Lambda)}\bar{\Delta}(0)\Delta(0) \\
&\quad = \frac{1}{U(\Lambda)}\bar{\Delta}(0)\Delta(0) - \sum_{\omega_n}\sum_{\boldsymbol{k}}{}' \ln\det \begin{bmatrix} -\mathrm{i}\omega_n + \xi_{\boldsymbol{k}} & \mathrm{i}\frac{1}{\sqrt{\beta V}}\Delta(0) \\ \mathrm{i}\frac{1}{\sqrt{\beta V}}\bar{\Delta}(0) & -\mathrm{i}\omega_n - \xi_{\boldsymbol{k}} \end{bmatrix} \\
&\quad = \frac{1}{U(\Lambda)}\bar{\Delta}(0)\Delta(0) - \sum_{\omega_n}\sum_{\boldsymbol{k}}{}' \ln\left[-\omega_n^2 - \xi_{\boldsymbol{k}}^2 + \frac{1}{\beta V}\bar{\Delta}(0)\Delta(0)\right] + \text{const.}
\end{aligned} \tag{5.1.18}$$

Here, we defined $\sum_{\boldsymbol{k}}' = \sum_{\boldsymbol{k}, \Lambda - d\Lambda < |\xi_{\boldsymbol{k}}| < \Lambda}$. Expanding the logarithm with respect to $1/(\beta V)\bar{\Delta}(0)\Delta(0)$ and comparing the coefficients of $\bar{\Delta}(0)\Delta(0)$, we obtain

$$\frac{1}{U(\Lambda - \mathrm{d}\Lambda)} = \frac{1}{U(\Lambda)} + \frac{1}{\beta V} \sum_{\omega_n} \sum_{k}{}' \frac{1}{\omega_n^2 + \xi_k^2} \ . \tag{5.1.19}$$

Assuming $k_{\mathrm{B}}T \ll \Lambda$, and writing an integral instead of the sum in ω_n, we obtain

$$\frac{1}{\beta V} \sum_{\omega_n} \sum_{k}{}' \frac{1}{\omega_n^2 + \xi_k^2} \cong \frac{1}{V} \sum_{k}{}' \int_{-\infty}^{\infty} \frac{\mathrm{d}\omega}{2\pi} \frac{1}{\omega^2 + \xi_k^2} = \frac{1}{2V} \sum_{k}{}' \frac{1}{|\xi_k|} = \rho_0 \frac{\mathrm{d}\Lambda}{\Lambda} \ . \tag{5.1.20}$$

Here, ρ_0 is the (density of states/spin×volume) on the Fermi surface. We therefore obtain

$$\frac{1}{U(\Lambda - \mathrm{d}\Lambda)} = \frac{1}{U(\Lambda)} + \rho_0 \frac{\mathrm{d}\Lambda}{\Lambda} \tag{5.1.21}$$

or

$$\frac{\mathrm{d}}{\mathrm{d}\ln\Lambda} \left[\frac{1}{U(\Lambda)} \right] = -\rho_0 \ . \tag{5.1.21$'$}$$

We observe that the effective repulsive potential $U(\Lambda)$ becomes smaller for lower energies. This is due to the fact that because of the repulsive potential, other electrons no longer come close, and they feel the interaction less strongely. Integrating (5.1.21′) from $\Lambda = \Lambda_0$ down to $\Lambda = \omega_{\mathrm{D}}$, we obtain

$$\frac{1}{U(\omega_{\mathrm{D}})} - \frac{1}{U(\Lambda_0)} = -\rho_0 \ln \frac{\omega_{\mathrm{D}}}{\Lambda_0} \ , \tag{5.1.22}$$

and because $U(\Lambda_0) = U$, we obtain

$$U(\omega_{\mathrm{D}}) = \frac{U}{1 + \rho_0 U \ln \dfrac{\Lambda_0}{\omega_{\mathrm{D}}}} \ . \tag{5.1.22$'$}$$

With $\Lambda_0 \sim \varepsilon_{\mathrm{F}} \sim 10^4\,\mathrm{K}$ and $\omega_D \sim 10^2\,\mathrm{K}$, we obtain $\ln \varepsilon_{\mathrm{F}}/\omega_{\mathrm{D}} \sim 5$. Writing the repulsive force dimensionless, $\mu^* = \rho_0 U(\omega_{\mathrm{D}})$, using $\mu_0 = \rho_0 U(\omega_{\mathrm{D}})$ (this is not the chemical potential) we obtain

$$\mu^* = \rho_0 U(\omega_{\mathrm{D}}) = \frac{\rho_0 U}{1 + \rho_0 U \ln \dfrac{\Lambda_0}{\omega_{\mathrm{D}}}} = \frac{\mu}{1 + \mu \ln \dfrac{\Lambda_0}{\omega_{\mathrm{D}}}} \leq \frac{1}{\ln \dfrac{\Lambda_0}{\omega_{\mathrm{D}}}} \ . \tag{5.1.23}$$

μ^* becomes even smaller than 1/5. As a result, the attractive force of weak phonons may dominate the repulsive force. In this case, for low energies ($|\omega| < \omega_{\mathrm{D}}$) an effective attractive force acts between the electrons, justifying the Hamiltonian of the BCS theory.

In a very direct manner, this can be interpreted as follows. An electron with energy of about ε_{F} moving with high speed has an average rest time in the zone around one ion of about $1/\varepsilon_{\mathrm{F}}$. On the other hand, the motion of the ion (phonon) has the time scale of about $1/\omega_D$, which is (compared with the electrons) very slow. When an electron arrives at an ion, the energy of the nearby nucleus is relaxed, and therefore a good relation with the guest (the electron) at this place will be established. However, although the electron is moving away in the time $1/\varepsilon_F$, even without a guest the ion provides good accommodation for electrons during the time $1/\omega_D$. Now, let us assume that another traveller (electron) comes its way. He will be attracted by the place where good accommodation is provided. Because the electrons do not meet each other, although they do not have a good relation to each other, both feel attracted by the same host. By this mechanism, the attractive force in the BCS theory arises, leading to superconductivity. Before we enter into details, we give an overview of superconductivity. In the following discussion, we set $\hbar = c = 1$.

When we have to explain in a word what superconductivity means, the answer might be that the quantum mechanical phase of the electron system becomes "solid" as if it were a rigid body, and gains rigidity. In this sense, this phenomenon is analogous to superfluidity described in Sect. 4.2. The difference between both is that electrons are charged and coupled through the electromagnetic field, but superfluids (for example, helium) are neutral. As a result, in superconductors remarkable electromagnetic phenomena emerge.

We consider a sample of length L where the phase of the wave function $\varphi_i(\boldsymbol{r})$ fulfils the boundary condition that it differs on both sides only by $\Delta\theta$. Therefore, after performing the gauge transformation

$$\varphi_i(\boldsymbol{r}) = \mathrm{e}^{\mathrm{i}(\Delta\theta/L)x}\bar{\varphi}_i(\boldsymbol{r}) \tag{5.1.24}$$

we might require just the normal periodic boundary conditions for $\bar{\varphi}_i$. Corresponding to (5.1.24), the differential operator in the Hamiltonian becomes

$$\nabla\varphi_i(\boldsymbol{r}) = \nabla\left[\mathrm{e}^{\mathrm{i}(\Delta\theta/L)x}\bar{\varphi}_i(\boldsymbol{r})\right] = \mathrm{e}^{\mathrm{i}(\Delta\theta/L)x}\left[\nabla + \mathrm{i}\frac{\Delta\theta}{L}\hat{\boldsymbol{e}}_x\right]\bar{\varphi}_i(\boldsymbol{r}) \ , \tag{5.1.25}$$

which just corresponds to the presence of a vector potential $\boldsymbol{A} = (\Delta\theta/eL)\hat{\boldsymbol{e}}_x$. Here, $\hat{\boldsymbol{e}}_x = (1,0,0)$ is the unit vector in the x direction. Concerning the twisting angle $\Delta\theta$, when the degree of freedom of the phase behaves like a solid body, there should arise a "macroscopic" growth $K(\Delta\theta)^2/2$ of the free energy, with K being the coefficient of rigidity. Correspondingly, owing to the vector potential, the energy increase is just

$$\Delta G = \frac{e^2\rho_\mathrm{s}}{2m}\int \boldsymbol{A}^2\,\mathrm{d}V \ . \tag{5.1.26}$$

Here, ρ_s is the so-called superfluid density, becoming smaller with increasing temperature and zero at the phase transition temperature T_c.

Because the current density $\boldsymbol{J}$ is the derivative with respect to $\boldsymbol{A}$ of the free energy, we obtain from (5.1.26)

$$J = -\frac{\delta \Delta G}{\delta A(r)} = -\frac{e^2 \rho_s}{m} A(r) . \tag{5.1.27}$$

This is nothing but the London equation. More precisely, using the Maxwell equation $\mathrm{rot}\, \boldsymbol{H} = \mathrm{rot}\ \mathrm{rot}\, \boldsymbol{A} = \mathrm{grad}(\mathrm{div}\, \boldsymbol{A}) - \Delta \boldsymbol{A}$, we obtain the London equation:

$$\begin{aligned} \mathrm{rot}\, \boldsymbol{H} &= \mathrm{rot}\,\mathrm{rot}\, \boldsymbol{A} = \mathrm{grad}(\mathrm{div}\, \boldsymbol{A}) - \Delta \boldsymbol{A} \\ &= -\Delta \boldsymbol{A} = -\frac{4\pi e^2 \rho_s}{m} \boldsymbol{A} . \end{aligned} \tag{5.1.28}$$

Here, the Coulomb gauge condition $\mathrm{div} \boldsymbol{A} = 0$ has been applied for the vector potential. From (5.1.28) we conclude that the penetration depth λ of the magnetic field is given by $(4\pi e^2 \rho_s / m)^{-1/2}$. The magnetic flux does not reach deeper inside, which corresponds to the well-known Meissner effect. Tracing it back to its origin, we come back to the rigidity expressed in (5.1.26). Here, we discuss more in detail the significance of (5.1.26).

First, it is important to notice that (5.1.26) is the increase in free energy, and describes a thermal equilibrium state. That is, no energy dissipation occurs. This is due to the fact that only the electric field $\boldsymbol{E}$ can do work in the system, but not the magnetic field $\boldsymbol{H}$. Because we are thinking of time-independent $\boldsymbol{A}$ at the moment, the electric field $\boldsymbol{E}$ can be expressed only in terms of the scalar potential ϕ:

$$\boldsymbol{E} = -\nabla \phi . \tag{5.1.29}$$

The energy dissipation becomes

$$Q = \int \mathrm{d}V\, \boldsymbol{J} \cdot \boldsymbol{E} = \int \mathrm{d}V \frac{1}{4\pi\lambda^2} \boldsymbol{A} \cdot \boldsymbol{\nabla} \phi = -\int \mathrm{d}V \frac{1}{4\pi\lambda^2} \phi\, \mathrm{div}\, \boldsymbol{A} = 0 \tag{5.1.30}$$

and therefore vanishes. Here, we used the Gauss theorem, and the Coulomb gauge condition $\mathrm{div}\, \boldsymbol{A} = 0$. Therefore, $\boldsymbol{A}$ describes a magnetic field, and the current given in (5.1.27) expresses the corresponding induced diamagnetic current density.

So, what is the difference from the diamagnetic current density occurring in a non-superconducting material? The difference is that the vector potential itself appears in (5.1.26). In a normal conductive material, the presence of a magnetic field $\boldsymbol{H} = \mathrm{rot}\, \boldsymbol{A}$ leads also to an increase in the free energy, but in any case the expression contains the term $\boldsymbol{H} = \mathrm{rot}\, \boldsymbol{A}$; the vector potential $\boldsymbol{A}$ without derivatives does not appear. Explicitely, one can write

$$\Delta G = \int \frac{1}{2} \chi \boldsymbol{H}^2 \, \mathrm{d}V = \int \frac{1}{2} \chi (\mathrm{rot}\, \boldsymbol{A})^2 \, \mathrm{d}V . \tag{5.1.31}$$

Here, χ is Landau's diamagnetic susceptibility. The magnetization $\boldsymbol{M}$ is given by the derivative of ΔG with respect to $\boldsymbol{H}$:

$$\boldsymbol{M} = -\frac{\delta \Delta G}{\delta \boldsymbol{H}} = -\chi \boldsymbol{H} \ . \tag{5.1.32}$$

On the other hand, the diamagnetic current density is a derivative of (5.1.31) with respect to $\boldsymbol{A}$:

$$\boldsymbol{J} = -\frac{\delta \Delta G}{\delta \boldsymbol{A}} = -\chi \, \mathrm{rot} \, \mathrm{rot} \, \boldsymbol{A} = \mathrm{rot} \, \boldsymbol{M} \tag{5.1.33}$$

leading to a very well-known equation. Now, we consider the total diamagnetic current flow I through one intersection of a sample. Introducing a surface S that is large enough, containing this intersection and reaching also the vacuum outside of the sample, we conclude from Stokes theorem for the current I:

$$I = \int_S \boldsymbol{J} \cdot \mathrm{d}\boldsymbol{S} = \int_S \mathrm{rot} \, \boldsymbol{M} \cdot \mathrm{d}\boldsymbol{S} = \int_{\partial S} \boldsymbol{M} \cdot \mathrm{d}\boldsymbol{l} \ . \tag{5.1.34}$$

Here, ∂S is the boundary of the surface, and because the magnetization $\boldsymbol{M}$ is zero outside of the sample, the integral vanishes. In this manner, the diamagnetic current in a normal conductive material must cancel out and become zero in total. On the other hand, the superconductive diamagnetic current (5.1.27) is a macroscopic current which can be observed. Roughly speaking, (5.1.27) is the diamagnetic current (5.1.33) that becomes macroscopically visible and, correspondingly, until the magnetic flux is totally repelled, a strong diamagnetic current (perfect diamagnetism) emerges.

As mentioned above, the vanishing resistance for superconductors signifies that the superconducting diamagnetic current also flows when the electric potential difference is zero. The notion of the limit of a small electrical resistance ρ easily leads to misunderstandings. Rather, a physical picture is that the electrons moving chaotically in the normal conduction phase take each others' hands in the superconductor phase to rebuff the enemy (magnetic field) as a united community. Saying it in another way, the current density $\boldsymbol{J}$ consists of a paramagnetic current $\boldsymbol{J}_{\mathrm{p}}$ and a diamagnetic current $\boldsymbol{J}_{\mathrm{d}}$. The paramagnetic current $\boldsymbol{J}_{\mathrm{p}}$ will be defined later on in (5.1.70) as a contribution due to the strain of the wave function under the electromagnetic field. On the other hand, the diamagnetic current $\boldsymbol{J}_{\mathrm{d}}$ is given by (5.1.27) when replacing ρ_{s} by the electron density $\sum_\sigma \bar{\psi}_\sigma(x)\psi_\sigma(x)$. In the normal conducting phase $\boldsymbol{J}_{\mathrm{p}}$ and $\boldsymbol{J}_{\mathrm{d}}$ almost cancel each other, and only the small Landau diamagnetic current (5.1.33) remains. In the superconducting phase, as has been mentioned several times, the wave function gains rigidity and is no longer influenced by the electromagnetic field. As a result, the cancellation no longer occurs, and (5.2.27) describes the perfect diamagnetic current density.

So, what might be the microscopic mechanism that leads to a "rigid phase"? The BCS theory gives an answer to this question, as will be explained in what follows. The phenomenon that the quantum mechanical phase reaches macroscopic dimensions (this is called coherency) is realized for the Bose condensation, as was described in Sect. 4.2. However, because electrons are fermions, they cannot all condensate in the ground state. But when pairs of two electrons attracting each other (called Cooper pairs) occur, these bosons consisting of two fermions do condensate. Because the Cooper pair carries the charge $-2e$, the system becomes superconducting. However, this picture is not exactly correct, because the separation of the bound state energy for a pair of electrons from the continuum would require an enormous attractive force of the order of ε_F (some eV). Considering the fact that the fundamental interaction between the electrons is the repulsive Coulomb force, this seems to be impossible. In a realistic metal, the electron states in the vicinity of the Fermi surface reorganize themselves in such a way that bosonic fields, which are the fields $\hat{\Phi}^\dagger$ and $\hat{\Phi}$ given in (5.1.1) and (5.1.2), do condensate.

To describe this feature, we start with the following BCS Hamiltonian:

$$H = \int \mathrm{d}\boldsymbol{r}\, \hat{\psi}^\dagger_\sigma(\boldsymbol{r}) \left[-\frac{1}{2m}\nabla^2 - \mu \right] \hat{\psi}_\sigma(\boldsymbol{r}) - g \int \mathrm{d}\boldsymbol{r}\, \hat{\psi}^\dagger_\uparrow(\boldsymbol{r}) \hat{\psi}^\dagger_\downarrow(\boldsymbol{r}) \hat{\psi}_\downarrow(\boldsymbol{r}) \hat{\psi}_\uparrow(\boldsymbol{r})$$
$$+ \frac{1}{2} \int \mathrm{d}\boldsymbol{r}\, \mathrm{d}\boldsymbol{r}'\, \hat{\psi}^\dagger_\sigma(\boldsymbol{r}) \hat{\psi}^\dagger_{\sigma'}(\boldsymbol{r}') \frac{e^2}{|\boldsymbol{r}-\boldsymbol{r}'|} \hat{\psi}_{\sigma'}(\boldsymbol{r}') \hat{\psi}_\sigma(\boldsymbol{r}) \ . \tag{5.1.35}$$

As has already been mentioned, this is the effective Hamiltonian describing the low-energy states in a width of order ω_D in the vicinity of the Fermi surface. The attractive force g in the second term is given by

$$\lambda = \rho_0 g = \lambda_0 - \mu^* \ , \tag{5.1.36}$$

with g_0 being the attractive force due to the phonons (written dimensionless with $\lambda_0 = \rho_0 g_0$).

Using again the Stratonovich–Hubbard transformation, the partition function can be written as

$$Z = \int \mathcal{D}\bar{\psi}_\sigma \mathcal{D}\psi_\sigma \mathcal{D}\bar{\Delta} \mathcal{D}\Delta \mathcal{D}\varphi\, \mathrm{e}^{-S(\bar{\psi}_\sigma, \psi_\sigma, \bar{\Delta}, \Delta, \varphi)} \ , \tag{5.1.37}$$

$$S(\bar{\psi}_\sigma, \psi_\sigma, \bar{\Delta}, \Delta, \varphi) = \int_0^\beta \mathrm{d}\tau \int \mathrm{d}\boldsymbol{r} \left\{ \frac{1}{g} \bar{\Delta}(\boldsymbol{r},\tau) \Delta(\boldsymbol{r},\tau) + \frac{1}{8\pi} [\nabla \varphi(\boldsymbol{r},\tau)]^2 \right.$$
$$+ \bar{\psi}_\sigma(\boldsymbol{r},\tau) \left[\partial_\tau - \frac{\nabla^2}{2m} \mu + \mathrm{i} e \varphi(\boldsymbol{r},\tau) \right] \psi_\sigma(\boldsymbol{r},\tau)$$
$$\left. + \bar{\Delta}(\boldsymbol{r},\tau) \psi_\downarrow(\boldsymbol{r},\tau) \psi_\uparrow(\boldsymbol{r},\tau) + \Delta(\boldsymbol{r},\tau) \bar{\psi}_\uparrow(\boldsymbol{r},\tau) \bar{\psi}_\downarrow(\boldsymbol{r},\tau) \right\} . \tag{5.1.38}$$

Comparing this with (5.1.15), notice that the complex unit i is not multiplied by $\bar{\Delta}$ and Δ. This is due to the minus sign before the g in (5.1.35), which signals the attractive force.

The derivation of the differential equation (5.1.22) can be adopted to the case $\Lambda < \omega_D$ almost without change, with the result

$$\frac{\mathrm{d}}{\mathrm{d}\ln\Lambda}\left[\frac{1}{g(\Lambda)}\right]=\rho_0 \ , \tag{5.1.39}$$

leading with the initial condition $g(\omega_{\mathrm{D}})=g$ to

$$g(\Lambda)=\frac{g}{1-\rho_0 g\ln\dfrac{\omega_{\mathrm{D}}}{\Lambda}} \ . \tag{5.1.40}$$

This scaling law signifies that the attractive force becomes stronger for lower energies. Furthermore, from (5.1.40) we conclude that for $\Lambda=\Lambda_{\mathrm{c}}=\omega_{\mathrm{D}}\exp[-1/(\rho_0 g)]$, $g(\Lambda)$ diverges, which means that some kind of instability occurs. [Essentially, because (5.1.40) has been derived under the assumption that the interaction g is small, the above theory should be regarded as being qualitative.] When the particles come closer to each other due to an attractive force, they will feel the attraction even more strongly.

Now, the instability itself is just the superconductivity. In order to see this, we integrate out the fermions $\bar{\psi}_\sigma$ and ψ_σ in (5.1.37) and (5.1.38) and deduce an effective theory for $\bar{\Delta}$, Δ and φ. The lowest-order term in φ, i.e. the quadratic term, has already been derived in Sect. 4.1. It has been possible to terminate the expansion at the second order because in this framework the fluctuation around φ has been stable, that is, the potential had a minimum at $\varphi=0$. However, things are different for $\bar{\Delta}$ and Δ. Especially for $q=(\boldsymbol{q},\omega_m)=0$, we obtain for the static component when deducing the potential for $\bar{\Delta}(0)$ and $\Delta(0)$ in the same manner as in (5.1.19)

$$\begin{aligned} S_0 &= a(T)\bar{\Delta}(0)\Delta(0) \\ &= \left(\frac{1}{g}-\frac{1}{\beta V}\sum_{\omega_n}\sum_{\boldsymbol{k}}\frac{1}{\omega_n^2+\xi_{\boldsymbol{k}}^2}\right)\bar{\Delta}(0)\Delta(0) \ . \end{aligned} \tag{5.1.41}$$

Two points are different from (5.1.19). (i) The second term has a minus sign. As mentioned above, this is due to the attractiveness of the force. (ii) The summation in $\boldsymbol{k}$ has to be performed in the whole region $|\xi_{\boldsymbol{k}}|<\omega_{\mathrm{D}}$. This time, we perform the summation $\sum_{\omega_n}$ arising due to the finite temperature exactly and obtain

$$\begin{aligned} a(T) &= \frac{1}{g}-\frac{1}{V}\sum_{\boldsymbol{k}}\frac{f(-\xi_{\boldsymbol{k}})-f(\xi_{\boldsymbol{k}})}{2\xi_{\boldsymbol{k}}} \\ &= \frac{1}{g}-\rho_0\int_{-\omega_{\mathrm{D}}}^{\omega_{\mathrm{D}}}\mathrm{d}\xi\,\frac{1-2f(\xi)}{2\xi} \ . \end{aligned} \tag{5.1.42}$$

The temperature T_{c} where $a(T)$ changes its sign, that it, the temperature fulfilling $a(T_{\mathrm{c}})=0$, is the transition temperature of the BCS theory to the superconducting phase. It is given by

$$k_{\mathrm{B}}T_{\mathrm{c}}=1.14\,\omega_{\mathrm{D}}\exp\left(-\frac{1}{\rho_0 g}\right) \ . \tag{5.1.43}$$

In what follows the Boltzmann constant is set to one, $k_B = 1$. In the vicinity of T_c, the following expansion holds:

$$\begin{aligned} a(T) = a(T) - a(T_c) &= \rho_0 \int_{-\omega_D}^{\omega_D} \frac{f(\xi, T) - f(\xi, T_c)}{\xi} \, d\xi \\ &\cong \rho_0 \int_{-\omega_D}^{\omega_D} \frac{1}{\xi} \frac{\partial f(\xi, T)}{\partial T} \bigg|_{T=T_c} d\xi \cdot (T - T_c) \\ &= -\rho_0 \int_{-\omega_D}^{\omega_D} \frac{1}{\xi} \frac{\partial f(\xi, T)}{\partial \xi} \bigg|_{T=T_c} \frac{\xi}{T_c} \, d\xi \cdot (T - T_c) \\ &\cong \rho_0 \frac{T - T_c}{T_c} \; . \end{aligned} \tag{5.1.44}$$

Because $a(T) > 0$ for $T > T_c$, the potential for $\Delta_0^2 = \bar{\Delta}(0)\Delta(0)$ has a stable minimum at $\Delta_0 = 0$, as shown in Fig. 5.1. On the other hand, for $T < T_c$, $a(T) < 0$, and therefore $\Delta_0 = 0$ becomes an unstable maximum. In this case, higher order terms in $\bar{\Delta}$ and Δ may stabilize the potential and, as indicated by dotted lines in Fig. 5.1, a minimum at a finite Δ_0 may arise. Indeed, using (4.1.36) for the expansion up to Δ_0^4, its coefficient becomes positive. When $|a(T)|$ is small, that is for $|T - T_c| \ll T_c$, it should be possible to stop the expansion there. This is the basic idea of the so-called Ginzburg–Landau theory.

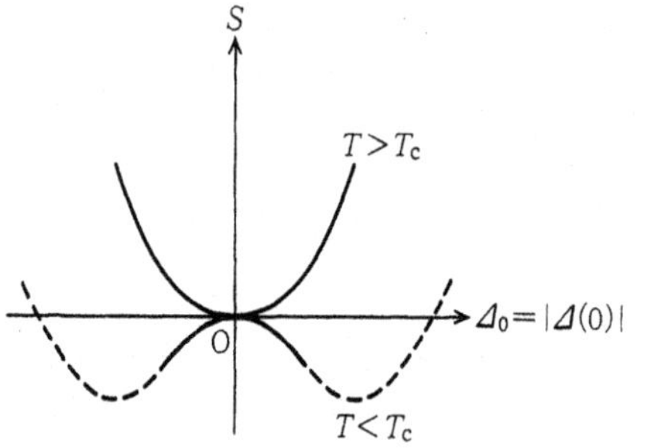

Fig. 5.1. The 'potential' of the order parameter of superconductivity

When restricting the discussion just to the static component, the action

$$S = \beta F\left(\{\bar{\Delta}(\boldsymbol{r})\}, \{\Delta(\boldsymbol{r})\}\right) \tag{5.1.45}$$

can be expressed as a functional of the free energy $F(\{\bar{\Delta}(\boldsymbol{r})\}, \{\Delta(\boldsymbol{r})\})$. As has been discussed in Fig. 3.6a, b of Sect. 3.2, the restriction to $\omega_m = 0$ means that the quantum fluctuations of the fields $\bar{\Delta}$ and Δ are ignored. Therefore, (5.1.45) could be called the classical theory of superconductivity. However, this does not mean that quantum mechanics is totally ignored; the quantum aspects of the electronic system are, in principle, relevant when deriving $F(\{\bar{\Delta}\}, \{\Delta\})$. We might compare this to magnetism, where even when in principle quantum effects are responsible, a classical theory for the magnetization $M(\boldsymbol{r})$ is developed.

Next, we examine explicitly what $F(\{\bar{\Delta}(\boldsymbol{r})\},\{\Delta(\boldsymbol{r})\})$ might be. We assume that $\bar{\Delta}(\boldsymbol{r})$ and $\Delta(\boldsymbol{r})$ change slowly in space and take into account only terms up to second order in $\nabla\bar{\Delta}(\boldsymbol{r})$ and $\nabla\Delta(\boldsymbol{r})$. In $\bar{\Delta}$ and Δ, we will use all orders up to infinity. We write approximately

$$F\left(\{\bar{\Delta}(\boldsymbol{r})\},\{\Delta(\boldsymbol{r})\}\right) \cong \sum_{\boldsymbol{q}} \left[a(\boldsymbol{q})-a(\boldsymbol{0})\right]\bar{\Delta}(\boldsymbol{q})\Delta(\boldsymbol{q}) + F(\bar{\Delta},\Delta) \ . \quad (5.1.46)$$

In the above equation, $\sum_{\boldsymbol{q}} a(\boldsymbol{q})\bar{\Delta}(\boldsymbol{q})\Delta(\boldsymbol{q})$ is the second-order term, and $F(\bar{\Delta},\Delta)$ is the free energy for constant $\bar{\Delta}$ and Δ. The calculation of $a(\boldsymbol{q})$ is a slight generalization of (5.1.41), leading to

$$a(\boldsymbol{q}) = \frac{1}{g} - \frac{1}{V}\sum_{\boldsymbol{k}} \frac{1-f(\xi_{\boldsymbol{k}+\boldsymbol{q}/2})-f(\xi_{\boldsymbol{k}-\boldsymbol{q}/2})}{\xi_{\boldsymbol{k}+\boldsymbol{q}/2}+\xi_{\boldsymbol{k}-\boldsymbol{q}/2}} \ . \quad (5.1.47)$$

(The derivation is left to the reader as an exercise). In the same manner as in (4.1.47) and afterwards, we obtain up to order $O(|\boldsymbol{q}|^2)$:

$$a(\boldsymbol{q})-a(\boldsymbol{0}) \cong \frac{1}{V}\sum_{\boldsymbol{k}} \left\{ \frac{1-2f(\xi_{\boldsymbol{k}})}{2\xi_{\boldsymbol{k}}} - \frac{1-2f(\xi_{\boldsymbol{k}}) - \dfrac{\partial f(\xi_{\boldsymbol{k}})}{\partial \xi_{\boldsymbol{k}}}\dfrac{|\boldsymbol{q}|^2}{4m} - \dfrac{\partial^2 f(\xi_{\boldsymbol{k}})}{\partial \xi_{\boldsymbol{k}}^2}\left(\dfrac{\boldsymbol{k}\cdot\boldsymbol{q}}{2m}\right)^2}{2\xi_{\boldsymbol{k}} + \dfrac{|\boldsymbol{q}|^2}{4m}} \right\} .$$

Performing the variable substitution $\xi_{\boldsymbol{k}}+|\boldsymbol{q}|^2/8m \to \xi_{\boldsymbol{k}}$ in the $|\boldsymbol{k}|$-integration, we obtain

$$\begin{aligned} a(\boldsymbol{q})-a(\boldsymbol{0}) &\cong \frac{1}{V}\sum_{\boldsymbol{k}} \frac{1}{2\xi_{\boldsymbol{k}}}\frac{\partial^2 f(\xi_{\boldsymbol{k}})}{\partial \xi_{\boldsymbol{k}}^2}\left(\frac{\boldsymbol{k}\cdot\boldsymbol{q}}{2m}\right)^2 \\ &= \rho_0 \int_{-\omega_{\mathrm{D}}}^{\omega_{\mathrm{D}}} \frac{\mathrm{d}\xi}{2\xi}\frac{\partial^2 f(\xi)}{\partial \xi^2}\frac{1}{3}\frac{\mu+\xi}{2m}|\boldsymbol{q}|^2 \ . \end{aligned} \quad (5.1.48)$$

Here, $\partial f(\xi)/\partial\xi$ and $\partial^2 f(\xi)/\partial\xi^2$ become large in the region $|\xi| \le T$. Therefore, for $T \simeq T_{\mathrm{c}} \ll \omega_{\mathrm{D}}$, we obtain

$$a(\boldsymbol{q})-a(\boldsymbol{0}) \cong \rho_0 \int_{-\infty}^{\infty} \mathrm{d}\xi \frac{1}{\xi}\frac{\partial^2 f(\xi)}{\partial \xi^2}\frac{\mu}{12m}|\boldsymbol{q}|^2 \ . \quad (5.1.49)$$

Using the ζ-function $\zeta(\sigma) = \sum_{n=1}^{\infty} 1/n^{\sigma}$, the integral can be calculated as

$$\begin{aligned} \int_{-\infty}^{\infty} \mathrm{d}\xi \frac{1}{\xi}\frac{\partial^2 f(\xi)}{\partial \xi^2} &= \int_{-\infty}^{\infty} \mathrm{d}\xi \frac{2}{\xi^3}\tilde{f}(\xi) = 2\frac{2\pi}{\beta}\sum_{n=0}^{\infty}\left[\frac{\beta}{\pi(2n+1)}\right]^3 \\ &= \frac{4\beta^2}{\pi^2}\frac{7}{8}\zeta(3) = \frac{7\beta^2}{2\pi^2}\zeta(3) \ , \end{aligned} \quad (5.1.50)$$

where we defined $\tilde{f}(\xi) \equiv f(\xi) - 1/2 + (1/4)\beta\xi$. Using $\mu = \frac{1}{2}mv_F^2$, (5.1.49) becomes

$$a(\boldsymbol{q}) - a(\boldsymbol{0}) \cong \frac{7\zeta(3)}{48} \frac{\rho_0 v_{\mathrm{F}}^2}{(\pi T)^2} |\boldsymbol{q}|^2 \ . \tag{5.1.51}$$

Next, we determine the second term on the right-hand side of (5.1.46). For the calculation, it is sufficient to set $\bar{\Delta}(\boldsymbol{r},\tau) = \bar{\Delta}$ and $\Delta(\boldsymbol{r},\tau) = \Delta$ in (5.1.38):

$$Z = \mathrm{e}^{-\beta F(\bar{\Delta},\Delta)} = \int \mathcal{D}\bar{\psi}_\sigma \mathcal{D}\psi_\sigma \, \mathrm{e}^{-S(\bar{\psi}_\sigma,\psi_\sigma,\bar{\Delta},\Delta)} \ , \tag{5.1.52}$$

$$\begin{aligned} &S(\bar{\psi}_\sigma, \psi_\sigma, \bar{\Delta}, \Delta) \\ &\quad = \beta V \frac{\bar{\Delta}\Delta}{g} + \sum_{\omega_n}\sum_{\boldsymbol{k}} {}^{\mathrm{t}}\!\begin{bmatrix} \bar{\psi}_\uparrow(k) \\ \psi_\downarrow(-k) \end{bmatrix} \begin{bmatrix} -\mathrm{i}\omega_n + \xi_{\boldsymbol{k}} & \Delta \\ \bar{\Delta} & -\mathrm{i}\omega_n - \xi_{\boldsymbol{k}} \end{bmatrix} \begin{bmatrix} \psi_\uparrow(k) \\ \bar{\psi}_\downarrow(-k) \end{bmatrix} \ ; \end{aligned} \tag{5.1.53}$$

therefore, we obtain

$$\begin{aligned} -\beta F(\bar{\Delta},\Delta) &= \sum_{\omega_n}\sum_{\boldsymbol{k}} \ln\det \begin{bmatrix} -\mathrm{i}\omega_n + \xi_{\boldsymbol{k}} & \Delta \\ \bar{\Delta} & -\mathrm{i}\omega_n - \xi_{\boldsymbol{k}} \end{bmatrix} - \frac{\beta V}{g}\bar{\Delta}\Delta \\ &= \sum_{\omega_n}\sum_{\boldsymbol{k}} \ln(-\omega_n^2 - \xi_{\boldsymbol{k}}^2 - \bar{\Delta}\Delta) - \frac{\beta V}{g}\bar{\Delta}\Delta \\ &= \sum_{\omega_n}\sum_{\boldsymbol{k}} \ln(-\omega_n^2 - \xi_{\boldsymbol{k}}^2) + \sum_{\omega_n}\sum_{\boldsymbol{k}} \ln\left(1 + \frac{\bar{\Delta}\Delta}{\omega_n^2 + \xi_{\boldsymbol{k}}^2}\right) - \frac{\beta V}{g}\bar{\Delta}\Delta \ . \end{aligned} \tag{5.1.54}$$

Here,

$$\begin{aligned} \Delta F(\bar{\Delta},\Delta) &= F(\bar{\Delta},\Delta) - F(0,0) \\ &= -\frac{1}{\beta}\sum_{\omega_n}\sum_{\boldsymbol{k}} \ln\left(1 + \frac{\bar{\Delta}\Delta}{\omega_n^2 + \xi_{\boldsymbol{k}}^2}\right) + \frac{V}{g}\bar{\Delta}\Delta \end{aligned} \tag{5.1.55}$$

is a function only of $\Delta_0^2 = \bar{\Delta}\Delta$. For S_0 in (5.1.41) being the first term (that is, the term quadratic in Δ_0) from the above expression, we can successively expand higher order terms in Δ_0^2. The fourth-order term becomes

$$\begin{aligned} \Delta F^{(4)} &= \frac{1}{2\beta}\sum_{\omega_n}\sum_{\boldsymbol{k}} \frac{(\bar{\Delta}\Delta)^2}{(\omega_n^2 + \xi_{\boldsymbol{k}}^2)^2} \\ &= \frac{V}{2\beta}\sum_{\omega_n} \rho_0 \int_{-\omega_{\mathrm{D}}}^{\omega_{\mathrm{D}}} \frac{\mathrm{d}\xi}{(\omega_n^2 + \xi^2)^2} (\bar{\Delta}\Delta)^2 \\ &\cong \frac{\rho_0 V}{4} \frac{\pi}{\beta} \sum_{\omega_n} \frac{1}{|\omega_n|^3} (\bar{\Delta}\Delta)^2 \end{aligned}$$

$$= \frac{\rho_0 V}{4} \frac{\pi}{\beta} \left(\frac{\beta}{\pi}\right)^3 \cdot 2 \cdot \frac{7}{8}\zeta(3)(\bar{\Delta}\Delta)^2$$

$$= \frac{7\zeta(3)}{16} \frac{\rho_0 V}{(\pi T)^2} (\bar{\Delta}\Delta)^2 \ . \tag{5.1.56}$$

Putting the pieces together, from (5.1.44), (5.1.51) and (5.1.56), and setting $\bar{\Delta}(\boldsymbol{r}) = \Delta^*(\boldsymbol{r})$, we obtain

$$\Delta F \cong \int \mathrm{d}\boldsymbol{r}\, \rho_0 \left[\frac{T - T_\mathrm{c}}{T_\mathrm{c}} |\Delta(\boldsymbol{r})|^2 + \frac{7\zeta(3)}{48} \left(\frac{v_\mathrm{F}}{\pi T}\right)^2 |\nabla \Delta(\boldsymbol{r})|^2 + \frac{7\zeta(3)}{16} \frac{1}{(\pi T)^2} |\Delta(\boldsymbol{r})|^4 \right] \ . \tag{5.1.57}$$

This is the free energy of the GL theory.

Under what conditions might it be justified to stop the expansion of (5.1.55) in $\Delta_0^2 = \bar{\Delta}\Delta$ after the fourth order? The general term $\Delta F^{(2m)}$ can be estimated by

$$\Delta F^{(2m)} = \frac{1}{m\beta} \sum_{\omega_n, \boldsymbol{k}} \frac{(-\bar{\Delta}\Delta)^m}{(\omega_n^2 + \xi_{\boldsymbol{k}}^2)^m}$$

$$\sim \frac{V}{m}\rho_0 \left(\frac{\beta}{\pi}\right)^{2m-2} \sum_n \frac{1}{(2n+1)^{2m-1}} (-\bar{\Delta}\Delta)^m \ ,$$

therefore we see that

$$\Delta F^{(2m+2)} / \Delta F^{(2m)} \sim \beta^2 \bar{\Delta}\Delta \tag{5.1.58}$$

is the dimensionless expansion coefficient.

Next, we discuss the meaning of the action (5.1.53). We introduce the new Grassman numbers $\alpha(k)$ and $\beta(k)$ by

$$\begin{bmatrix} \psi_\uparrow(k) \\ \bar{\psi}_\downarrow(-k) \end{bmatrix} = \begin{bmatrix} \cos\theta_{\boldsymbol{k}} & -\sin\theta_{\boldsymbol{k}} \\ \sin\theta_{\boldsymbol{k}} & \cos\theta_{\boldsymbol{k}} \end{bmatrix} \begin{bmatrix} \alpha(k) \\ \bar{\beta}(-k) \end{bmatrix} \ . \tag{5.1.59}$$

We choose $\cos\theta_{\boldsymbol{k}}$ and $\sin\theta_{\boldsymbol{k}}$ in such a manner that (5.1.53) becomes diagonal with respect to $\alpha(k)$ and $\beta(k)$. Therefore, we set

$$\cos 2\theta_{\boldsymbol{k}} = \frac{\xi_{\boldsymbol{k}}}{\sqrt{\xi_{\boldsymbol{k}}^2 + \Delta_0^2}} \ , \qquad \sin 2\theta_{\boldsymbol{k}} = \frac{\Delta_0}{\sqrt{\xi_{\boldsymbol{k}}^2 + \Delta_0^2}} \ .$$

Here, we set $\bar{\Delta} = \Delta = \Delta_0$. Then, (5.1.53) becomes

$$S(\bar{\psi}_\sigma, \psi_\sigma, \Delta_0) = S(\bar{\alpha}, \alpha, \bar{\beta}, \beta, \Delta_0)$$

$$= \beta V \frac{\Delta_0^2}{g} + \sum_{\mathrm{i}\omega_n} \sum_{\boldsymbol{k}} {}^{\mathrm{t}}\begin{bmatrix} \bar{\alpha}(k) \\ \beta(-k) \end{bmatrix} \begin{bmatrix} -\mathrm{i}\omega_n + \sqrt{\xi_{\boldsymbol{k}}^2 + \Delta_0^2} & 0 \\ 0 & -\mathrm{i}\omega_n - \sqrt{\xi_{\boldsymbol{k}}^2 + \Delta_0^2} \end{bmatrix} \begin{bmatrix} \alpha(k) \\ \bar{\beta}(-k) \end{bmatrix}$$

$$= \beta V \frac{\Delta_0^2}{g} + \sum_{\mathrm{i}\omega_n} \sum_{\boldsymbol{k}} \left(-\mathrm{i}\omega_n + \sqrt{\xi_{\boldsymbol{k}}^2 + \Delta_0^2} \right) \left(\bar{\alpha}(k)\alpha(k) + \bar{\beta}(k)\beta(k) \right) \ .$$

This is nothing but the action corresponding to the Hamiltonian

$$\mathcal{H} = \sum_{\boldsymbol{k}} E_{\boldsymbol{k}} \left(\alpha^\dagger(\boldsymbol{k})\alpha(\boldsymbol{k}) + \beta^\dagger(\boldsymbol{k})\beta(\boldsymbol{k}) \right) \ . \tag{5.1.60}$$

Owing to (5.1.60), we conclude that $\alpha^\dagger$, α, $\beta^\dagger$ and β are the operators of elementary excitations occurring in the superconductor, and their energy dispersion is given by $\pm E_{\boldsymbol{k}} = \pm\sqrt{\xi_{\boldsymbol{k}}^2 + \Delta_0^2}$. The physical meaning of $\Delta_0 = \sqrt{\bar{\Delta}\Delta}$ is the presence of a $2\Delta_0$ energy gap. Therefore, it should be clear that the ratio of this Δ_0 and the average thermal energy T is an important scale in the theory. For $\Delta_0 \ll T$, more and more particles become excited and surpass the gap; on the other hand, for $\Delta_0 \gg T$, thermal excitation becomes almost impossible. Determining the average value of Δ_0 from the expansion up to fourth order in (5.1.57), for $T < T_\mathrm{c}$ we obtain

$$\Delta_0 = \sqrt{\frac{8\pi^2}{7\zeta(3)} \frac{T^2}{T_\mathrm{c}} (T_\mathrm{c} - T)} \ . \tag{5.1.61}$$

Therefore, we can conclude from (5.1.58) that (5.1.57) is a good expansion for $|T_\mathrm{c} - T|/T_\mathrm{c} \ll 1$.

So far, we introduced the so-called GL theory. However, the BCS mean field theory corresponds to direct determination of the minimum of ΔF in (5.1.55):

$$\frac{1}{V}\frac{\partial \Delta F}{\partial \Delta_0^2} = \frac{1}{g} - \frac{1}{\beta}\sum_{\omega_n}\frac{1}{V}\sum_{\boldsymbol{k}}\frac{1}{\omega_n^2 + \xi_{\boldsymbol{k}}^2 + \Delta_0^2} = 0 \ , \tag{5.1.62}$$

where the summation of the ω_n can be performed immediately, and from (5.1.62) the BCS mean field equation is regained:

$$\frac{1}{g} = \rho_0 \int_{-\omega_\mathrm{D}}^{\omega_\mathrm{D}} \mathrm{d}\xi \frac{1}{2\sqrt{\xi^2 + \Delta_0^2}} \left[1 - 2f\left(\sqrt{\xi^2 + \Delta_0^2}\right) \right] \ . \tag{5.1.63}$$

As is well known, a solution $\Delta_0 = \Delta_0(T)$ of this equation similar to (5.1.61) is recovered, and for $T = 0$ the value

$$\Delta_0(T = 0) \cong 2\omega_\mathrm{D} \exp\left(-\frac{1}{\rho_0 g} \right) \tag{5.1.64}$$

is obtained. Here, the assumption $\rho_0 g \ll 1$ (weak binding) has been made.

Above, we introduced the so-called classical theory of superconductivity. Because the discussion has been limited to the static mode ($\omega_m = 0$) only, it has been possible to regard φ and $\bar{\Delta}, \Delta$ as independent degrees of freedom. This is due to the fact that canonical variables appear in the action as $\mathrm{i}p\dot{x}$, and this term vanishes for $\omega_m = 0$. This is similar to the fact that the x-integration and the p-integration can be done independently in classical statistical mechanics. In the present case, the canonical fields of φ and $\bar{\Delta}, \Delta$

are $\rho = \bar{\psi}_\sigma \psi_\sigma$ and $\psi_\downarrow \psi_\uparrow, \bar{\psi}_\uparrow \bar{\psi}_\downarrow$, respectively. When proceeding to quantum dynamics, that is, $\omega_m \neq 0$, it becomes necessary to take both together into account.

When the notation of $\bar{\Delta}$ and Δ was introduced, it was stressed that $\bar{\Delta}$ is not necessarily the complex conjugate of Δ and both are independent integration variables. In what follows, however, we choose integration paths that fulfil the relation $\bar{\Delta}(\boldsymbol{r}, \tau) = \Delta^*(\boldsymbol{r}, \tau)$. We now write $x = (\boldsymbol{r}, \tau)$ for short. We split the degrees of freedom of the amplitude and the phase of $\Delta(x)$ and $\Delta^*(x)$ and write

$$\Delta(x) = |\Delta(x)| \, \mathrm{e}^{2\mathrm{i}\theta(x)} \ , \qquad \Delta^*(x) = |\Delta(x)| \, \mathrm{e}^{-2\mathrm{i}\theta(x)} \ . \qquad (5.1.65)$$

The mean field equation depends only on the absolute value $|\Delta|$, and even for the uniform variation of the amplitude (that is, the variation $|\Delta(x)| = |\Delta| + \delta\Delta$ being independent of x) the action (energy) of the system will be altered. Therefore, the fluctuation modes of the amplitude have a finite frequency. On the other hand, a uniform variation of the phase $\theta \to \theta + \delta\theta$ leaves the action invariant. Therefore, a Goldstone mode with frequency approaching zero for $\boldsymbol{q} \to 0$, as introduced in Sect. 3.2, is expected to exist for this kind of phase fluctuation.

Now, we fix the amplitude $|\Delta(x)|$ of the effective action describing the low-energy dynamics by setting $|\Delta| = |\Delta^*| = \Delta_0$, and focus only on the fluctuations of the phase $\theta(x)$ and the potential $\varphi(x)$. Correspondingly, transforming the Grassmann fields $\psi_\sigma(x)$ and $\bar{\psi}_\sigma(x)$ as

$$\psi_\sigma(x) = \tilde{\psi}_\sigma(x) \, \mathrm{e}^{\mathrm{i}\theta(x)} \ , \qquad \bar{\psi}_\sigma(x) = \bar{\tilde{\psi}}_\sigma(x) \, \mathrm{e}^{-\mathrm{i}\theta(x)} \ , \qquad (5.1.66)$$

the action (5.1.38) becomes up to constants

$$S = S_0 + S_1 + S_2 \ , \qquad (5.1.67\mathrm{a})$$

$$S_0 = \sum_\sigma \int \mathrm{d}x \, \bar{\tilde{\psi}}_\sigma(x) \left[\frac{\partial}{\partial \tau} - \frac{\nabla^2}{2m} - \mu \right] \tilde{\psi}_\sigma(x)$$
$$+ \Delta_0 \int \mathrm{d}x \left\{ \tilde{\psi}_\downarrow(x) \tilde{\psi}_\uparrow(x) + \bar{\tilde{\psi}}_\uparrow(x) \bar{\tilde{\psi}}_\downarrow(x) \right\} \ , \qquad (5.1.67\mathrm{b})$$

$$S_1 = \mathrm{i} \sum_\sigma \int \mathrm{d}x \, \bar{\tilde{\psi}}_\sigma(x) \left[\partial_\tau \theta(x) + e\varphi(x) \right] \tilde{\psi}_\sigma(x)$$
$$+ \int \mathrm{d}x \, \nabla\theta(x) \cdot \frac{1}{2m\mathrm{i}} \sum_\sigma \left[\bar{\tilde{\psi}}_\sigma(x) \nabla \tilde{\psi}_\sigma(x) - \nabla \bar{\tilde{\psi}}_\sigma(x) \tilde{\psi}_\sigma(x) \right] \ , \qquad (5.1.67\mathrm{c})$$

$$S_2 = \sum_\sigma \int \mathrm{d}x \frac{1}{2m} \left(\nabla\theta(x) \right)^2 \bar{\tilde{\psi}}_\sigma(x) \tilde{\psi}_\sigma(x) \ . \qquad (5.1.67\mathrm{d})$$

It is important to notice that (5.1.67) is independent of the constant part of $\theta(x)$, because θ is always accompanied by a derivative, that is $\partial_\tau \theta$ or $\nabla\theta(x)$. From (5.1.67) we need to derive an effective action up to second order in $\varphi(x)$,

$\partial_\tau \theta(x)$ and $\nabla\theta(x)$. In order to do so, we use the matrix $\tilde{M}_0$ (see Sect. 4.1 and (5.1.53)) and the matrices V_1 and V_2 corresponding to S_1 and S_2 to write

$$-\operatorname{Tr}\ln\left(1+\tilde{M}_0^{-1}(V_1+V_2)\right) \ .$$

It is sufficient to expand this expression up to second order in V_1 and first-order in V_2. The first-order contribution in V_2 is the most simple one, corresponding to replacing $\sum_\sigma \bar{\tilde{\psi}}_\sigma(x)\tilde{\psi}_\sigma(x)$ by its average value with respect to S_0, i.e. the average electron density n.

Concerning S_1, we first remark that it is possible to rewrite (5.1.67c) in the following manner:

$$S_1 = \mathrm{i}\int \mathrm{d}x\, \rho(x)\,[\partial_\tau\theta(x)+e\varphi(x)] + \int \mathrm{d}x\, \nabla\theta(x)\cdot \boldsymbol{J}_\mathrm{p}(x) \ . \qquad (5.1.68)$$

Here

$$\rho(x) = \sum_\sigma \bar{\tilde{\psi}}_\sigma(x)\tilde{\psi}_\sigma(x) = \sum_\sigma \bar{\psi}_\sigma(x)\psi_\sigma(x) \qquad (5.1.69)$$

is the electron density, and

$$\boldsymbol{J}_\mathrm{p}(x) = \frac{1}{2m\mathrm{i}}\sum_\sigma \left[\bar{\tilde{\psi}}_\sigma(x)\nabla\tilde{\psi}_\sigma(x) - \nabla\bar{\tilde{\psi}}_\sigma(x)\tilde{\psi}_\sigma(x)\right] \qquad (5.1.70)$$

is the ferromagnetic current, which we have already mentioned. The first-order contribution of S_1 (that is, V_1) vanishes because of $\langle \boldsymbol{J}_\mathrm{p}(x)\rangle = \mathbf{0}$ and $\int \mathrm{d}x\,[\partial_\tau\theta + e\varphi] = 0$.

Next, we calculate the second-order term in $S_1(V_1)$. Finally, this can be reduced to calculating the correlation functions $\langle\rho(x)\rho(0)\rangle$, $\langle J_\mathrm{p}^\alpha(x) J_\mathrm{p}^\beta(0)\rangle$ and $\langle\rho(x)\boldsymbol{J}_\mathrm{p}(0)\rangle$ in the superconducting phase ($\Delta_0 > 0$). From symmetry arguments we immediately understand that $\langle\rho(x)\boldsymbol{J}_\mathrm{p}(0)\rangle = \mathbf{0}$, and no cross terms from the first and the second terms in (5.1.67c) arise. Let us now turn to explicit calculations. For $k = (\boldsymbol{k}, \mathrm{i}\omega_n)$ fixed, $\tilde{M}_0$ and $\tilde{G}_0 = -\tilde{M}_0^{-1}$ are 2×2 matrices:

$$\begin{aligned}\tilde{G}_0(k) &= \begin{bmatrix} \mathrm{i}\omega_n - \xi_{\boldsymbol{k}} & -\Delta_0 \\ -\Delta_0 & \mathrm{i}\omega_n + \xi_{\boldsymbol{k}} \end{bmatrix}^{-1} \\ &= -\frac{1}{\omega_n^2+\xi_{\boldsymbol{k}}^2+\Delta_0^2}\begin{bmatrix} \mathrm{i}\omega_n + \xi_{\boldsymbol{k}} & +\Delta_0 \\ +\Delta_0 & \mathrm{i}\omega_n - \xi_{\boldsymbol{k}} \end{bmatrix} \ . \end{aligned} \qquad (5.1.71)$$

The corresponding Tr is a summation with respect to k and the trace tr over the 2×2 matrix. Corresponding to (5.1.67c), the matrix elements are given by

$$\begin{aligned}(V_1)_{k,k'} &= \frac{1}{(\beta V)^{1/2}}\left[(\omega_n - \omega_{n'})\theta(k-k') + \mathrm{i}e\varphi(k-k')\right]\tau_3 \\ &+ \frac{1}{(\beta V)^{1/2}}\mathrm{i}\frac{(\boldsymbol{k}-\boldsymbol{k}')\cdot(\boldsymbol{k}+\boldsymbol{k}')}{2m}\theta(k-k')\hat{1} \ . \end{aligned} \qquad (5.1.72)$$

Here, we defined

$$\tau_3 = \begin{bmatrix} 1 & 0 \\ 0 & -1 \end{bmatrix} , \quad \hat{1} = \begin{bmatrix} 1 & 0 \\ 0 & 1 \end{bmatrix} .$$

Therefore, the second-order term with respect to V_1 is given by

$$\begin{aligned} \frac{1}{2}\operatorname{Tr}\left[\tilde{G}_0 V_1 \tilde{G}_0 V_1\right] &= \frac{1}{2}\sum_{k,q} \operatorname{tr}\left[\tilde{G}_0(k+q)(V_1)_{k+q,k}\tilde{G}_0(k)(V_1)_{k,k+q}\right] \\ &= \frac{1}{2\beta V}\sum_{k,q}(\omega_m\theta(q) + \mathrm{i}e\varphi(q)) \\ &\quad \times (-\omega_m\theta(-q) + \mathrm{i}e\varphi(-q))\operatorname{tr}\left[\tilde{G}_0(k+q)\tau_3\tilde{G}_0(k)\tau_3\right] \\ &\quad + \frac{1}{2\beta V}\sum_{k,q} q_i q_j \frac{(k_i+q_i/2)(k_j+q_j/2)}{m^2}\theta(q)\theta(-q)\operatorname{tr}\left[\tilde{G}_0(k+q)\tilde{G}_0(k)\right] . \end{aligned} \tag{5.1.73}$$

Here, we introduced $k - k' = q = (\boldsymbol{q}, i\omega_m)$ to rewrite the sum in k and k' into a sum in k and q. Using the density correlation function $\pi^{(0)}(q)$ and the paramagnetic current density correlation function $\pi^{(\perp)}(q)$ that will be defined in what follows, (5.1.73) can be expressed as

$$\begin{aligned} (5.1.73) &= \frac{1}{2}\sum_q \pi^{(0)}(q)(\omega_m\theta(q) + \mathrm{i}e\varphi(q))(-\omega_m\theta(-q) + \mathrm{i}e\varphi(-q)) \\ &\quad + \frac{1}{2}\sum_q q_i q_j \pi_{ij}^{(\perp)}(q)\theta(q)\theta(-q) . \end{aligned} \tag{5.1.74}$$

Here, we defined

$$\begin{aligned} \pi^{(0)}(q) &= \frac{1}{\beta V}\sum_{\boldsymbol{k},\omega_n} \operatorname{tr}\left[\tilde{G}_0(k+q)\tau_3\tilde{G}_0(k)\tau_3\right] \\ &= \frac{-2}{\beta V}\sum_{\boldsymbol{k},\omega_n} \frac{\Delta_0^2 - \xi_{\boldsymbol{k}}\xi_{\boldsymbol{k}+\boldsymbol{q}} + \omega_n(\omega_n+\omega_m)}{\left[\omega_n^2 + E_{\boldsymbol{k}}^2\right]\left[(\omega_n+\omega_m)^2 + E_{\boldsymbol{k}+\boldsymbol{q}}^2\right]} \end{aligned} \tag{5.1.75a}$$

and

$$\begin{aligned} \pi_{ij}^{(\perp)} &= \frac{1}{\beta V}\sum_{\boldsymbol{k},\omega_n} \frac{(k_i+q_i/2)(k_j+q_j/2)}{m^2}\operatorname{tr}\left[\tilde{G}_0(k+q)\tilde{G}_0(k)\right] \\ &= \frac{2}{\beta V}\sum_{\boldsymbol{k},\omega_n} \frac{(k_i+q_i/2)(k_j+q_j/2)}{m^2} \\ &\quad \times \frac{\Delta_0^2 + \xi_{\boldsymbol{k}}\xi_{\boldsymbol{k}+\boldsymbol{q}} - \omega_n(\omega_n+\omega_m)}{\left[\omega_n^2 + E_{\boldsymbol{k}}^2\right]\left[(\omega_n+\omega_m)^2 + E_{\boldsymbol{k}+\boldsymbol{q}}^2\right]} . \end{aligned} \tag{5.1.75b}$$

Important in (5.1.75) is the fact that owing to the presence of the $2\Delta_0$ gap, it is possible to expand both $\pi^{(0)}(q)$ and $\pi^{(\perp)}(q)$ in the limit $|\omega_m| \to 0$, $|\boldsymbol{q}| \to 0$. Restricting the discussion to zero temperature and by setting $q = 0$, we obtain

$$\begin{aligned}\pi^{(0)}(0) &= -2\rho_0 \int \mathrm{d}\xi_{\boldsymbol{k}} \int_{-\infty}^{\infty} \frac{\mathrm{d}\omega}{2\pi} \frac{\Delta_0^2 - \xi_{\boldsymbol{k}}^2 + \omega^2}{(\omega^2 + E_{\boldsymbol{k}}^2)^2} \\ &= -\rho_0 \int \mathrm{d}\xi \frac{\Delta_0^2}{[\xi^2 + \Delta_0^2]^{3/2}} = -2\rho_0 \ ,\end{aligned} \tag{5.1.76a}$$

$$\begin{aligned}\pi_{ij}^{(\perp)} &= 2 \int \frac{\mathrm{d}\boldsymbol{k}}{(2\pi)^3} \frac{(k_i + q_i/2)(k_j + q_j/2)}{m^2} \int_{-\infty}^{\infty} \frac{\mathrm{d}\omega}{2\pi} \frac{\Delta_0^2 + \xi_{\boldsymbol{k}}^2 - \omega^2}{(\omega^2 + E_{\boldsymbol{k}}^2)^2} \\ &= 0 \ .\end{aligned} \tag{5.1.76b}$$

The expression (5.1.76a) agrees with the result (4.1.53) in the normal phase in the limit $i\omega_m = 0$, $\boldsymbol{q} \to \boldsymbol{0}$, whereas the result (5.1.76b) is characteristic for superconductivity. That is, the paramagnetic current $\boldsymbol{J}_\mathrm{p}$ is the contribution to the current density caused by the strain of the wave function due to the external field. Owing to the presence of Δ_0, the wave function becomes "solid" and no strain occurs, therefore $\boldsymbol{J}_\mathrm{p}$ does not emerge. Saying it in terms of the twist $\nabla\theta$ of the phase in the action, only the contribution $(n/2m)(\nabla\theta)^2$ of the action S_2 discussed earlier remains (corresponding to the diamagnetic current density). This is a manifestation of the rigidity of the phase, i.e. the Meissner effect. At finite temperature, $\pi_{ij}^{(\perp)}(0)$ obtains a finite value caused by quasi-particles that surpass the gap due to thermal excitation. For $T < T_\mathrm{c}$, this is not sufficient to cancel the diamagnetic current, and n in $(n/2m)(\nabla\theta)^2$ is simply replaced by ρ_S (the superconducting current density), being smaller than n.

Putting things together, the $|\omega_m| \ll \Delta_0$ and $v_\mathrm{F}\,|\boldsymbol{q}| \ll \Delta_0$ low-energy, long wavelength effective action is given by

$$\begin{aligned}S_{\text{eff}} = \sum_q \Bigg[&\frac{\boldsymbol{q}^2}{8\pi} \varphi(q)\varphi(-q) + \rho_0(e\varphi(q) - \mathrm{i}\omega_m\theta(q))(e\varphi(-q) + \mathrm{i}\omega_m\theta(-q)) \\ &+ \frac{\rho_\mathrm{s}}{2m}\boldsymbol{q}^2\theta(q)\theta(-q)\Bigg] \ .\end{aligned} \tag{5.1.77}$$

Integrating out φ in (5.1.77) (by completing the square), we obtain for θ

$$S_{\text{eff}}(\{\theta\}) = \sum_q \left[\frac{\rho_0 \boldsymbol{q}^2/8\pi}{\boldsymbol{q}^2/8\pi + e^2\rho_0}\omega_m^2 + \frac{\rho_\mathrm{s}\boldsymbol{q}^2}{2m} \right] \theta(q)\theta(-q) \ . \tag{5.1.78}$$

As is clear from this equation, the long-range characteristics of the Coulomb force, that is, ω_m^2 multiplied by $\boldsymbol{q}^2$, has $\boldsymbol{q}^2$ as a common factor with the second term, and therefore for $\boldsymbol{q} \to 0$ a Goldstone mode with $\omega_q \to 0$ does not exist. (If the interaction between the electrons were short range, then the coefficient of ω_m^2 would reach a constant value, and a mode with the dispersion relation of an acoustic wave $\omega_{\boldsymbol{q}} \sim |\boldsymbol{q}|$ would exist. This corresponds to the Bogoliubov mode that appeared in the Bogoliubov theory of Bose condensation, as discussed in Sect. 4.2.) This "missing Goldstone mode" can be found in the plasma excitations that exist at higher frequencies. (5.1.77) is only the correct description in the limit $|\omega_m| \ll \Delta_0$, $v_F\,|\boldsymbol{q}| \ll \Delta_0$ and does not take into account plasma excitations. For $|\omega_m| \gg \Delta_0$, although the temperature is finite, the contribution of $\pi^{(\perp)}(q)$ is small, and ρ_S in (5.1.77) should approach n.

Replacing ρ_0 in (5.1.77) again by $-\frac{1}{2}\pi(q)$ for general q and this time performing the integration with respect to θ, we obtain for φ

$$S_{\text{eff}}(\{\varphi\}) = \sum_q \frac{\dfrac{n\boldsymbol{q}^2}{16\pi m} - \dfrac{1}{2}\pi^{(0)}(q)\left[\dfrac{ne^2}{2m} + \dfrac{\omega_m^2}{8\pi}\right]}{\dfrac{n\boldsymbol{q}^2}{2m} - \dfrac{1}{2}\pi^{(0)}(q)\omega_m^2}\, \boldsymbol{q}^2 \varphi(q)\varphi(-q) \ . \qquad (5.1.79)$$

Because for $\boldsymbol{q} \to \boldsymbol{0}$, $\pi^{(0)}(q)$ reaches a value different from zero, in this limit we obtain

$$S_{\text{eff}} = \sum_{|\boldsymbol{q}|:\,\text{small}} \sum_{i\omega_m} \frac{\boldsymbol{q}^2}{8\pi} \left(1 + \frac{\omega_{\text{p}}^2}{\omega_m^2}\right) \varphi(q)\varphi(-q) \ . \qquad (5.1.80)$$

Here, $\omega_{\text{p}} = (4\pi ne^2/m)^{1/2}$ is the plasma frequency. As for the normal conduction phase, (5.1.80) is the action of the scalar potential with respect to plasma excitations [that is, leading to the same Green function as (4.1.60)]. Obviously, the behaviour at high energies ω_{p} ($\gg \Delta_0$) is not strongly affected by the superconductivity.

5.2 Macroscopic Quantum Effects and Dissipation: The Josephson Junction

As was discussed in the last section, in superconductors the quantal phase θ becomes coherent over the whole crystal, and therefore the quantum properties of the electron wave become visible at a macroscopic level. The Josephson effect that will now be introduced is a striking phenomenon related to this quantal phase. Recently, there has been increasing interest in quantum effects at a macroscopic level. In such systems, due to the fact that the microscopic degrees of freedom have been integrated over, the effect of irreversibility, i.e. dissipation, becomes important.

Now, we consider a system where two superconductors A and B are "weakly" coupled (Fig. 5.2). Explicitly, it is possible to contact them through an insulating layer of thickness of about 10^{-7}cm, or to install a point contact

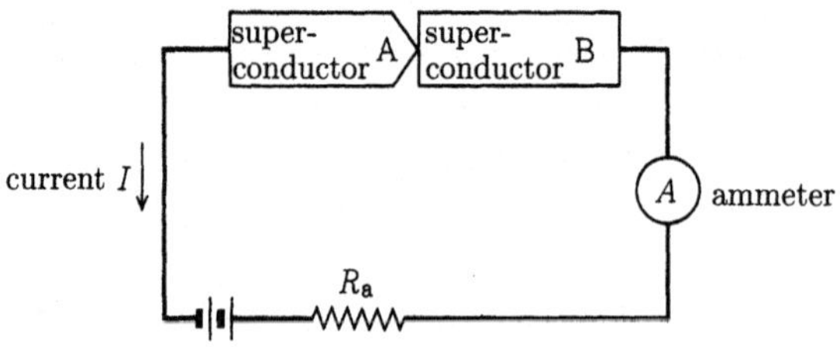

Fig. 5.2. A circuit containing a Josephson junction

between them. Because A and B are both superconductors, as mentioned in Sect. 5.1, the quantum mechanical phase becomes solid and uniform [being the phase $2\theta(\boldsymbol{r},\tau)$ of $\Delta(\boldsymbol{r},\tau)$, which will be called $\phi(\boldsymbol{r},\tau)$ in what follows]. Therefore, it can be treated as a macroscopic variable. That is, for both superconductors A and B we can introduce the functions $\phi(\boldsymbol{r},\tau) = 2\theta(\boldsymbol{r},\tau) = \phi_{\mathrm{A}}(\tau)$ and $\phi_{\mathrm{B}}(\tau)$.

Next, we consider (5.1.68). The first term, with ρ being the electron density and φ being the potential, reads

$$\mathrm{i}\int \mathrm{d}\boldsymbol{r}\,\mathrm{d}\tau\,\rho(\boldsymbol{r},\tau)\,[\partial_\tau\theta(\boldsymbol{r},\tau) + e\varphi(\boldsymbol{r},\tau)] \quad . \tag{5.2.1}$$

Considering only the $\boldsymbol{r}$-independent contribution for A and B, respectively, we obtain

$$\mathrm{i}\int \mathrm{d}\tau\,N_\alpha(\tau)\,[\partial_\tau\theta_\alpha(\tau) + e\varphi_\alpha(\tau)] \qquad (\alpha = \mathrm{A},\mathrm{B}) \quad . \tag{5.2.2}$$

Setting $N = N_{\mathrm{A}} + N_{\mathrm{B}}$, $\Delta N = N_{\mathrm{A}} - N_{\mathrm{B}}$, $\bar{\theta} = \theta_{\mathrm{A}} + \theta_{\mathrm{B}}$, $\Delta\theta = \theta_{\mathrm{A}} - \theta_{\mathrm{B}}$, $\tilde{\varphi} = \varphi_{\mathrm{A}} + \varphi_{\mathrm{B}}$ and $\Delta\varphi = \varphi_{\mathrm{A}} - \varphi_{\mathrm{B}}$, then equation (5.2.2) for both indices A and B reads

$$\frac{1}{2}\mathrm{i}N\int \mathrm{d}\tau\,\left[\partial_\tau\bar{\theta} + e\bar{\varphi}\right] + \frac{1}{2}\mathrm{i}\int \mathrm{d}\tau\,\Delta N\,[\partial_\tau\Delta\theta + e\Delta\varphi] \quad . \tag{5.2.3}$$

Here, we set $N = N_{\mathrm{A}} + N_{\mathrm{B}}$ constant. Then, the "centre of mass" degree of freedom becomes irrelevant, and only the second term in (5.2.3), representing the degree of freedom of the phase difference, becomes important. Of course, if both superconductors were totally independent, there would be no physical difference, whatever the value of the uniform phases $\phi_{\mathrm{A}} = 2\theta_{\mathrm{A}}$ and $\phi_{\mathrm{B}} = 2\theta_{\mathrm{B}}$ might be. Introducing a weak coupling, an additional energy $\Delta E(\phi)$ depending on the phase difference $\Delta\phi = \phi_{\mathrm{A}} - \phi_{\mathrm{B}}$ is added.

Explicitly, this can be expressed as

$$\begin{aligned}\Delta E(\phi) &= -2U\cos(\phi_{\mathrm{A}} - \phi_{\mathrm{B}})\\ &= -2U\cos(\Delta\phi) \quad .\end{aligned} \tag{5.2.4}$$

Combining (5.2.3) and (5.2.4), we obtain the action describing the degree of freedom of the phase difference:

$$S_{\text{diff}} = \frac{\mathrm{i}}{4}\int \mathrm{d}\tau \left\{\Delta N(\tau)\left[\partial_\tau(\Delta\phi) + 2e\Delta\varphi\right] - 8U\cos(\Delta\phi)\right\} \ . \tag{5.2.5}$$

Let us now consider the equation of motion derived by this action. It is sufficient to require $\delta S_{\text{diff}} = 0$:

$$\begin{aligned}\delta S_{\text{diff}} &= \frac{\mathrm{i}}{4}\int \mathrm{d}\tau\, \delta(\Delta N)\left[\partial_\tau(\Delta\phi) + 2e\Delta\varphi\right] \\ &\quad + \int \mathrm{d}\tau\, \delta(\Delta\phi)\left[-\frac{\mathrm{i}}{4}\partial_\tau(\Delta N) + 2U\sin(\Delta\phi)\right] \\ &= 0 \ .\end{aligned} \tag{5.2.6}$$

We therefore obtain

$$\mathrm{i}\frac{\mathrm{d}\Delta N}{\mathrm{d}\tau} = 8U\sin(\Delta\phi) \tag{5.2.7}$$

$$\mathrm{i}\frac{\mathrm{d}\Delta\phi}{\mathrm{d}\tau} = -2\mathrm{i}e\Delta\varphi \ . \tag{5.2.8}$$

Replacing the imaginary time τ by the real time t by writing $\tau = it$ (remember that in this moment the scalar potential φ must also be replaced by $-\mathrm{i}\varphi$), because the current $I = I_{\mathrm{B}\to\mathrm{A}}$ from B to A can be expressed as $(-e)\times\dot{N}_{\mathrm{A}} = (-e)\times(\Delta\dot{N}/2)$, and using the fact that the potential difference $\Delta\varphi$ is the electric voltage V between A and B, from (5.2.7) and (5.2.8) we obtain

$$I = -\frac{4eU}{\hbar}\sin(\Delta\phi) \equiv -I_0\sin(\Delta\phi) \tag{5.2.9}$$

$$\frac{\mathrm{d}\Delta\phi}{\mathrm{d}t} = -\frac{2e}{\hbar}V \ , \tag{5.2.10}$$

where $\hbar$ has been reintroduced. These two equations are the basic equations of the Josephson junction.

These equations describe in principle the same physics as the Josephson equations of superfluids, namely (4.2.32), (4.2.33) and (4.2.34), discussed in Sect. 4.2. Different is the fact that because the Josphson junction U is weak, the phase difference $\Delta\phi$ need not necessarily be small compared with 2π. Therefore, (5.2.9) is the generalization of (4.2.32). (5.2.10) is the discretized form of (4.2.33).

When a constant current I ($< I_0$) flows, in the circuit shown in Fig. 5.2 it follows from (5.2.9) that the phase difference $\Delta\phi$ reaches a fixed value independent of time, and therefore due to (5.2.10) the voltage V becomes zero. Therefore, the potential difference is caused totally by the external resistance R_{a}, fulfilling the relation $V_{\mathrm{a}} = R_{\mathrm{a}}I$. In such a way, the resistance of the system of the two superconductors A and B, including the junction region, is zero. This is the so-called dc-Josephson effect. On the other hand, from (5.2.10) we deduce that for the case when there is a finite potential difference between A and B, $\Delta\phi$ increases in time, and due to (5.2.9) the current $I(t)$ behaves in time as

$$I(t) = +I_0 \sin\left[\frac{2e}{\hbar}Vt + \alpha\right] \quad . \tag{5.2.11}$$

This is the so-called ac-Josephson effect.

In this manner, the Josephson effect arises because of the tendency of the phases between two different superconductors to become uniform, as is described by the energy in (5.2.4). We can imagine one superconductor to be fictitiously separated into many segments and define for every segment a phase θ_i and think of Josephson junctions (not necessarily weak) between each of them. Then, corresponding to (5.2.4), for every neighbouring pair $\langle ij \rangle$, we describe the energy of the system by

$$\Delta E = -2U \sum_{\langle ij \rangle} \cos(\phi_i - \phi_j) \quad . \tag{5.2.12}$$

For the case when U is strong enough, the phase difference between neighbouring phases $\phi_i - \phi_j$ is small and the cos can be expanded. In the continuum limit, with a being the length of one side of every microscopic system and d the dimension of the space, we can write

$$\Delta E = Ua^{2-d} \int \mathrm{d}\boldsymbol{r}\, (\nabla\phi(\boldsymbol{r}))^2 \quad . \tag{5.2.13}$$

This corresponds to the last term in (5.1.77). We conclude that the Josephson junction that seeks to equalize the phase difference is at the heart of superconductivity.

So far, the dissipation effect has not been included in the discussion. We will now discuss this effect following Caldeira and Leggett.

We start by considering the classical dynamics of a dissipative macroscopic system. The model is shown in Fig. 5.3, where two superconductors are joined by a connection with capacity C, resistance R_S and Josephson coupling U. In what follows, we set $\hbar = 1$. With $\Delta\phi$ being the phase difference, the current flowing from A to B is given by

$$I = C\frac{\mathrm{d}V}{\mathrm{d}t} - I_0 \sin\Delta\phi + \frac{1}{R_\mathrm{S}}V \quad . \tag{5.2.14}$$

Using (5.2.10)

$$V = -\frac{1}{2e}\frac{\mathrm{d}\Delta\phi}{\mathrm{d}t} \equiv -\frac{\mathrm{d}\Delta\tilde{\phi}}{\mathrm{d}t} \qquad \left(\Delta\tilde{\phi} \equiv \frac{1}{2e}\Delta\phi\right)$$

for (5.2.14) we obtain

$$C\frac{\mathrm{d}^2\Delta\tilde{\phi}}{\mathrm{d}t^2} + \frac{1}{R_\mathrm{S}}\frac{\mathrm{d}\Delta\tilde{\phi}}{\mathrm{d}t} + I_0 \sin\left[2e\Delta\tilde{\phi}\right] = -I \quad . \tag{5.2.15}$$

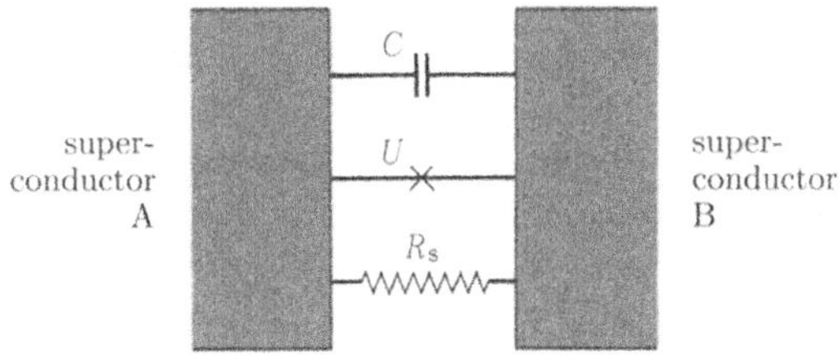

Fig. 5.3. Model of the Josephson junction

When regarding $\Delta\tilde{\phi}$ as the position of some point-like particle, then the equation just describes the Newtonian equation of motion of a particle with mass C moving in a potential $V(\Delta\tilde{\phi}) = -(1/2e)I_0\cos(2e\Delta\tilde{\phi}) + I\Delta\phi$ while being damped by a force proportional to the inverse of R_S. Then, I is an external field that is inclining toward the "wash board potential" $\cos(\Delta\phi)$.

Caldeira and Leggett asked what might be the microscopic quantum mechanical structure leading to the behaviour described by the classical equation (5.2.15). They concluded that it is sufficient to describe the interaction of the microscopic variables $\Delta\phi$ to the microscopic degrees of freedom by rewriting them as an ensemble of harmonic oscillators and couple them linearly. Explicitly, we choose the Lagrangian to be

$$L = \frac{1}{2}C\left(\frac{\mathrm{d}\Delta\tilde{\phi}}{\mathrm{d}t}\right)^2 - V(\Delta\tilde{\phi}) + \sum_\alpha \lambda_\alpha x_\alpha \Delta\tilde{\phi} + \frac{1}{2}\sum_\alpha m_\alpha\left[\left(\frac{\mathrm{d}x_\alpha}{\mathrm{d}t}\right)^2 - \omega_\alpha^2 x_\alpha^2\right] - \sum_\alpha \frac{\lambda_\alpha^2}{2m_\alpha\omega_\alpha^2}(\Delta\tilde{\phi})^2 \ . \quad (5.2.16)$$

The last term in the above equation has been introduced to cancel out the change in the adiabatic potential of $\Delta\tilde{\phi}$ that arises due to the coupling with a system of harmonic oscillators (heat bath). It can be determined explicitly by the requirement that for time independent $\Delta\tilde{\phi}$, the heat bath does not influence the system.

The equations of motion following from this Lagrangian are given by

$$\frac{\mathrm{d}}{\mathrm{d}t}\left(\frac{\partial L}{\partial \Delta\dot{\tilde{\phi}}}\right) = \frac{\partial L}{\partial \Delta\tilde{\phi}} \longrightarrow C\frac{\mathrm{d}^2\Delta\tilde{\phi}}{\mathrm{d}t^2} = -\frac{\partial V(\Delta\tilde{\phi})}{\partial\Delta\tilde{\phi}} + \sum_\alpha \lambda_\alpha x_\alpha - \sum_\alpha \frac{\lambda_\alpha^2}{m_\alpha\omega_\alpha^2}\Delta\tilde{\phi} \ , \quad (5.2.17)$$

$$\frac{\mathrm{d}}{\mathrm{d}t}\left(\frac{\partial L}{\partial \dot{x}_\alpha}\right) = \frac{\partial L}{\partial x_\alpha} \longrightarrow m_\alpha\frac{\mathrm{d}^2 x_\alpha}{\mathrm{d}t^2} = -m_\alpha\omega_\alpha^2 x_\alpha + \lambda_\alpha\Delta\tilde{\phi} \ . \quad (5.2.18)$$

Performing a Fourier transformation of the time variable (that is, setting $\Delta\tilde{\phi}(t) = \mathrm{e}^{-\mathrm{i}\omega t}\Delta\tilde{\phi}(\omega)$) from (5.2.18) we obtain

$$x_\alpha(\omega) = \frac{\lambda_\alpha \Delta\tilde{\phi}(\omega)}{m_\alpha(\omega_\alpha^2 - \omega^2)} \ . \tag{5.2.19}$$

Inserting this expression into the second term of the right-hand side of (5.2.17), we obtain

$$\sum_\alpha \frac{\lambda_\alpha^2}{m_\alpha(\omega_\alpha^2 - \omega^2)} \Delta\tilde{\phi}(\omega) \ . \tag{5.2.20}$$

λ_α, m_α and ω_α should be determined in such a way that this term becomes $-(1/R_\mathrm{S})\Delta\dot{\phi}(t)$ (or $(\mathrm{i}\omega/R_\mathrm{S})\Delta\dot{\tilde{\phi}}(\omega)$ when Fourier transformed). When the energy denominator (5.2.20) becomes zero, it is necessary to give ω a little complex part ($\omega \to \omega + \mathrm{i}\varepsilon, \varepsilon > 0$). Physically, this manipulation corresponds to the boundary condition $\Delta\phi(t) = 0$ for $t \to \infty$ when solving (5.2.17) and (5.2.18). Then, we write $\mathrm{i}J(\omega)$ for the complex part of the coefficient of $\Delta\tilde{\phi}(\omega)$. $J(\omega)$ is the spectral function that represents the interaction with the heat bath and is given by

$$J(\omega) = \pi \sum_\alpha \frac{\lambda_\alpha^2}{2m_\alpha\omega_\alpha} \delta(\omega - \omega_\alpha) \ . \tag{5.2.21}$$

As will become clear later on, all the information concerning the action of the macroscopic degrees of freedom of the heat bath is contained herein. For $J(\omega) = (1/R_\mathrm{S})\omega$, (5.2.17) and (5.2.18) do reproduce (5.2.15).

After the Lagrangian L is found to be (5.2.16), it must be quantized. Introducing the imaginary time formalism, the partition function becomes

$$Z = \int \prod_\alpha \mathcal{D}x_\alpha(\tau)\mathcal{D}\Delta\tilde{\phi}(\tau) \exp[-S(\{\Delta\tilde{\phi}(\tau)\}, \{x_\alpha(\tau)\})] \ . \tag{5.2.22}$$

The action is given by

$$\begin{aligned} S = S_0 + \int_0^\beta \mathrm{d}\tau \Bigg\{ & \sum_\alpha \frac{1}{2} m_\alpha \left[\left(\frac{\mathrm{d}x_\alpha(\tau)}{\mathrm{d}\tau} \right)^2 + \omega_\alpha^2 x_\alpha(\tau)^2 \right] \\ & + \sum_\alpha \lambda_\alpha x_\alpha(\tau)\Delta\tilde{\phi}(\tau) + \sum_\alpha \frac{\lambda_\alpha^2}{2m_\alpha\omega_\alpha^2} (\Delta\tilde{\phi}(\tau))^2 \Bigg\} \ . \end{aligned} \tag{5.2.23}$$

Here, S_0 is the action that describes the system without interaction with the heat bath

$$S_0 = \int_0^\beta \mathrm{d}\tau \left\{ \frac{1}{2} C \left(\frac{\mathrm{d}\Delta\tilde{\phi}(\tau)}{\mathrm{d}\tau} \right)^2 + V(\Delta\tilde{\phi}(\tau)) \right\} \ . \tag{5.2.24}$$

Because the $\int \prod_\alpha \mathcal{D}x_\alpha(\tau)$ integral in (5.2.22) is a Gaussian integral, it can be performed. With the Fourier transformation

$$x_\alpha(\tau) = \frac{1}{\sqrt{\beta}} \sum_{\mathrm{i}\omega_n} x_\alpha(\mathrm{i}\omega_n)\, \mathrm{e}^{-\mathrm{i}\omega_n\tau} \tag{5.2.25}$$

the functional integral becomes

$$\int \prod_{\alpha} \mathcal{D}x_\alpha(\tau) \to \int \prod_{\alpha} \prod_{\mathrm{i}\omega_n:\mathrm{half}} \mathrm{d}[\mathrm{Re}\, x_\alpha(\mathrm{i}\omega_n)]\, \mathrm{d}[\mathrm{Im}\, x_\alpha(\mathrm{i}\omega_n)] \ . \quad (5.2.26)$$

Because the action can be expressed as

$$\begin{aligned} S = S_0 + \sum_{\mathrm{i}\omega_n} \sum_{\alpha} \Big\{ & \frac{1}{2} m_\alpha [\omega_n^2 + \omega_\alpha^2] x_\alpha(\mathrm{i}\omega_n) x_\alpha(-\mathrm{i}\omega_n) \\ & + \frac{1}{2} \lambda_\alpha \left[x_\alpha(\mathrm{i}\omega_n) \Delta\tilde{\phi}(-\mathrm{i}\omega_n) + x_\alpha(-\mathrm{i}\omega_n) \Delta\tilde{\phi}(\mathrm{i}\omega_n) \right] \\ & + \frac{\lambda_\alpha^2}{2 m_\alpha \omega_\alpha^2} \Delta\tilde{\phi}(\mathrm{i}\omega_n) \Delta\tilde{\phi}(-\mathrm{i}\omega_n) \Big\} \ , \end{aligned} \quad (5.2.27)$$

when shifting the origin of $x_\alpha(\mathrm{i}\omega_n)$ and performing the integration, for $\Delta\tilde{\phi}$ we obtain the effective action

$$S_{\mathrm{eff}} = S_0 + \sum_{\mathrm{i}\omega_n} K(\mathrm{i}\omega_n) \Delta\tilde{\phi}(\mathrm{i}\omega_n) \Delta\tilde{\phi}(-\mathrm{i}\omega_n) \ , \quad (5.2.28)$$

where we define

$$\begin{aligned} K(\mathrm{i}\omega_n) &= \sum_{\alpha} \frac{\lambda_\alpha^2}{2 m_\alpha} \left(\frac{1}{\omega_\alpha^2} - \frac{1}{\omega_n^2 + \omega_\alpha^2} \right) \\ &= \int_0^\infty \frac{\mathrm{d}\omega}{\pi} J(\omega) \left(\frac{1}{\omega} - \frac{\omega}{\omega_n^2 + \omega^2} \right) \ . \end{aligned} \quad (5.2.29)$$

Expressing (5.2.28) in imaginary time, we obtain

$$S_{\mathrm{eff}} = S_0 + \int_0^\beta \mathrm{d}\tau \int_0^\beta \mathrm{d}\tau' \, K(\tau - \tau') \Delta\tilde{\phi}(\tau) \Delta\tilde{\phi}(\tau') \ , \quad (5.2.30)$$

with the kernel $K(\tau)$ given by

$$K(\tau) = \frac{1}{\beta} \sum_{\mathrm{i}\omega_n} K(\mathrm{i}\omega_n)\, \mathrm{e}^{-\mathrm{i}\omega_n \tau} \ . \quad (5.2.31)$$

Inserting $K(\mathrm{i}\omega_n)$ (5.2.29), the result is

$$\begin{aligned} K(\tau) &= \int_0^\infty \frac{\mathrm{d}\omega}{\pi} \frac{J(\omega)}{\omega} \frac{1}{\beta} \sum_{\mathrm{i}\omega_n} \mathrm{e}^{-\mathrm{i}\omega_n \tau} - \int_0^\infty \frac{\mathrm{d}\omega}{\pi} \omega J(\omega) \frac{1}{\beta} \sum_{\mathrm{i}\omega_n} \frac{\mathrm{e}^{-\mathrm{i}\omega_n \tau}}{\omega_n^2 + \omega^2} \\ &= \int_0^\infty \frac{\mathrm{d}\omega}{\pi} \frac{J(\omega)}{\omega} \cdot \sum_{m} \delta(\tau - m\beta) - \int_0^\infty \frac{\mathrm{d}\omega}{\pi} \omega J(\omega) \frac{\cosh[\omega(|\tau| - \beta/2)]}{2\omega \sinh(\beta\omega/2)} \ . \end{aligned} \quad (5.2.32)$$

This must be inserted into (5.2.30), and using the equation

$$\sum_{m=-\infty}^{\infty} \mathrm{e}^{-\omega|\tau+m\beta|} = \frac{\cosh[\omega(|\tau|-\beta/2)]}{\sinh(\beta\omega/2)} \tag{5.2.33}$$

and the fact that $\Delta\tilde{\phi}(\tau)$ satisfies the periodic boundary conditions $\Delta\tilde{\phi}(\tau+\beta) = \Delta\tilde{\phi}(\tau)$, we obtain

$$\begin{aligned} S_{\text{eff}} = S_0 &+ \int_0^\beta \mathrm{d}\tau' \left(\int_0^\infty \frac{\mathrm{d}\omega}{\pi} \frac{J(\omega)}{\omega} \right) (\Delta\tilde{\phi}(\tau'))^2 \\ &- \int_{-\infty}^{\infty} \mathrm{d}\tau \int_0^\beta \mathrm{d}\tau' \left(\int_0^\infty \frac{\mathrm{d}\omega}{2\pi} J(\omega)\, \mathrm{e}^{-\omega|\tau-\tau'|} \right) \Delta\tilde{\phi}(\tau)\Delta\tilde{\phi}(\tau') \ . \end{aligned} \tag{5.2.34}$$

Using the identity

$$\frac{1}{\omega} = \frac{1}{2} \int_{-\infty}^{\infty} \mathrm{d}\tau\, \mathrm{e}^{-\omega|\tau-\tau'|} \ , \tag{5.2.35}$$

(5.2.34) can be written as

$$S_{\text{eff}} = S_0 + \int_{-\infty}^{\infty} \mathrm{d}\tau \int_0^\beta \mathrm{d}\tau'\, \hat{K}(|\tau-\tau'|)(\Delta\tilde{\phi}(\tau) - \Delta\tilde{\phi}(\tau'))^2 \ . \tag{5.2.36}$$

Here, $\hat{K}(|\tau-\tau'|)$ is the kernel of equation (5.2.34):

$$\hat{K}(|\tau-\tau'|) = \int_0^\infty \frac{\mathrm{d}\omega}{4\pi} J(\omega)\, \mathrm{e}^{-\omega|\tau-\tau'|} \ . \tag{5.2.37}$$

Especially for $J(\omega) = \omega/R_{\mathrm{S}}$, we obtain

$$\hat{K}(|\tau-\tau'|) = \frac{1}{4\pi R_{\mathrm{S}}} \frac{1}{|\tau-\tau'|^2} \ . \tag{5.2.38}$$

Finally, the effective action becomes

$$\begin{aligned} S_{\text{eff}} &= \int_0^\beta \mathrm{d}\tau \left\{ \frac{1}{2} C \left(\frac{\mathrm{d}\Delta\tilde{\phi}}{\mathrm{d}\tau} \right)^2 + V(\Delta\tilde{\phi}) \right\} \\ &\quad + \int_{-\infty}^{\infty} \mathrm{d}\tau \int_0^\beta \mathrm{d}\tau' \frac{1}{4\pi R_{\mathrm{S}}} \left(\frac{\Delta\tilde{\phi}(\tau) - \Delta\tilde{\phi}(\tau')}{\tau-\tau'} \right)^2 \\ &= \int_0^\beta \mathrm{d}\tau \left\{ \frac{1}{2} \left(\frac{\hbar}{2e} \right)^2 \left(\frac{\mathrm{d}\Delta\phi}{\mathrm{d}\tau} \right)^2 - \frac{I_0}{2e} \cos(\Delta\phi) + I\Delta\phi \right\} \\ &\quad + \int_{-\infty}^{\infty} \mathrm{d}\tau \int_0^\beta \mathrm{d}\tau' \frac{\hbar}{16\pi e^2 R_{\mathrm{S}}} \left(\frac{\Delta\phi(\tau) - \Delta\phi(\tau')}{\tau-\tau'} \right)^2 \ . \end{aligned} \tag{5.2.39}$$

Here, we reintroduced $\hbar$ and used the relation $\Delta\tilde{\phi} = (\hbar/2e)\Delta\phi$. This action describes two different states, depending on the strength of the dissipation. We will discuss this point in what follows.

The action (5.2.39) describes the damped movement of a particle in the periodic cos-potential (inclined for the case where a current I is present). Returning to (5.2.28), the term describing the dissipation can be expressed as

$$S_{\text{eff}} = S_0 + \sum_{\mathrm{i}\omega_n} \frac{\alpha}{4\pi} |\omega_n| \Delta\phi(\mathrm{i}\omega_n)\Delta\phi(-\mathrm{i}\omega_n) \ . \tag{5.2.40}$$

Here, we introduced $\alpha = R_{\mathrm{Q}}/R_{\mathrm{S}}$, where R_{Q} is the so-called quantum resistance $R_{\mathrm{Q}} \equiv h/4e^2 = \pi\hbar/2e^2$. In what follows, the action (5.2.40) will be used. For simplicity, we discuss the case $I = 0$. Furthermore, because for low energy, $|\omega_n|$ is larger than the "kinetic energy" ω_n^2, the latter is ignored. Instead, a cut-off Λ of the magnitude of α/C is introduced. For $\alpha \simeq 1$, Λ can be considered as the approximate "band width" when no dissipation occurs. In this case, only the action S_0,

$$S_0 = -\frac{I_0}{2e}\int_0^\beta \cos(\Delta\phi)\,\mathrm{d}\tau = -y\int_0^\beta \cos(\Delta\phi)\,\mathrm{d}\tau \ , \tag{5.2.41}$$

has to be taken into account. y is the height of the potential; the limit of a weak potential ($|y| \ll \Lambda$) and the limit of a strong potential ($|y| \gg \Lambda$) will now be discussed separately.

We start by discussing the case $|y| \ll \Lambda$, where perturbation theory in y can be applied. However, because the free propagator $G_0(\mathrm{i}\omega_n)$ behaves like $\propto 1/|\omega_n|$, simple perturbation theory cannot be applied. In such cases, the renormalization group becomes a powerful tool. Its principal idea is to integrate out little by little the high-frequency components, to implement them in a renormalized, effective action for the low-frequency modes and to pursue the change in the effective action. Here, we restrict our discussion to zero temperature. We split $\Delta\phi(\mathrm{i}\omega)$

$$\Delta\phi(\mathrm{i}\omega) = \Delta\phi_1(\mathrm{i}\omega)\theta(|\omega| < \Lambda - \mathrm{d}\Lambda) + \Delta\phi_2(\mathrm{i}\omega)\theta(\Lambda - \mathrm{d}\Lambda < |\omega| < \Lambda) \tag{5.2.42}$$

into the low-frequency part $\Delta\phi_1$ and the high-frequency part $\Delta\phi_2$ and integrate out only ϕ_2. θ is a function that is equal to 1 when the condition in its argument is satisfied, otherwise it is zero.

$$Z = \int \mathcal{D}\Delta\phi_1 \mathcal{D}\Delta\phi_2\, \mathrm{e}^{-S_{\text{eff}}(\Delta\phi_1+\Delta\phi_2)} = \int \mathcal{D}\Delta\phi_1\, \mathrm{e}^{-\tilde{S}_{\text{eff}}(\Delta\phi_1)} \ . \tag{5.2.43}$$

The question of what the new action $\tilde{S}_{\text{eff}}(\Delta\phi_1)$ of $\Delta\phi_1$ might be is in general a difficult one. For small y, $\tilde{S}_{\text{eff}}(\Delta\phi_1)$ can be determined perturbatively with respect to y. Explicitly, we write

$$\begin{aligned}\tilde{S}_{\text{eff}}(\Delta\phi_1) = {} & \frac{\alpha}{4\pi}\sum_{|\omega|<\Lambda-\mathrm{d}\Lambda} |\omega|\Delta\phi_1(\mathrm{i}\omega)\Delta\phi_1(-\mathrm{i}\omega) \\ & - \ln\left\langle \exp\left[-y\int \cos(\Delta\phi_1 + \Delta\phi_2)\,\mathrm{d}\tau\right]\right\rangle_{\Delta\phi_2} \ .\end{aligned} \tag{5.2.44}$$

The averaging is defined by

$$\langle A\rangle_{\Delta\phi_2} \propto \int \mathcal{D}\Delta\phi_2 \exp\left\{-\frac{\alpha}{4\pi}\sum_{\Lambda-\mathrm{d}\Lambda<|\omega|<\Lambda} |\omega|\Delta\phi_2(\mathrm{i}\omega)\Delta\phi_2(-\mathrm{i}\omega)\right\} A\ . \tag{5.2.45}$$

Expanding the exponential up to first order in y, we obtain

$$\left\langle 1 - y\int \cos(\Delta\phi_1 + \Delta\phi_2)\,\mathrm{d}\tau + \cdots\right\rangle_{\Delta\phi_2}\ . \tag{5.2.46}$$

The first term equals 1, the second term is a Gaussian average with respect to $\Delta\phi_2$, and therefore we obtain

$$-y\int \mathrm{d}\tau\,\mathrm{e}^{-1/2\langle(\Delta\phi_2(\tau))^2\rangle}\cos\Delta\phi_1(\tau)\ . \tag{5.2.47}$$

Because of

$$\langle(\Delta\phi_2(\tau))^2\rangle = \frac{2}{\alpha}\frac{\mathrm{d}\Lambda}{\Lambda} \tag{5.2.48}$$

(5.2.47) becomes

$$-y\,\mathrm{e}^{(-1/\alpha)(\mathrm{d}\Lambda/\Lambda)}\int \mathrm{d}\tau\cos\Delta\phi_1(\tau)\ . \tag{5.2.49}$$

Therefore, the new problem for $\Delta\phi_1$ has as cut-off $\Lambda - \mathrm{d}\Lambda$ instead of Λ, and y is replaced by $\tilde{y} = y\exp[-(1/\alpha)(\mathrm{d}\Lambda/\Lambda)]$. The relation between the potential y and the "band width" Λ determines whether the particles are extended or localized. Asking how this ratio $y_0 \equiv y/\Lambda$ is altered, we obtain

$$y_0(\Lambda) \to y_0(\Lambda - \mathrm{d}\Lambda) = \frac{y\,\mathrm{e}^{(-1/\alpha)(\mathrm{d}\Lambda/\Lambda)}}{\Lambda - \mathrm{d}\Lambda} = y_0(\Lambda)\left[1 + \left(1 - \frac{1}{\alpha}\right)\frac{\mathrm{d}\Lambda}{\Lambda}\right] \tag{5.2.50}$$

and the following differential equation for $y_0(\Lambda)$ as a function of the cut-off Λ:

$$\Lambda\frac{\mathrm{d}y_0(\Lambda)}{\mathrm{d}\Lambda} = \Lambda\frac{y_0(\Lambda) - y_0(\Lambda - \mathrm{d}\Lambda)}{\mathrm{d}\Lambda} = \left(\frac{1}{\alpha} - 1\right)y_0(\Lambda)\ . \tag{5.2.51}$$

We conclude that for small Λ, that is, in the limit of low energy, $y_0(\Lambda)$ increases for $\alpha > 1$ and decreases for $\alpha < 1$. Therefore, $\alpha = 1$ is the boundary, where for $\alpha < 1$ the potential effectively has no influence and the particle wave function spreads out as a wave and becomes an extended state. On the other hand, for $\alpha > 1$ owing to the dissipation effect the kinetic energy is suppressed and the particle comes to rest in one of the minima of the potential.

This has been an analysis of the limit of a weak potential $|y| \ll \Lambda$. Also for $|y| \gg \Lambda$ at the boundary value $\alpha = 1$ the transition from local to non-local behaviour occurs. We derive this fact by looking at the theory after having performed a duality transformation.

In the limit $|y| \gg \Lambda$, the probability that tunnelling between the minima of the potential occurs will be very small. In Sect. 2.1, we understood that in the imaginary time formulation of the path integral, the tunnelling effect is represented by instantons. The action of one instanton S_{inst} is given roughly by $\sqrt{yC}$. When in total n instantons or anti-instantons are present, the contribution is multiplied by the factor z^n. With $z \propto \mathrm{e}^{-S_{\text{inst}}}$, Fig. 5.4 shows the configuration of a field containing instantons. Assuming that the instantons are dilute enough, we can ignore the interaction between the instantons which comes from S_0 in (5.2.40). Then, the partition function Z reads

$$Z = \sum_{n=0}^{\infty} z^n \sum_{\{e_i\}} \int_0^{\beta} \mathrm{d}\tau_n \int_0^{\tau_n} \mathrm{d}\tau_{n-1} \cdots \int_0^{\tau_2} \mathrm{d}\tau_1 \exp[-S_{\text{dissip.}}] \ . \quad (5.2.52)$$

Here, τ_i is the location of the ith instanton, and e_i is the topological charge distinguishing the instanton ($e_i = +1$) from the anti-instanton ($e_i = -1$). $S_{\text{dissip.}}$ is the second term of (5.2.40), where the instanton configuration has been inserted.

As is evident from Fig. 5.4, when one (anti-)instanton is coming in, the value of $\Delta\phi(\tau)$ before and after changes by 2π, and the effect goes over a large range in τ. In order to localize it to the vicinity of the (anti-)instanton, it is sufficient to consider the derivative

$$\frac{\mathrm{d}(\Delta\phi(\tau))}{\mathrm{d}\tau} = \sum_i e_i h(\tau - \tau_i) \ . \quad (5.2.53)$$

Here, $h(\tau - \tau_1)$ is the derivative at time τ of the one-instanton configuration. The Fourier transformation of (5.2.53) reads

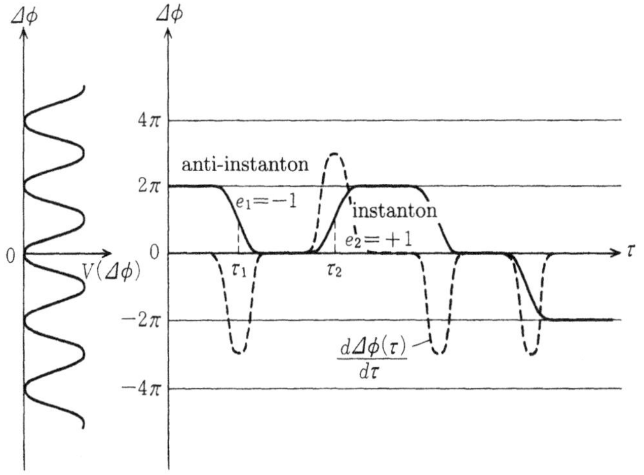

Fig. 5.4. Configuration containing instantons and anti-instantons

$$-\mathrm{i}\omega\Delta\phi(\mathrm{i}\omega) = \sum_i e_i \hat{h}(\mathrm{i}\omega)\,\mathrm{e}^{\mathrm{i}\omega\tau_i} \ , \tag{5.2.54}$$

where $\hat{h}(\mathrm{i}\omega)$ is the Fourier transformation of $h(\tau)$:

$$\hat{h}(\mathrm{i}\omega) = \frac{1}{\sqrt{\beta}} \int \mathrm{d}\tau\,\mathrm{e}^{\mathrm{i}\omega\tau} h(\tau) \ . \tag{5.2.55}$$

For $\omega \to 0$, it becomes the integral of $h(\tau)/\sqrt{\beta}$, i.e. $2\pi/\sqrt{\beta}$. Inserting this into the second term at the right-hand side of (5.2.40), we obtain

$$\sum_{\mathrm{i}\omega} \frac{\alpha}{4\pi}|\omega|\frac{1}{\omega^2}\sum_{ij} e_i e_j \hat{h}(\mathrm{i}\omega)\hat{h}(-\mathrm{i}\omega)\,\mathrm{e}^{\mathrm{i}\omega(\tau_i-\tau_j)} \underset{\omega:\,\mathrm{small}}{\cong} \sum_{ij} F(\tau_i-\tau_j)e_i e_j \ . \tag{5.2.56}$$

Here, $F(\tau_i - \tau_j)$ is given by

$$F(\tau_i-\tau_j) = \frac{1}{\beta}\sum_{\mathrm{i}\omega}\frac{\pi\alpha}{|\omega|}\,\mathrm{e}^{\mathrm{i}\omega(\tau_i-\tau_j)} \ . \tag{5.2.57}$$

For large $|\tau_i - \tau_j|$, it behaves like $\ln|\tau_i - \tau_j|$, signifying the presence of a "non-local" interaction between the instantons.

Therefore, (5.2.52) can be written as

$$Z = \sum_{n=0}^{\infty} z^n \sum_{\{e_i\}} \int_0^{\beta} \mathrm{d}\tau_n \ldots \int_0^{\tau_2} \mathrm{d}\tau_1 \exp\left[-\sum_{ij} F(\tau_i-\tau_j)e_i e_j\right] \ . \tag{5.2.58}$$

Performing one more step by expressing the interaction between e_i and e_j using the functional integral over the new field $q(\tau)$

$$\begin{aligned}
&\int \mathcal{D}q(\tau)\exp\left[-\sum_{\mathrm{i}\omega}\frac{|\omega|}{4\pi\alpha}q(\mathrm{i}\omega)q(-\mathrm{i}\omega) + \mathrm{i}\sum_i e_i q(\tau_i)\right] \\
&= \int \mathcal{D}q(\tau)\exp\left[-\sum_{\mathrm{i}\omega}\left\{\frac{|\omega|}{4\pi\alpha}q(\mathrm{i}\omega)q(-\mathrm{i}\omega) + \frac{\mathrm{i}}{\sqrt{\beta}}\sum_i e_i\,\mathrm{e}^{-\mathrm{i}\omega\tau_1}q(\mathrm{i}\omega)\right\}\right] \\
&= \exp\left[-\frac{1}{\beta}\sum_{ij}\sum_{\mathrm{i}\omega}\frac{\pi\alpha}{|\omega|}\,\mathrm{e}^{\mathrm{i}\omega(\tau_i-\tau_j)}e_i e_j\right] \ ,
\end{aligned} \tag{5.2.59}$$

the partition function becomes

$$\begin{aligned}
Z = &\int \mathcal{D}q(\tau) \sum_{n=0}^{\infty} z^n \sum_{\{e_i\}} \int_0^{\beta} \mathrm{d}\tau_n \ldots \int_0^{\tau_2} \mathrm{d}\tau_1 \\
&\times \exp\left[-\sum_{\mathrm{i}\omega}\frac{|\omega|}{4\pi\alpha}q(\mathrm{i}\omega)q(-\mathrm{i}\omega) + \mathrm{i}\sum_i e_i q(\tau_i)\right]
\end{aligned}$$

$$= \int \mathcal{D}q(\tau) \exp\left[-\sum_{\mathrm{i}\omega} \frac{|\omega|}{4\pi\alpha} q(\mathrm{i}\omega) q(-\mathrm{i}\omega)\right] \sum_{n=0}^{\infty} \frac{z^n}{n!} \left(2\int_0^\beta \mathrm{d}\tau \cos q(\tau)\right)^n$$

$$= \int \mathcal{D}q(\tau) \exp\left[-\sum_{\mathrm{i}\omega} \frac{|\omega|}{4\pi\alpha} q(\mathrm{i}\omega) q(-\mathrm{i}\omega) + 2z\int_0^\beta \mathrm{d}\tau \cos q(\tau)\right] . \quad (5.2.60)$$

and surprisingly, we finally regained the same model as described in (5.2.40) and (5.2.41). Notice the correspondence $\alpha \leftrightarrow 1/\alpha$, $y \leftrightarrow z$. In both models, large and small dissipations, potential and tunnelling are dual to each other.

The analysis based on the renormalization group is also valid for (5.2.60): for $1/\alpha < 1$, the effect of z becomes irrelevant, leading to a localized wave function; for $1/\alpha > 1$, the effect of z becomes relevant and the system becomes non-localized. Combining this with the analysis of the weak potential limit, the whole phase diagram can be given by Fig. 5.5. It is natural to assume that the phase boundary is independent of y and becomes rectilinear.

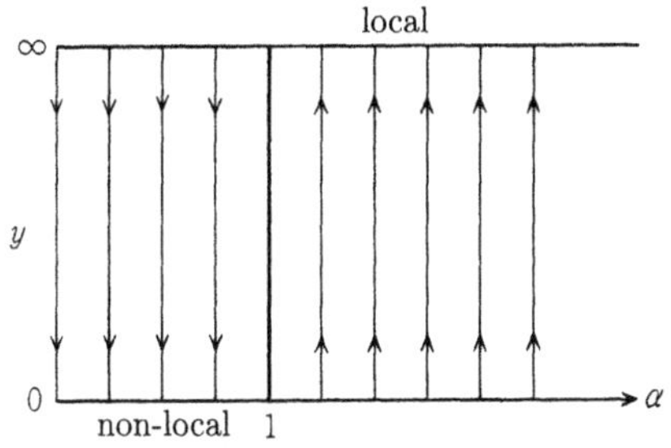

Fig. 5.5. Phase diagram expected for the dissipative system

We will now discuss the significance of the result obtained for the Josephson junction. Owing to the relation $\alpha = R_{\mathrm{Q}}/R_{\mathrm{S}}$, at the critical value R_{Q} of the junction resistance R_{S}, which is the universal resistance, phase transition occurs. It may be surprising that for smaller junction resistances R_{S} the dissipation α becomes larger. Currently, we are discussing the dynamics of the phase difference $\Delta\phi$, and because it is the canonical conjugate degree of freedom of the particle number ΔN and its time derivative, the current, the effect of R_{S} will be inverse for them. For large R_{S}, α becomes small, and therefore the wave function becomes extended with respect to the phase difference $\Delta\phi$; on the other hand the particle number ΔN becomes localized, and as a result superconductivity breaks down. For a thin superconducting film that can be regarded as being two dimensional, the striking phenomenon of an insulator–superconductor transition has been observed at the critical resistance nearby R_{Q}. Models based on the phase transition occurring at one Josephson junction have been proposed for the theoretical analysis; however, the problem is not that simple, and it may be necessary to consider a system of many Josephson junctions.

In the next section, this problem will be discussed in more detail.

5.3 The Superconductor–Insulator Phase Transition in Two Dimensions and the Quantum Vortices

In some sense, the content of this section is a repetition of Sect. 3.3, where we understood that the lattice structure allows the creation of vortices, and as a result the Kosterlitz–Thouless transition occurs. In a superconductor, there is the degree of freedom of the amplitude of the order parameter, and when it becomes zero, the phase becomes singular, that is, the creation of vortices becomes possible. In this section, we discuss the superconductor in two dimensions using the continuum theory, focusing on the physical results.

We begin by discussing the fundamental properties of the superconductor in two dimensions. In Sect. 5.1, the BCS theory of superconductors has been developed from the point of view of functional integrals; however, because in two dimensions the effect of fluctuations having been neglected in the BCS theory become relevant, a more detailed discussion is necessary. Explicitly, in the case of finite temperature, thermal excitations that do not depend on the imaginary time are dominant, and we will use the framework of the Ginzburg–Landau (GL) theory presented in Sect. 5.1 for the discussion. With standard notation, the functional of the free energy is given by ($\psi(\boldsymbol{r})$ corresponds to $\Delta(\boldsymbol{r})$ of Sect. 5.1)

$$F[\psi(\boldsymbol{r}),\psi^*(\boldsymbol{r}),\boldsymbol{A}(\boldsymbol{r})] = \int \mathrm{d}^2\boldsymbol{r}\left[\frac{1}{2m}\left|\left(-\mathrm{i}\hbar\nabla + \frac{2e}{c}\boldsymbol{A}\right)\psi(\boldsymbol{r})\right|^2 \right.$$
$$\left. + a(T)\,|\psi(\boldsymbol{r})|^2 + \frac{b}{2}\,|\psi(\boldsymbol{r})|^4 + \frac{1}{8\pi}(\nabla\times\boldsymbol{A})^2\right] \quad (5.3.1)$$

and the partition function is given by

$$Z = \int D\psi(\boldsymbol{r})\exp\left\{-\beta F[\psi(\boldsymbol{r}),\psi^*(\boldsymbol{r}),\boldsymbol{A}(\boldsymbol{r})]\right\} . \quad (5.3.2)$$

In a narrow sense, the GL theory (classic approximation) corresponds to the saddle-point solution of the functional integral of $F[\psi(\boldsymbol{r})\psi^*(\boldsymbol{r}),\boldsymbol{A}(\boldsymbol{r})]$. The condition $\delta F/\delta\psi^*(\boldsymbol{r}) = 0$ in the functional integral leads to

$$a(T)\psi + b|\psi|^2\psi + \frac{1}{2m}\left(-\mathrm{i}\hbar\nabla + \frac{2e\boldsymbol{A}}{c}\right)^2\psi = 0 \quad (5.3.3)$$

and from $\delta F/\delta\boldsymbol{A}(\boldsymbol{r}) = \boldsymbol{0}$, we obtain

$$\boldsymbol{\nabla}\times\boldsymbol{H} = \frac{4\pi}{c}\boldsymbol{j}_{\mathrm{s}} . \quad (5.3.4)$$

Here, the superconducting current $\boldsymbol{j}_s$ is given by

$$\boldsymbol{j}_{\mathrm{s}}(\boldsymbol{r}) = -\frac{2\hbar}{\mathrm{i}m}(\psi^*\nabla\psi - \psi\nabla\psi^*) - \frac{4e^2}{mc}\psi^*\psi\boldsymbol{A} \quad (5.3.5)$$

and the magnetic field $\boldsymbol{H}$ is given by $\boldsymbol{H} = \nabla \times \boldsymbol{A}$. Equations (5.3.3), (5.3.4) and (5.3.5) are the fundamental equations of the GL theory. The superfluid density ρ_s is defined as the ratio of the macroscopic average of the superconductor velocity (with $\theta(\boldsymbol{r})$ being the phase of $\psi(\boldsymbol{r})$):

$$\boldsymbol{v}_s(\boldsymbol{r}) \equiv \frac{\hbar}{m}\left[\nabla\theta(\boldsymbol{r}) + \frac{2e}{c\hbar}\boldsymbol{A}\right] \tag{5.3.6}$$

and the macroscopic superconducting current $\boldsymbol{j}_s(\boldsymbol{r})$ ($\langle \boldsymbol{j}_s \rangle = -2e\rho_s\langle v \rangle_s$). Setting $\psi(\boldsymbol{r}) = |\psi(\boldsymbol{r})|\, e^{i\theta(\boldsymbol{r})}$ and inserting this expression into (5.3.5), we obtain

$$\boldsymbol{j}_s(\boldsymbol{r}) = -\frac{2e\hbar}{m}|\psi(\boldsymbol{r})|^2\left(\nabla\theta(\boldsymbol{r}) + \frac{2e}{c\hbar}\boldsymbol{A}\right) . \tag{5.3.7}$$

In a homogeneous, infinite system, for $\psi_0(\boldsymbol{r})$ (when no magnetic field is present, $\boldsymbol{A} = \boldsymbol{0}$) a space-independent solution ψ_0 can be chosen. With $a(T) = a'(T - T_c)$, the solution is given by $\psi_0 = 0$ for $T > T_c$, and by $\psi_0 = \sqrt{-a(T)/b}$ for $T < T_c$.

Next, we improve this approximation in two directions. For simplicity, we ignore the fluctuations of the electromagnetic field and set $\boldsymbol{A} = \boldsymbol{0}$ (this corresponds to an infinite penetration depth of the magnetic field, in this case (5.3.3) also describes the superfluidity of ^{4}He).

First, we consider Gaussian fluctuations around ψ_c for $T < T_c$. Because low-energy fluctuations are fluctuations of the phase of $\psi_c(\boldsymbol{r})$, we set $\psi_0(\boldsymbol{r}) = \psi_0\, e^{i\theta(\boldsymbol{r})}$ and insert this into (5.3.1). Up to constants, the result is

$$\beta\Delta F = \frac{1}{2}K\int d^2\boldsymbol{r}\,(\nabla\theta(\boldsymbol{r}))^2 . \tag{5.3.8}$$

Here, K is given by $K = \hbar^2\psi_0^2/mk_BT$. Then, (5.3.8) has the same form as the energy in the spin wave approximation of the two-dimensional XY model in the low-energy limit. The analysis of Sect. 3.2 can be applied as it stands, and we obtain

$$\begin{aligned}\langle\psi(\boldsymbol{r})\rangle &= \psi_0\langle\exp[i\theta(\boldsymbol{r})]\rangle \\ &= \psi_0\exp[-\langle\theta(\boldsymbol{r})^2\rangle/2] = 0 ,\end{aligned} \tag{5.3.9a}$$

$$\begin{aligned}\langle\psi^*(\boldsymbol{r})\psi(\boldsymbol{r}')\rangle &= \psi_0^2\langle\exp[-i(\theta(\boldsymbol{r}) - \theta(\boldsymbol{r}'))]\rangle \\ &\sim |\boldsymbol{r} - \boldsymbol{r}'|^{-1/2\pi K} .\end{aligned} \tag{5.3.9b}$$

It is important to notice that although the superconductive density ρ_s and the average value of the order parameter $\langle\psi(\boldsymbol{r})\rangle$ are not totally independent, they are physically absolutely different quantities. That is, even when $\langle\psi(\boldsymbol{r})\rangle$ is zero, ρ_s can have a finite value. In order to see this, consider a two-dimensional superconductor, or superfluid that is wrapped around a cylinder, with a velocity field $\boldsymbol{v}$ of the superfluid flowing in the direction of the arrows as indicated in Fig. 5.6.

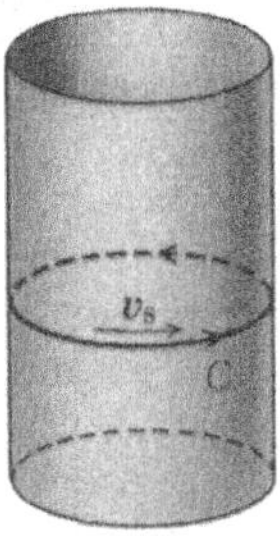

Fig. 5.6. Two-dimensional superconductor with the topology of a cylinder

The system is a superconductor or a superfluid ($\rho_s \neq 0$) when $\boldsymbol{v}_s$ is nonzero and does not decrease, that is, when $\boldsymbol{v}_s \neq 0$ is a stable state of the system.

We now consider the line integral $I = \oint_C \boldsymbol{v}_s \cdot d\boldsymbol{r}$ of $\boldsymbol{v}_s$ around the cylinder following the path C. From (5.3.6) for the case $\boldsymbol{A} = \boldsymbol{0}$ it follows that

$$I = \oint_C \boldsymbol{v}_s \cdot d\boldsymbol{r} = \frac{\hbar}{m}[\theta]_C \ . \tag{5.3.10}$$

The change in the phase $[\theta]_C$ is quantized because it must be an integer multiplied by 2π. Therefore, I also is quantized and must be an integer multiplied by $2\pi\hbar/m$. In such a way, I is a topological quantity that counts the winding. Under a continuous transformation, it does not change continuously. Therefore, as long as $\theta(\boldsymbol{r})$ is a single-valued continuous function of the two-dimensional position vector $\boldsymbol{r}$, I and therefore $\boldsymbol{v}_s$ cannot decrease to zero. Indeed, directly from the free energy (5.3.8), we obtain $\rho_s = \psi_0^2$. The line integral I only changes when the path C reaches a singularity where $\theta(\boldsymbol{r})$ cannot be defined. In such a case, $[\theta]_C$ changes discontinuously about $\pm 2\pi$. Singular points where $\theta(\boldsymbol{r})$ cannot be defined are points where the amplitude ψ becomes zero, that is, in the centre of a vortex.

In the functional integral, besides the homogeneous solution ψ_0, a vortex contributes as an additional space-dependent saddle-point solution. Far from the centre of the vortex $\boldsymbol{r}_0$ compared with the correlation length ξ, this solution behaves as $\psi(\boldsymbol{r}) \sim \psi_0\, e^{\pm i\phi(\boldsymbol{r}-\boldsymbol{r}_0)}$ with $\phi(\boldsymbol{r})$ being the angle. Therefore, similar to the XY model, the energy of one vortex is logarithmically divergent. Furthermore, when thinking about a neutral vortex–antivortex system, a two-dimensional classical Coulomb gas model with logarithmic interaction is obtained. Around the vortex $\boldsymbol{v}_s$ a whirl emerges, and in the Coulomb gas model the chirality corresponds to the sign of the charge.

We consider this correspondence further. A Coulomb charge creates an electric field $\boldsymbol{E}$ around itself. Because the field lines are straight lines coming from the origin, they are obtained by rotating the whirl $\boldsymbol{v}_s$ by the angle $\pi/2$. Mathematically, this can be expressed as

$$\boldsymbol{E}(\boldsymbol{r}) \equiv -\frac{2\pi K m}{\hbar}\hat{\boldsymbol{z}} \times \boldsymbol{v}_s(\boldsymbol{r}) \ . \tag{5.3.11a}$$

Here, we set $\hat{\boldsymbol{z}} = (0, 0, 1)$. Then, on the other hand, $\boldsymbol{v}_\mathrm{s}$ itself can be expressed as

$$\boldsymbol{v}_\mathrm{s}(\boldsymbol{r}) = \frac{\hbar}{2\pi K m} \hat{\boldsymbol{z}} \times \boldsymbol{E}(\boldsymbol{r}) \ . \tag{5.3.11b}$$

Now, we split the superfluid velocity into two parts and write $\boldsymbol{v}_\mathrm{s} = \boldsymbol{v}_0 + \boldsymbol{v}_1$. Here, $\boldsymbol{v}_0$ is the contribution of the vortex fulfilling $\nabla \cdot \boldsymbol{v}_0 = 0$, and $\boldsymbol{v}_1$ is the contribution fulfilling $\nabla \times \boldsymbol{v}_1 = 0$ which can be written as the gradient of a single valued function θ_1. This is nothing but the well-known vector decomposition theorem in vector analysis. Were are at the moment only interested in the whirl and therefore set $\boldsymbol{v}_1 = \boldsymbol{0}$. Then, we get $\nabla \cdot \boldsymbol{v}_\mathrm{s} = \nabla \cdot \boldsymbol{v}_0 = 0$, and in the interpretation of (5.3.11a), this translates into

$$\nabla \times \boldsymbol{E} = \boldsymbol{0} \ . \tag{5.3.12}$$

Introducing the vortex density $N(\boldsymbol{r})$ ($\boldsymbol{r}_i$ is the position of the ith vortex, e_i is its direction ± 1)

$$N(\boldsymbol{r}) = \sum_i e_i \delta(\boldsymbol{r} - \boldsymbol{r}_i) \tag{5.3.13}$$

we obtain

$$\oint_{\partial S} \boldsymbol{v}_0(\boldsymbol{r}) \cdot \mathrm{d}\boldsymbol{r} = \frac{2\pi\hbar}{m} \iint_S \mathrm{d}S \, N(\boldsymbol{r}) \ . \tag{5.3.14}$$

In differential form, this corresponds to

$$\nabla \times \frac{2\pi K m}{\hbar} \boldsymbol{v}_\mathrm{s}(\boldsymbol{r}) = \nabla \cdot \boldsymbol{E}(\boldsymbol{r}) \hat{\boldsymbol{z}} = 4\pi^2 K N(\boldsymbol{r}) \hat{\boldsymbol{z}} \ . \tag{5.3.15}$$

We conclude that the electric field and the superfluid velocity around one vortex at the origin are given by

$$\boldsymbol{E}(\boldsymbol{r}) = \frac{2\pi K \boldsymbol{r}}{|\boldsymbol{r}|^2} \ , \tag{5.3.16a}$$

$$\boldsymbol{v}_\mathrm{s}(\boldsymbol{r}) = \frac{\hbar}{m} \frac{\hat{\boldsymbol{z}} \times \boldsymbol{r}}{|\boldsymbol{r}|^2} \ . \tag{5.3.16b}$$

The idea of the correspondence between superfluid velocity and electric field, and between vortex density $N(\boldsymbol{r})$ and electric charge, will also be applied in a more general model which will be introduced later on, where the quantum fluctuations will also be taken into account. The XY model being defined on a lattice, and the GL theory describing the degree of freedom of the phase, are different; however, in both cases the KT phase transition occurs, and the properties of the phase transition are similar for both. That is, in both cases a phase transition ocurs between a high-temperature phase where free (anti-) vortices are present, and a low-temperature phase where all vortices are combined to neutral pairs. Because the decrease in the line integral I of the superfluid velocity field that has been discussed before does not arise when a neutral pair passes the path C, but only for the case when free (anti-) vortices are present, the phase transition occurring in the two-dimensional superconductor is a KT transition. Below the transition temperature, the Meissner effect occurs (in the case of helium, superfluidity arises). As should be clear from the analysis in Sect. 3.3, where the phase transition temperature is approached from the low-temperature side, using the spin wave approximation, we conclude

$$\lim_{T\to T_c-0} \frac{mk_BT}{\hbar^2\rho_s} = \frac{\pi}{2} , \tag{5.3.17}$$

because of $K = 2/\pi$ at the transition point.

On the other hand, when approaching the transition temperature from above, because $\rho_s = 0$ all the time, ρ_s jumps at $T = T_c$ to a finite value, that is the universal value given in (5.3.17). This is the so-called universal jump discovered by Nelson and Kosterlitz. However, the above discussion is not totally correct. Taking into account the contribution of bound vortex–antivortex pairs and small fluctuations in the amplitude, K is altered and at a macroscopic scale it becomes a renormalized K_R. However, when K is replaced by K_R, the above theory can be applied as it stands. Therefore, for the macroscopic ρ_s, the conclusion (5.3.17) will not be modified.

The above discussion did not take into account the electromagnetic field. Including the electromagnetic field in the discussion, we note that the vector potential $\boldsymbol{A}$ behaves such that $\nabla\theta(\boldsymbol{r}) + (2e/\hbar c)\boldsymbol{A}$ decreases faster than $1/|\boldsymbol{r}|$ far away from the vortex. (The other way round, this effect can be seen as screening of the electromagnetic field by the superconductor. In practice, a solution that is mutually consistent is determined.) As a result, logarithmic divergence no longer appears in the vortex energy, and at finite temperature free vortices always are thermally excited. In a narrow sense, no superconductivity can occur.

However, in practice the ratio of the relevant length scale with the screening length mentioned above has to be regarded. Considering a thin superconducting layer with film thickness d, because compared with a three-dimensional superconductor, the screening ability to the electromagnetic field is smaller, and the penetration depth becomes longer, $\lambda_{\text{eff}} = \lambda^2/d$ compared with the value λ in three dimensions. Vortices at distances smaller than λ_{eff} can be thought of as interacting logarithmically. Restricting the discussion to the physics at this scale, the above considerations can be applied. When the linear dimension R of the sample is smaller than λ_{eff}, the energy of the vortex is given by

$$U_1 = \frac{\phi_0^2}{16\pi\lambda_{\text{eff}}} \ln\left(\frac{R}{\xi}\right) , \tag{5.3.18}$$

with ϕ_0 being the flux quantum $hc/2e$ ([24]). ξ is the correlation length of the superconductor. Notice that λ and $\lambda_{\text{eff}} = \lambda^2/d$ are temperature dependent and become divergent for $T \to T_{\text{BCS}}$.

Following the BCS theory including the effect of impurities, with l being the mean free path of the electron, λ is given by (in the contamination limit)

$$\begin{aligned}\lambda^2 &= \lambda(0)^2 \cdot \frac{\xi_0}{l} \cdot \left[\frac{\Delta(T)}{\Delta(0)} \cdot \tanh\left(\frac{1}{2}\beta\Delta(T)\right)\right]^{-1} \\ &\equiv \lambda(0)^2 \cdot \frac{\xi_0}{l} \cdot f^{-1}(T) \ . \end{aligned} \tag{5.3.19}$$

Here, $\lambda(0) = \sqrt{mc^2/4\pi ne^2}$ is the penetration depth at zero temperature for the case when no contaminations are present, and $\Delta(T)$ is the BCS gap at temperature T. At zero temperature, the correlation length ξ_0 is given by $\hbar v_{\text{F}}/\pi\Delta(0) = 0.18\hbar v_{\text{F}}/k_{\text{B}}T_{\text{BCS}}$, and using the resistivity ρ, the resistance R_{2D} of the two-dimensional surface is given by $\rho/d = mv_{\text{F}}/ne^2ld$, and we finally obtain

$$\lambda_{\text{eff}} = 0.0143R_{\text{2D}} \cdot \frac{\hbar c^2}{k_{\text{B}}T_{\text{BCS}}} \cdot f^{-1}(T) \ . \tag{5.3.20}$$

Inserting this into (5.3.18), U_1 becomes

$$U_1 = 4.36 \cdot \frac{\hbar/e^2}{R_{\text{2D}}} \cdot k_{\text{B}}T_{\text{BCS}} \cdot f(T) \ln\left(\frac{R}{\xi}\right) \ . \tag{5.3.21}$$

It is possible to determine the KT phase transition temperature T_{KT} by requiring that at this temperature, U_1 and the entropy become equal [Sect. 3.3]:

$$2.18 \cdot \frac{\hbar/e^2}{R_{\text{2D}}} \cdot f(T_{\text{KT}}/T_{\text{BCS}}) = T_{\text{KT}}/T_{\text{BCS}} \ . \tag{5.3.22}$$

Here, we denoted explicitly that the temperature dependence in $f(T)$ appears in the form T/T_{BCS}.

Owing to (5.3.22), we conclude that $T_{\text{KT}}/T_{\text{BCS}}$ is a function only of $R_{\text{2D}}/(\hbar/e^2)$, and the curve should follow the universal behaviour indicated by the full line in Fig. 5.7. This was first indicated by Beasley, and the experiments are qualitatively in agreement. A more precise experiment (the dotted line in Fig. 5.7) clarified that in the vicinity of $R_{\text{2D}} \cong h/(2e)^2 \cong 6.5\,\text{k}\Omega$, T_{KT} slips off from the curve mentioned above and becomes zero. That is, no superconductivity occurs at zero temperature. This means that two effects that have not been included so far in the theory, namely Anderson localization and the Coulomb charging effect are important, and it is necessary to take into account the quantum fluctuations, that is, the (imaginary) time dependence of the fields. However, because this problem is extremely difficult and the studies are still going on, here we will discuss a simplified model.

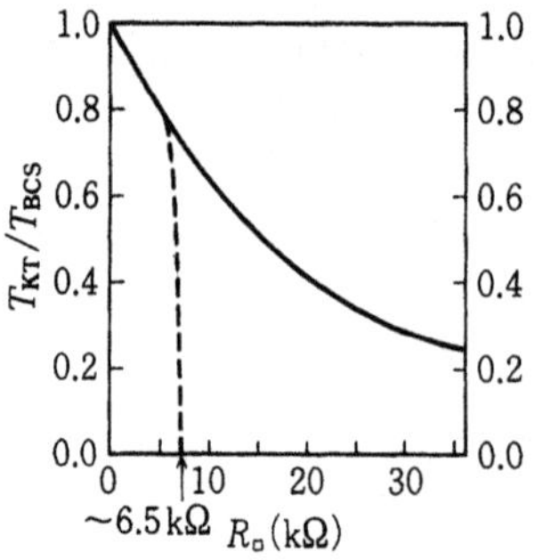

Fig. 5.7. The KT transition temperature of a two-dimensional superconductor [32]. The solid line is the solution (5.3.22), and the dotted line corresponds to the experiments (schematic)

We start with the Hamiltonian of the Josephson junction model:

$$H = \sum_i \frac{n_i^2}{2C} - \sum_{i,j} J_{ij} \cos(\theta_i - \theta_j) \ . \tag{5.3.23}$$

Here, we use the units where $2e = \hbar = c = 1$. C is the capacity of every single microscopic part, and the first term expresses the Coulomb charging effect. The number of Cooper pairs (bosons) n_i is the canonical conjugate variable to the Josephson phase θ_i, fulfilling the commutation relation

$$[n_i, \theta_j] = \mathrm{i}\delta_{ij} \ . \tag{5.3.24}$$

Therefore, (5.3.23) can be interpreted as follows. Owing to $n_i = \mathrm{i}\partial/\partial\theta_i$, the first term on the right-hand side corresponds to the kinetic energy of the phase with mass corresponding to C. The kinetic energy becomes small in a state where θ_i is extended; on the other hand, the potential energy represented by the second term becomes small when θ_i reaches a fixed value. Owing to (5.3.24), the uncertainty is given by $\Delta n \Delta\theta > 2\pi$, leading to a competition between them, and the physics is determined by the ratio of both. When the potential wins, the system is a superconductor, and when C becomes small and the kinetic energy wins, the system becomes an insulator. This is a brief outline of the story, which will be made more precise now.

The partition function of the system is given by

$$Z = \int \prod_i \mathcal{D}\theta_i(\tau) \exp\left[-\int_0^\beta \mathrm{d}\tau \left(\sum_i \frac{C}{2} \dot{\theta}_i(\tau)^2 - \sum_{i,j} J_{ij} \cos(\theta_i(\tau) - \theta_j(\tau)) \right) \right] . \tag{5.3.25}$$

With a being the lattice spacing, the continuum limit is given by

$$Z = \int \mathcal{D}\theta(\boldsymbol{r},\tau) \exp\left[-\int_0^\beta \mathrm{d}\tau \int \mathrm{d}\boldsymbol{r} \left(\frac{C}{2a^2} (\partial_\tau \theta(\boldsymbol{r},\tau))^2 + \frac{J}{2} (\nabla\theta(\boldsymbol{r},\tau))^2 \right) \right] . \tag{5.3.26}$$

Substituting $(C/Ja^2)^{1/2}\boldsymbol{r}$ by $\boldsymbol{r}$, we obtain

$$Z = \int \mathcal{D}\theta(\boldsymbol{r},\tau) \exp\left[-\int_0^\beta \mathrm{d}\tau \int \mathrm{d}\boldsymbol{r} \frac{J}{2}\left[(\partial_\tau \theta(\boldsymbol{r},\tau))^2 + (\nabla\theta(\boldsymbol{r},\tau))^2\right]\right] . \tag{5.3.27}$$

We combine the space and time components in $x = (\boldsymbol{r}, \tau)$ and the corresponding ∇ and ∂_τ in a three-dimensional gradient vector $\nabla_{3\mathrm{D}} = (\nabla, \partial_\tau)$. With this notation, the action (5.3.27) can be written as

$$S = \int \frac{J}{2}(\nabla_{3\mathrm{D}}\theta(x))^2 \,\mathrm{d}^3x = \int \frac{J}{2}(\partial_\mu \theta(x))^2 \,\mathrm{d}^3x . \tag{5.3.28}$$

Because this action is quadratic in the scalar field $\theta(x)$, the reader might think that no phase transition can occur. Indeed, this is the case when $\theta(x)$ is a steady, differentiable field. However, assuming the presence of vortices, this is not correct. $\theta(x)$ is singular at the centre of a vortex, and when describing a circle around, the value jumps about $\pm 2\pi$. That is, $\theta(x)$ becomes a multi-valued function.

This feature can be described by splitting $\theta(x)$ into a single-valued part $\theta_0(x)$ and a multi-valued part $\theta_\mathrm{v}(x)$:

$$\theta(x) = \theta_0(x) + \theta_\mathrm{v}(x) . \tag{5.3.29}$$

$\theta_0(x)$ corresponds to the normal phonon contribution in (5.3.28), and $\theta_\mathrm{v}(x)$ to the vortices. That is, when at time τ a vortex with centre $\boldsymbol{R}(\tau) = (X_1(\tau), X_2(\tau))$ is present, we obtain

$$\begin{aligned} \theta_\mathrm{v}(x) = \theta_\mathrm{v}(\boldsymbol{r},\tau) &= a(\boldsymbol{r} - \boldsymbol{r}(\tau)) \\ &\equiv \tan^{-1}\left(\frac{x_2 - X_2(\tau)}{x_1 - X_1(\tau)}\right) . \end{aligned} \tag{5.3.30}$$

Here, $\alpha(\boldsymbol{r} - \boldsymbol{R}(\tau))$ is the angle measured from the centre $\boldsymbol{R}(\tau)$ between the space coordinate $\boldsymbol{r}$ and the x axis, being multi-valued with multiples of 2π.

However, the derivative becomes

$$\nabla_r \alpha(\boldsymbol{r} - \boldsymbol{R}(\tau)) = \frac{1}{|\boldsymbol{r} - \boldsymbol{R}(\tau)|^2}(-(x_2 - X_2(\tau)), x_1 - X_1(\tau)) \tag{5.3.31}$$

and is not multi-valued. At the centre of the vortex $\boldsymbol{r} = \boldsymbol{R}(\tau)$, a singularity arises, and as should be clear from the first definition, for an arbitrary loop C encircling $\boldsymbol{R}(\tau)$, we obtain

$$\oint_C \mathrm{d}\boldsymbol{r} \cdot \nabla_r \alpha(\boldsymbol{r} - \boldsymbol{R}(\tau)) = 2\pi . \tag{5.3.32}$$

From this equation, we deduce that

$$\nabla_r \times \nabla_r \alpha(\boldsymbol{r} - \boldsymbol{R}(\tau)) = 2\pi\delta(\boldsymbol{r} - \boldsymbol{R}(\tau))\hat{\boldsymbol{z}} \tag{5.3.33}$$

or

$$(\partial_1\partial_2 - \partial_2\partial_1)\alpha(\boldsymbol{r} - \boldsymbol{R}(\tau)) = 2\pi\delta(\boldsymbol{r} - \boldsymbol{R}(\tau)) \ . \tag{5.3.34}$$

Having this relation in mind, we define the "vortex density" current $j_\mu(x)$ as

$$j_\mu(x) = \frac{1}{2\pi}\varepsilon_{\mu\nu\lambda}\partial_\nu\partial_\lambda\theta_{\mathrm{v}}(x) \ . \tag{5.3.35}$$

Here, μ, ν, λ runs over $0, 1, 2$, and $\varepsilon_{\mu\nu\lambda}$ is the totally antisymmetric tensor with $\varepsilon_{012} = 1$. For example, for $\mu = 0$, we obtain

$$\begin{aligned} j_0(x) &= \frac{1}{2\pi}\varepsilon_{0\alpha\beta}\partial_\alpha\partial_\beta\theta_{\mathrm{v}}(x) \\ &= \frac{1}{2\pi}(\partial_1\partial_2 - \partial_2\partial_1)\theta_{\mathrm{v}}(\boldsymbol{r}, \tau) \ , \end{aligned} \tag{5.3.36}$$

and when $\theta_{\mathrm{v}}(\boldsymbol{r}, \tau)$ is inserted into (5.3.30), due to (5.3.34) we obtain

$$j_0(\boldsymbol{r}, \tau) = \delta(\boldsymbol{r} - \boldsymbol{R}(\tau)) \tag{5.3.37}$$

and therefore $j_0(\boldsymbol{r}, \tau)$ indeed represents the vortex density. On the other hand, for the space components $j_\alpha(x)$ $(\alpha = 1, 2)$

$$\begin{aligned} j_\alpha(x) &= \frac{1}{2\pi}\varepsilon_{\alpha 0\beta}(\partial_0\partial_\beta - \partial_\beta\partial_0)\theta_{\mathrm{v}}(x) \\ &= -\frac{1}{2\pi}\varepsilon_{\alpha\beta}(\partial_0\partial_\beta - \partial_\beta\partial_0)\theta_{\mathrm{v}}(x) \ . \end{aligned} \tag{5.3.38}$$

If we insert (5.3.30), we obtain

$$\boldsymbol{j}(\boldsymbol{r}, \tau) = [\partial_\tau \boldsymbol{R}(\tau)]\,\delta(\boldsymbol{r} - \boldsymbol{R}(\tau)) \ . \tag{5.3.39}$$

Interpreting the centre of a vortex as a particle, $\boldsymbol{j}(\boldsymbol{r}, \tau)$ becomes the particle current density. When the same considerations are made for an antivortex, the sign of j_0 and $\boldsymbol{j}$ will change. Therefore, in terms of the "vortex charge", assigning $+1$ to a vortex, and -1 to an antivortex, j_0 becomes the vortex charge density, and $\boldsymbol{j}$ the vortex charge current density.

After these preliminaries, we perform the Stratonovich–Hubbard transformation in (5.3.28). The sum of states becomes

$$Z = \int \mathcal{D}\theta_0 \mathcal{D}\theta_{\mathrm{v}} \mathcal{D}J_\mu \, \mathrm{e}^{-S(\theta_0, \theta_{\mathrm{v}}, J_\mu)} \ , \tag{5.3.40}$$

$$S(\theta_0, \theta_{\mathrm{v}}, J_\mu) = \int \mathrm{d}^3x \left\{ \frac{1}{2J} J_\mu(x)^2 + \mathrm{i}J_\mu(x)(\partial_\mu\theta_0 + \partial_\mu\theta_{\mathrm{v}}) \right\} \ . \tag{5.3.41}$$

Because θ_0 is a normal scalar field it can immediately be integrated out:

$$Z = \int \mathcal{D}\theta_{\mathrm{v}} \mathcal{D}J_\mu \, \mathrm{e}^{-S(\theta_{\mathrm{v}}, J_\mu)} \prod_{\tau, \boldsymbol{r}} \delta(\partial_\mu J_\mu(\boldsymbol{r}, \tau)) \ , \tag{5.3.42}$$

$$S(\theta_{\mathrm{v}}, J_\mu) = \int \mathrm{d}^3x \left\{ \frac{1}{2J} J_\mu(x)^2 + \mathrm{i}J_\mu(x)\partial_\mu\theta_{\mathrm{v}}(x) \right\} \ . \tag{5.3.43}$$

$J_\mu \propto J\partial_\mu\theta$ is the bosonic (Cooper pair) current.

The constraint condition

$$\partial_\mu J_\mu(x) = \nabla_{3D} \cdot \boldsymbol{J}_{3D}(x) = 0 \tag{5.3.44}$$

in (5.3.42) can be solved explicitly by introducing the vector potential $\boldsymbol{a}_{3D} = (a_\mu)$:

$$\boldsymbol{J}_{3D}(x) = \frac{1}{2\pi} \nabla_{3D} \times \boldsymbol{a}_{3D} \tag{5.3.45}$$

or

$$J_\mu = \frac{1}{2\pi} \varepsilon_{\mu\nu\lambda} \partial_\nu a_\lambda \ . \tag{5.3.46}$$

Using this, the second term S_{int} in (5.3.43) can be written as

$$\begin{aligned} S_{\text{int}} &= \mathrm{i} \int \mathrm{d}^3x \, J_\mu(x) \partial_\mu \theta_{\mathrm{v}}(x) = \mathrm{i} \int \mathrm{d}^3x \frac{1}{2\pi} \varepsilon_{\mu\nu\lambda} \partial_\nu a_\lambda(x) \partial_\mu \theta_{\mathrm{v}}(x) \\ &= -\mathrm{i} \int \mathrm{d}^3x \, a_\lambda(x) \frac{1}{2\pi} \varepsilon_{\mu\nu\lambda} \partial_\nu \partial_\mu \theta_{\mathrm{v}}(x) \\ &= \mathrm{i} \int \mathrm{d}^3x \, a_\lambda(x) \frac{1}{2\pi} \varepsilon_{\lambda\mu\nu} \partial_\mu \partial_\nu \theta_{\mathrm{v}}(x) = \mathrm{i} \int \mathrm{d}^3x \, a_\lambda(x) j_\lambda(x) \ . \end{aligned} \tag{5.3.47}$$

Finally, we arrived at a model where the centre of the vortex is considered as a particle, and where this particle interacts with the gauge field a_μ.

Putting things together, from (5.3.42) and (5.3.43) we obtain

$$Z = \int \mathcal{D}a_\mu \mathcal{D}j_\mu \, \mathrm{e}^{-S(j_\mu, a_\mu)} \ , \tag{5.3.48}$$

$$S(j_\mu, a_\mu) = A \sum_i \int \mathrm{d}s_i + \mathrm{i} \int a_\mu(x) j_\mu(x) \, \mathrm{d}^3x + \int \frac{(\nabla_{3D} \times \boldsymbol{a}_{3D}(x))^2}{8\pi^2 J} \mathrm{d}^3x \ . \tag{5.3.49}$$

The first term at the right-hand side of (5.3.49) corresponds to the energy cost at the core of the vortex, represented by the action A per unit length multiplied by the length of the world line $\int \mathrm{d}s_i$ of the ith vortex. This term is determined by the physics at small scale, and here it is added phenomenologically. The above transformation is called the duality transformation. It can be understood in a direct manner as follows.

We consider one flux line of the gauge field (magnetic flux). Let one charged particle go around this flux line. Then, due to the Aharonov–Bohm effect, the quantum mechanical phase changes by $\exp[\mathrm{i} \oint \boldsymbol{a} d\boldsymbol{r}]$. One is moving around the other, and the important thing to notice is that the physical result is not changed when the role of both is interchanged. In the original model, the Cooper pair (boson) was the charged particle and the vortex was the flux. After the transformation, the opposite is true, namely the vortex is the charged particle and the Cooper pair (boson) becomes the flux.

Regarding the world line of the vortex (the path described in the $(2+1)$-dimensional space) as the world line of a charged particle, the above formulation just corresponds to the path integral in first quantization. Now, we will write down the path integral for the corresponding quantum field theory in second quantization. As stated in the first term of the right-hand side of (5.3.49), the action of the world line of the vortex in the $(2+1)$-dimensional space is proportional to its length $t \equiv \int \mathrm{d}s$. Regarding a discrete space with lattice spacing 1, then the world line of the vortex has t segments. At zero temperature all world lines of vortices are loops. Choosing one specific loop, the partition function can be written as

$$Z_{\text{loop}} = \int_0^\infty \mathrm{d}t\, P(t)\, \mathrm{e}^{-At} \ . \tag{5.3.50}$$

Here, A is the action along one unit element, and $P(t)$ is the number of possible configurations of a closed loop consisting of t steps.

In order to determine $P(t)$, we introduce the probability distribution $p(x,x',t)$ for the probability of reaching the point x after t steps, having started at x'. $p(x,x',t)$ obeys the diffusion equation

$$\frac{\partial p(x,x',t)}{\partial t} = \frac{1}{2D}\nabla_x^2 p(x,x',t) \ , \tag{5.3.51}$$

where D is the dimension, and in our case is given by 3. $p(x,x',t)$ fulfils the initial condition $p(x,x',t) = \delta(x-x')$. Using the path integral, the solution of the diffusion equation can be expressed as

$$p(x,x',t) = N \int_{x(0)=x'}^{x(t)=x} \mathcal{D}x(t) \exp\left[-\int \mathrm{d}t \frac{D}{2}\dot{x}(t)^2\right] \ . \tag{5.3.52}$$

The normalization factor is determined such that

$$\int \mathrm{d}x\, p(x,x',t) = 1$$

holds.

Next, the total number of paths $\Gamma(x,x',t)$ going in t steps from x' to x is given by

$$\begin{aligned} \Gamma(x,x't) &= p(x,x',t)(2D)^t \\ &= N \int_{x(0)=x'}^{x(t)=x} \mathcal{D}x(t) \exp\left[-\int \mathrm{d}t \left(\frac{D}{2}\dot{x}(t)^2 - \ln(2D)\right)\right] \ . \end{aligned} \tag{5.3.53}$$

$\Gamma(x,x',t)$ can be expressed in the framework of the eigenvalue problem

$$\left(-\frac{1}{2D}\nabla_x^2 - \ln(2D)\right)\Phi_m(x) = E_m \Phi_m(x) \tag{5.3.54}$$

by using the eigenstates $\Phi_m(x)$ and the eigenvalues E_m:

$$\Gamma(x,x',t) = \sum_m \Phi_m(x)\Phi_m(x')\,\mathrm{e}^{-E_m t} \ . \tag{5.3.55}$$

Here, $\Phi_m(x)$ fulfils the normalization condition and completeness relation.

$$\int \mathrm{d}x\, \Phi_m(x)^2 = 1 \ , \qquad \sum_m \Phi_m(x)\Phi_m(x') = \delta(x-x') \ . \tag{5.3.56}$$

$\Gamma(x,x',t)$ and $P(t)$ are related by

$$\int \mathrm{d}x\, \Gamma(x,x,t) = tP(t) \ . \tag{5.3.57}$$

The explicit factor t appears because for one given loop the integral at the left-hand side, i.e. in the sum on the starting point of the lattice, can be performed at any of the t possible lattice points of the loop, and therefore one loop will be counted t times.

From the above equations, Z_{loop} can be expressed by

$$\begin{aligned} -\frac{\partial Z_{\mathrm{loop}}}{\partial A} &= \int_0^\infty \mathrm{d}t\, tP(t)\,\mathrm{e}^{-At} \\ &= \int_0^\infty \mathrm{d}t \int \mathrm{d}x\, \Gamma(x,x,t)\,\mathrm{e}^{-At} \\ &= \int_0^\infty \mathrm{d}t \sum_n \mathrm{e}^{-(E_n+A)t} = \sum_n \frac{1}{E_n+A} \\ &= \mathrm{Tr}\left(-\frac{1}{2D}\nabla^2 + A - \ln(2D)\right)^{-1} \ . \end{aligned} \tag{5.3.58}$$

Integrating (5.3.58) with respect to A, we obtain

$$\begin{aligned} Z_{\mathrm{loop}} &= -\,\mathrm{Tr}\ln\left(-\frac{1}{2D}\nabla^2 + A - \ln(2D)\right) + \text{const.} \\ &= -\ln\det\left(-\frac{1}{2D}\nabla^2 + A - \ln(2D)\right) + \text{const.} \end{aligned} \tag{5.3.59}$$

Assuming that the interaction between the loops can be ignored, the sum of states Z_{v} of the loop system (vortex system) can be written as

$$\begin{aligned} Z_{\mathrm{v}} &= \sum_n \frac{(Z_{\mathrm{loop}})^2}{n!} = \exp[Z_{\mathrm{loop}}] \\ &= \text{const.} \times \det{}^{-1}\left(-\frac{1}{2D}\nabla^2 + m^2\right) \ . \end{aligned} \tag{5.3.60}$$

Here, we set $m^2 = A - \ln(2D)$. The factor $1/n!$ removes the multiplicity caused by the fact that the configuration of the system does not change when the loops are interchanged.

Expressing the $\det^{-1}$ in the above equation as a functional of the complex fields φ and φ^*, we obtain

$$Z_{\mathrm{v}} = \int \mathcal{D}\varphi(x) \int \mathcal{D}\varphi^*(x) \exp\left[-\int \mathrm{d}x\, \varphi^* \left(-\frac{1}{2D}\nabla^2 + m^2\right)\varphi\right] . \quad (5.3.61)$$

What might the correlation function $\langle\varphi^*(x)\varphi(x')\rangle$ of the field theory of a complex field signify for the original vortex? In fact, we notice that this correlation function equals the average number $\int_0^\infty \Gamma(x,x',t)a^{-At}$ of vortices starting at x' and going to x. Recall that the action of a vortex of length t is just given by At, and therefore the probability becomes only about e^{-At} smaller. Integrating over all possible path lengths, the above equation is obtained. We will prove this relation. From (5.3.55), we obtain

$$\begin{aligned}
\int_0^\infty \mathrm{d}t\, \Gamma(x,x',t)\,\mathrm{e}^{-At} &= \sum_n \frac{\Phi_n(x)\Phi_n(x')}{E_n + A} \\
&= \sum_n \langle x|n\rangle \left\langle n \left| \frac{1}{-(1/2D)\nabla^2 + m^2} \right| n\right\rangle \langle n|x'\rangle \\
&= \left\langle x \left| \frac{1}{-(1/2D)\nabla^2 + m^2} \right| x'\right\rangle \\
&= \frac{1}{Z_{\mathrm{v}}} \int D\varphi D\varphi^* \cdot \varphi^*(x)\varphi(x') \\
&\quad \times \exp\left[-\int \mathrm{d}x\, \varphi^* \left(-\frac{1}{2D}\nabla^2 + m^2\right)\varphi\right] \\
&= \langle\varphi^*(x)\varphi(x')\rangle . \qquad (5.3.62)
\end{aligned}$$

Especially for $x = x'$, $\langle|\varphi(x)|^2\rangle$ is the local vortex density. On the other hand, for $|x - x'| \to \infty$ we obtain $\langle\varphi^*(x)\varphi(x')\rangle \to \langle\varphi^*\rangle\langle\varphi\rangle$, which is the probability for the appearance of infinitely large vortices.

It is possible to express the fact that more than one segments of vortices cannot occupy the same place by adding a term $\lambda|\varphi|^4$ to the action (5.3.61). Furthermore, in the first quantized world line picture, the vortex current density j_μ defined in (5.3.35) becomes

$$j_\mu(x) = \dot{x}_\mu(t)\delta(x - x(t)) . \quad (5.3.63)$$

Inserting this into S_{int} of (5.3.47), the result is

$$S_{\text{int}} = \int \mathrm{d}t\, a_\mu(x(t))\dot{x}_\mu(t) . \quad (5.3.64)$$

Performing the path integral (5.3.53) for this action, ∇ in (5.3.61) is replaced by $\nabla + \mathrm{i}a$. The action of the whole system becomes

$$S = \int \mathrm{d}x \left[\varphi^* \left(-\frac{1}{2D}(\nabla + \mathrm{i}a)^2 + m^2\right)\varphi + \lambda|\varphi|^4 + \frac{1}{16\pi^2 J} f_{\mu\nu}^2\right] . \quad (5.3.65)$$

This action coincides with the action of a scalar field interacting with a gauge field (scalar QED). This action can also be regarded as describing a (hypothetical) superconductor in three dimensions. By analogy, for $m^2 < 0$, $\langle\varphi\rangle$ and $\langle\varphi^*\rangle$ become finite, and owing to the Meissner effect, the gauge field becomes massive, i.e. exponentially damped in space, and therefore does not reach inside the sample. In terms of the original two-dimensional superconductor, vortex condensation occurs; that is, because vortices of infinite size exist, the condensation of the bosons, i.e. the Cooper pairs, disappear, and superconductivity is lost.

Next, we discuss the conductivity and the surface resistance of the system. We consider a two-dimensional sample $L \times L$ as shown in Fig. 5.8, where a current density $\boldsymbol{J}_0$ is flowing in the x direction. We assume also that a voltage is applied in the x direction. In the original bosonic problem $J_x = J_0$ and $E_x = E$ are related by $J_0 = \sigma E$. σ is the conductivity that can be determined by the current–current correlation of the bosons.

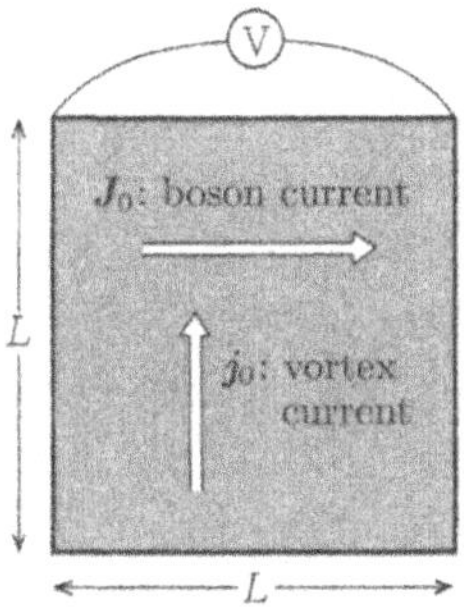

Fig. 5.8. Duality in the two-dimensional superconducting system

On the other hand we could think of the problem as a vortex system. The "electric field" $e_\alpha = \partial_\alpha a_0 - \partial_0 a_\alpha$ that is constructed with a_μ following (5.3.45) becomes $e_x = 0$, $e_y = 2\pi J_0$. The vortices that feel this "electric field" also flow only in the y direction with a current density $j_y = j = \tilde{\sigma} e_y = 2\pi\tilde{\sigma} J_0$. $\tilde{\sigma}$ is the vortex conductivity that can be calculated by the vortex current–current correlation function in the system with action (5.3.65). However, returning to the original problem, when a vortex current in the y direction is present, a voltage arises in the x direction, and following the Josephson equation (5.2.10), it is given by $V = 2\pi \times jL$. Because the resistance of the system is the ratio of V and the total current $I = LJ_0$, we obtain

$$R = \frac{R}{I} = \frac{LE}{LJ_0} = \frac{1}{\sigma} = \frac{2\pi jL}{LJ_0} = \frac{2\pi j}{J_0} = (2\pi)^2 \tilde{\sigma} \ . \tag{5.3.66}$$

We conclude that the conductivity of the model (5.3.28) and the model (5.3.65) fulfil the relation

$$\sigma\tilde{\sigma} = \frac{1}{(2\pi)^2} \ . \tag{5.3.67}$$

Reintroducing $2e$ and $\hbar$, and taking into account that in two dimensions the conductivity has the dimension $(2e)^2/\hbar$, (5.3.67) becomes

$$\sigma\tilde{\sigma} = \left(\frac{(2e)^2}{h}\right)^2 \quad . \tag{5.3.68}$$

Because in the superconducting phase the boson current is present, but not the vortex current, it corresponds to $\sigma = \infty$ and $\tilde{\sigma} = 0$. Conversely, in the insulator phase, the bosonic current disappears and only the vortex current is present, corresponding to $\sigma = 0$ and $\tilde{\sigma} = \infty$. Assuming that the phase transition between both is continuous, then exactly at the phase transition point, both values are expected to reach a finite value. Of course (5.3.28) and (5.3.65) are different models, and in general σ and $\tilde{\sigma}$ will gain different values at the phase transition point. However, when (in the low-energy region) both models become equal (this is called self-duality), the universal value

$$\sigma = \tilde{\sigma} = \frac{(2e)^2}{h} = (6.5\,\mathrm{k}\Omega)^{-1} \tag{5.3.69}$$

is obtained. It can be expected that even when self-duality is not exact, at the phase transition point (5.3.69) will not change drastically.

With the above considerations, based on the duality transformation, only the relation between two models has been clarified, but the model has not been solved. However, such a transformation often gives helpful insights into the physical phenomena. In the next chapter we will meet an example of an application.

6. Quantum Hall Liquid and the Chern–Simons Gauge Field

A two-dimensional electron system showing the quantum Hall effect when exposed to a strong magnetic field, the quantum Hall liquid, is a new type of quantum liquid. It has the astonishing property that the elementary excitations obey fractional statistics and that the charge is a fraction of e. In this chapter the quantum field theory of this quantum liquid is discussed, based on the statistical transmutation using the Chern–Simons gauge field.

6.1 Two-Dimensional Electron System

In solid state physics, at present, the two dimensional electronic system is one of the most extensively studied systems. Especially, a tremendous amount of work has been done on the 2D electron system under a strong magnetic field in relation to the quantum hall effect. It is impossible to give an overview over the whole scenario, therefore in the present section we will discuss the foundations, and in Sect. 6.2, one special point of view will be presented, being related to the Chern–Simons gauge field.

We start the discussion by considering a one electron problem. With m being the mass of the electron, and $-e$ (< 0) its charge, the Hamiltonian is given by

$$\mathcal{H}_1 = \frac{1}{2m}\left(-\mathrm{i}\hbar\nabla + \frac{e}{c}\boldsymbol{A}(\boldsymbol{r})\right)^2 + \frac{eB(\boldsymbol{r})}{2mc}\sigma_z \ . \tag{6.1.1}$$

The first term at the right-hand side describes the kinetic energy under the presence of a vector potential $\boldsymbol{A} = (A_x, A_y)$. In the second term, σ_z is (twice of) the z component of the spin of the electron, and the whole term represents the Zeemann energy due to the interaction of the spin with the z component $B = \partial_x A_y - \partial_y A_x$ of the magnetic field. Using the units $2m = \hbar = c = 1$, the system can simply be written as

$$\mathcal{H}_1 = (-\mathrm{i}\partial_x + eA_x)^2 + (-\mathrm{i}\partial_y + eA_y)^2 + eB\sigma_z \ . \tag{6.1.2}$$

We now introduce the combinations

$$\pi_\alpha = p_\alpha + eA_\alpha = -\mathrm{i}\partial_\alpha + eA_\alpha \tag{6.1.3}$$

and

$$\begin{aligned} \pi_- &= \pi_y - \mathrm{i}\sigma_z \pi_x \ , \\ \pi_+ &= \pi_-^\dagger = \pi_y + \mathrm{i}\sigma_z \pi_x \ . \end{aligned} \tag{6.1.4}$$

Owing to the relation

$$\begin{aligned} [\pi_x, \pi_y] &= e[p_x, A_y] + e[A_x, p_y] \\ &= -\mathrm{i}e(\partial_x A_y - \partial_y A_x) = -\mathrm{i}eB \end{aligned} \tag{6.1.5}$$

using (6.1.4) we can prove

$$\begin{aligned} \pi_+\pi_- &= (\pi_y + \mathrm{i}\sigma_z\pi_x)(\pi_y - \mathrm{i}\sigma_z\pi_x) \\ &= \pi_x^2 + \pi_y^2 + \mathrm{i}[\pi_x, \pi_y]\sigma_z = \pi_x^2 + \pi_y^2 + eB\sigma_z \ . \end{aligned} \tag{6.1.6}$$

Comparing (6.1.2) with (6.1.6), we obtain

$$\mathcal{H}_1 = \pi_+\pi_- = \pi_-^\dagger \pi_- \ . \tag{6.1.7}$$

Calling $|n\rangle$ the eigenstates of $\mathcal{H}_1$ and E_n the eigenenergy, from the above equation we obtain

$$\pi_-^\dagger \pi_- |n\rangle = E_n |n\rangle \ . \tag{6.1.8}$$

Multiplying both sides by $\langle n|$, we obtain

$$\langle n|\pi_-^\dagger \pi_-|n\rangle = \langle \pi_- n|\pi_- n\rangle = E_n \ . \tag{6.1.9}$$

We conclude that the energy E_n is non-negative and that the ground state $|0\rangle$ ($E_0 = 0$) fulfils $\pi_-|0\rangle = 0$. Explicitly for the wave function $f(\boldsymbol{r}) = {}^t[f_+(\boldsymbol{r}), f_-(\boldsymbol{r})]$ we obtain

$$[-\mathrm{i}\partial_y + eA_y - \mathrm{i}\sigma(-\mathrm{i}\partial_x + eA_x)] f_\sigma(\boldsymbol{r}) = 0 \ . \tag{6.1.10}$$

Here, we impose the Coulomb gauge condition for A_x and A_y:

$$\partial_x A_x + \partial_y A_y = 0 \ . \tag{6.1.11}$$

Introducing a scalar function ϕ

$$A_x = -\partial_y \phi \ , \qquad A_y = \partial_x \phi \tag{6.1.12}$$

in this gauge, for given $B(x, y)$, ϕ is determined by

$$\partial_x A_y - \partial_y A_x = \left(\partial_x^2 + \partial_y^2\right)\phi = B(x, y) \ . \tag{6.1.13}$$

With $\boldsymbol{r} = (x, y)$, the solution is given by

$$\phi(\boldsymbol{r}) = \frac{1}{2\pi} \int \ln \left| \frac{\boldsymbol{r} - \boldsymbol{r}'}{r_0} \right| \cdot B(\boldsymbol{r}')\, \mathrm{d}^2 \boldsymbol{r}' \ . \tag{6.1.14}$$

Using this $\phi(\boldsymbol{r})$, we set

$$f_\sigma = \mathrm{e}^{e\sigma\phi} g_\sigma \tag{6.1.15}$$

and, due to

$$\begin{aligned}[\partial_y + e\sigma A_x] f_\sigma &= \mathrm{e}^{e\sigma\phi}[\partial_y g_\sigma + e\sigma(\partial_y \phi + A_x) g_\sigma] \\ &= \mathrm{e}^{e\sigma\phi} \partial_y g_\sigma \ , \end{aligned} \tag{6.1.16}$$

$$\begin{aligned}[\partial_x - e\sigma A_y] f_\sigma &= \mathrm{e}^{e\sigma\phi}[\partial_x g_\sigma - e\sigma(-\partial_x \phi + A_y) g_\sigma] \\ &= \mathrm{e}^{e\sigma\phi} \partial_x g_\sigma \ , \end{aligned} \tag{6.1.17}$$

we obtain for (6.1.10)

$$\left(\frac{1}{\mathrm{i}}\partial_y - \sigma\partial_x\right) g_\sigma(x,y) = 0 \ . \tag{6.1.18}$$

We now introduce $z = x + \mathrm{i}y$, $\bar{z} = x - \mathrm{i}y$ and $\partial_x = \partial_z + \partial_{\bar{z}}$, $\partial_x = \partial_z - \partial_{\bar{z}}$ to express (6.1.18) as

$$[(1-\sigma)\partial_z - (1+\sigma)\partial_{\bar{z}}]\, g_\sigma(z,\bar{z}) = 0 \tag{6.1.19}$$

and

$$g_+(\boldsymbol{r}) = g_+(z) \ , \tag{6.1.20}$$

$$g_-(\boldsymbol{r}) = g_-(\bar{z}) \ . \tag{6.1.21}$$

On the other hand, in order that the integral over $|f_\sigma|^2$ is finite, with Φ being the total flux, it is necessary that

$$\lim_{|\boldsymbol{r}|\to\infty} |\boldsymbol{r}|^2 |f_\sigma(r)|^2 = \lim_{|\boldsymbol{r}|\to\infty} |\boldsymbol{r}|^2 \left(\frac{r_0}{|\boldsymbol{r}|}\right)^{-e\sigma\Phi/\pi} |g_\sigma(\boldsymbol{r})|^2 = 0 \tag{6.1.22}$$

holds. Here, we used that for $|\boldsymbol{r}| \to \infty$, (6.1.14) becomes

$$\begin{aligned}\phi(\boldsymbol{r}) &\underset{|\boldsymbol{r}|\to\infty}{\sim} \frac{1}{2\pi}\ln\left|\frac{\boldsymbol{r}}{r_0}\right| \int \mathrm{d}^2\boldsymbol{r}'\, B(\boldsymbol{r}') \\ &= \frac{\Phi}{2\pi}\ln\left|\frac{\boldsymbol{r}}{r_0}\right| \ . \end{aligned} \tag{6.1.23}$$

First, from (6.1.22) we deduce that $e\sigma\Phi < 0$ is necessary, therefore in the ground state only spin components opposite to the flux Φ contribute to the wave function of the state. Setting $\Phi < 0$ and $g_+(z) = z^m$ $(m \geq 0)$, (6.1.22) becomes

$$\lim_{|\boldsymbol{r}|\to\infty} |\boldsymbol{r}|^{2-e|\Phi|/\pi+2m} = 0 \ , \tag{6.1.24}$$

and with N being an integer, for

$$\frac{e|\Phi|}{2\pi} = N + \varepsilon \qquad (0 < \varepsilon < 1) \tag{6.1.25}$$

the N values $m = 0, 1, 2, \ldots, N-1$ are allowed for m. It is a very striking phenomenon that the total flux Φ of an arbitrary magnetic field $B(x, y)$ determines the characteristics to the ground state of the electron. Mathematically, this is one example for the Atiyah–Singer index theorem.

The above considerations are valid for an arbitrary magnetic field. Now, we will consider a homogeneous magnetic field. The magnetic field $-B(< 0)$ is applied in the z direction, and only the corresponding spin in the $+z$ direction is considered (that is, we set $\sigma_z = \sigma = 1$). The vector potential $\boldsymbol{A}$ is expressed in the symmetric gauge as

$$\boldsymbol{A} = (A_x, A_y) = \left(\tfrac{1}{2}By, -\tfrac{1}{2}Bx\right) \ . \tag{6.1.26}$$

Then π_- in (6.1.4) becomes

$$\pi_- = -\partial_x - \mathrm{i}\partial_y + \frac{eB}{2}(-\mathrm{i}y - x) = -2\partial_{\bar{z}} - \frac{eB}{2}z \ . \tag{6.1.27}$$

Here, we again used $z = x + iy$ and $\bar{z} = x - iy$. For the vector potential (6.1.26), the scalar potential ϕ (6.1.12) becomes

$$\phi = -\tfrac{1}{4}B(x^2 + y^2) = -\tfrac{1}{4}B\bar{z}z \ . \tag{6.1.28}$$

As is suggested by (6.1.15), for the wave function $\Psi_1(x, y) = \Psi_1(z, \bar{z})$, we make the ansatz (the index 1 signifies that a one electron state is considered)

$$\Psi_1(z, \bar{z}) = g(z, \bar{z})\, \mathrm{e}^{-(1/4)eB\bar{z}z} \ . \tag{6.1.29}$$

Then, we obtain

$$\pi_-\Psi_1 = -2\,\mathrm{e}^{-(1/4)eB\bar{z}z}\partial_{\bar{z}}g \ . \tag{6.1.30}$$

Because the lowest energy state fulfils $\pi_-\Psi_1 = 0$, we obtain $g(z, \bar{z}) = g(z)$, and therefore

$$\Psi_1(z, \bar{z}) = g(z)\, \mathrm{e}^{-(1/4)eB\bar{z}z} \tag{6.1.31}$$

$g(z)$ is analytic in z, and writing (up to normalization)

$$g_m(z) = z^m \qquad (m = 0, 1, 2, \ldots) \tag{6.1.32}$$

the number of allowed m, i.e. the degree of degeneracy G, can be calculated as follows.

For a round system with radius R, $\Psi_1^{(m)}(z, \bar{z}) = g_m(z) = 0$ must be zero for $|z| > R$. We consider $\rho_m(z, \bar{z}) = |\Psi_1^{(m)}(z)|^2$:

$$\rho_m(z, \bar{z}) = |z|^{2m}\, \mathrm{e}^{-(1/2)eB|z|^2} = f(|z|^2) \ . \tag{6.1.33}$$

As a function of $|z|^2$, $f(|z|^2)$ is maximal at $|z|^2 = 2m/eB$, and because this must be smaller than R^2,

$$0 \leqq \frac{2m}{eB} \leqq R^2 \qquad \text{or} \qquad 0 \leqq m \leqq \frac{eB}{2} R^2 \tag{6.1.34}$$

must hold. Therefore, G is given by

$$G = \frac{eB}{2} R^2 = \frac{eB}{2\pi} S \ . \tag{6.1.35}$$

Here, S is the surface πR^2 of the system. The G-times degenerate lowest energy state is the lowest $(n = 0)$ Landau level.

In order to clarify the meaning of m, we consider the z component of the angular momentum operator:

$$L_z = xp_y - yp_x = -\mathrm{i}(x\partial_y - y\partial_x) = z\partial_z - \bar{z}\partial_{\bar{z}} \ . \tag{6.1.36}$$

Calculating the commutator of π_- with L_z, we obtain

$$\begin{aligned} [\pi_-, L_z] &= \left[-2\partial_z - \frac{eB}{2} z, z\partial_z - \bar{z}\partial_{\bar{z}}\right] \\ &= +2\partial_{\bar{z}} + \frac{eB}{2} z = -\pi_- \end{aligned} \tag{6.1.37}$$

and in the same manner

$$[\pi_-^\dagger, L_z] = \pi_-^\dagger \ . \tag{6.1.38}$$

Therefore, we obtain

$$[H, L_z] = [\pi_-^\dagger \pi_-, L_z] = [\pi_-^\dagger, L_z]\pi_- + \pi_-^\dagger[\pi_-, L_z] = \pi_-^\dagger \pi_- + \pi_-^\dagger(-\pi_-) = 0 \tag{6.1.39}$$

and we conclude that L_z and H can be diagonalized simultaneously. Indeed, it can easily be confirmed that when acting with L_z on $\Psi_1^{(m)}(z, \bar{z})$, the result is

$$L_z \Psi_1^{(m)} = m \Psi_1^{(m)} \ . \tag{6.1.40}$$

Therefore, m is the z component of the angular momentum.

Furthermore, we consider the magnetic flux Φ_m that flows through the surface surrounding the orbit of $\Psi_1^{(m)}$. As was discussed in (6.1.33), the amplitude is maximal around the value $|z|^2 = 2m/eB$. Therefore, we obtain

$$\Phi_m = \pi |z|^2 B = \frac{2\pi m}{e} = m\phi_0 \ . \tag{6.1.41}$$

$\phi_0 = 2\pi/e$ is one magnetic flux quantum. We can assume that the state with angular momentum m describes a rotation around m magnetic flux quanta.

States with higher energy can be constructed by acting with $\pi_-^\dagger$. Using

$$\begin{aligned}[\pi_-, \pi_-^\dagger] &= [\pi_y - \mathrm{i}\pi_x, \pi_y + \mathrm{i}\pi_x] \\ &= -2\mathrm{i}[\pi_x, \pi_y] = 2eB = \text{const.} \ ,\end{aligned} \tag{6.1.42}$$

for a state Ψ fulfilling $H\Psi = E\Psi$ we obtain

$$\begin{aligned}H\pi_-^\dagger \Psi &= \pi_-^\dagger \pi_- \pi_-^\dagger \Psi = \pi_-^\dagger([\pi_-, \pi_-^\dagger] + H)\Psi \\ &= \pi_-^\dagger(2eB + E)\Psi = (2eB + E)\pi_-^\dagger \Psi\end{aligned} \tag{6.1.43}$$

and conclude that $\pi_-^\dagger \Psi$ again is an eigenstate of H, but with an energy raised by $2eB$. In such a manner, the Landau levels line up with an energy gap $2eB$ between each other and, as has been shown above, each level is G-times degenerate.

Up to now, we have discussed only the one electron problem. We now turn to the two electron problem. In what follows we assume that the magnetic field is strong and consider therefore only the lowest lying Landau level and only the spin up $\uparrow$ component. With $V(r_1 - r_2)$ being the interaction between the two electrons, the Hamiltonian $\mathcal{H}_2$ is given by

$$\mathcal{H}_2 = (-\mathrm{i}\nabla_1 + e\boldsymbol{A}(\boldsymbol{r}_1))^2 + (-\mathrm{i}\nabla_2 + e\boldsymbol{A}(\boldsymbol{r}_2))^2 + V(\boldsymbol{r}_1 - \boldsymbol{r}_2) \ . \tag{6.1.44}$$

Introducing the centre of mass coordinate $\boldsymbol{R} = (\boldsymbol{r}_1 + \boldsymbol{r}_2)/2$ and the relative coordinate $\boldsymbol{r} = \boldsymbol{r}_1 - \boldsymbol{r}_2$, and taking into account that $\boldsymbol{A}(\boldsymbol{r})$ as given in (6.1.26) is linear in x and y, and using the relations $\nabla_1 = \nabla_{\boldsymbol{R}}/2 + \nabla_{\boldsymbol{r}}$ and $\nabla_2 = \nabla_{\boldsymbol{R}}/2 - \nabla_{\boldsymbol{r}}$, $\mathcal{H}_2$ can be written as

$$\begin{aligned}\mathcal{H}_2 &= \left[\left(-\frac{\mathrm{i}}{2}\nabla_{\boldsymbol{R}} + e\boldsymbol{A}(\boldsymbol{R})\right) + \left(-\mathrm{i}\nabla_{\boldsymbol{r}} + \frac{e}{2}\boldsymbol{A}(\boldsymbol{r})\right)\right]^2 \\ &\quad + \left[\left(-\frac{\mathrm{i}}{2}\nabla_{\boldsymbol{R}} + e\boldsymbol{A}(\boldsymbol{R})\right) - \left(-\mathrm{i}\nabla_{\boldsymbol{r}} + \frac{e}{2}\boldsymbol{A}(\boldsymbol{R})\right)\right]^2 + V(\boldsymbol{r}) \\ &= \frac{1}{2}(-\mathrm{i}\nabla_{\boldsymbol{R}} + 2e\boldsymbol{A}(\boldsymbol{R}))^2 + 2\left(-\mathrm{i}\nabla_{\boldsymbol{r}} + \frac{e}{2}\boldsymbol{A}(\boldsymbol{r})\right)^2 + V(\boldsymbol{r}) \ .\end{aligned} \tag{6.1.45}$$

For a moment, we set $V(\boldsymbol{r}) = 0$. When the angular momentum of the centre of mass coordinate is zero, but the angular momentum of the relative coordinate is m, the wave function becomes

$$\Psi_2^{(m)}(z_1, \bar{z}_1, z_2, \bar{z}_2) = (z_1 - z_2)^m \exp\left[-\tfrac{1}{4}eB\left(|z_1|^2 + |z_2|^2\right)\right] \tag{6.1.46}$$

and $|z_1 - z_2|^2$ is of the order of $4m/eB$. That is, for larger m the rotation around each other takes place at a larger distance. Furthermore, notice that due to the anti-symmetry of the fermionic wave function, only odd numbers are allowed for m.

Next, we consider the N electron system. In this case, the relation between N and the degree of degeneracy G becomes important. The ratio

$$\nu = N/G \tag{6.1.47}$$

is called the filling factor of the Landau level. For $\nu = 1$, the wave function of the electronic system corresponds to a totally filled lowest Landau level and can be expressed using the Slater determinant as

$$\begin{aligned}
\Psi_n &= \begin{vmatrix} \Psi_1^{(0)}(r_1) & \Psi_1^{(1)}(r_1) & \ldots & \Psi_1^{(N-1)}(r_1) \\ \Psi_1^{(0)}(r_2) & \Psi_1^{(1)}(r_2) & \ldots & \Psi_1^{(N-1)}(r_2) \\ \cdots & \cdots & \cdots & \cdots \\ \Psi_1^{(0)}(r_N) & \Psi_1^{(1)}(r_N) & \ldots & \Psi_1^{(N-1)}(r_N) \end{vmatrix} \\
&= \begin{vmatrix} 1 & z_1 & z_1^2 & \ldots & z_1^{N-1} \\ 1 & z_2 & z_2^2 & \ldots & z_2^{N-1} \\ \cdots & \cdots & \cdots & \cdots & \cdots \\ 1 & z_N & z_N^2 & \ldots & z_N^{N-1} \end{vmatrix} \exp\left[-\frac{1}{4}eB\left(|z_1|^2 + \cdots + |z_n|^2\right)\right] \\
&= \prod_{i>j}(z_i - z_j)\exp\left[-\frac{1}{4}eB\left(|z_1|^2 + \cdots + |z_N|^2\right)\right] .
\end{aligned} \tag{6.1.48}$$

Recall that $\Psi_1^{(0)}, \Psi_1^{(1)}, \ldots, \Psi_1^{(N-1=G-1)}$ are the total number of states lying in the lowest Landau level.

Next, how can the state be expressed when ν is smaller than one? Of course, when there is no interaction between the electrons, N single particle states are occupied from G states, which gives rise to a huge degeneracy. When the repulsive Coulomb force acts between the electrons, this degeneracy is supposed to disappear. Especially for $\nu = 1/(2k+1)$ (k: integer), following Laughlin, to a very good approximation the wave function of the ground state is given by

$$\Psi_{\nu=1/(2k+1)} = \prod_{i>j}(z_i - z_j)^{2k+1}\exp\left[-\frac{1}{4}eB\left(|z_1|^2 + \cdots + |z_N|^2\right)\right] . \tag{6.1.49}$$

This state also describes an incompressible liquid. It had been proposed that a gap exists between this state and the excited states; this has now been established. In the next sections, this will be discussed further. Notice that (6.1.49) is the natural generalization of the formula (6.1.46) and therefore also includes the two electron correlation. Furthermore, setting $\nu = 1$, we regain equation (6.1.48).

6.2 Effective Theory of a Quantum Hall Liquid

In this section, using the statistical transmutation based on the Chern–Simons gauge field, the effective theory of a quantum Hall system will be discussed.

We start with the problem of an N-fermion system. We assume that owing to the Zeemann effect all spins are aligned up, and do not consider the spin degree of freedom further. In first quantization, the Hamiltonian is given by

$$\mathcal{H} = \sum_{i=1}^{N} \frac{1}{2m} \left[\boldsymbol{p}_i + e\boldsymbol{A}(\boldsymbol{r}_i)\right]^2 + \sum_i eA_0(\boldsymbol{r}_i) + \sum_{i<j} V(\boldsymbol{r}_i - \boldsymbol{r}_j) \ . \quad (6.2.1)$$

Here, $\boldsymbol{A}$ and A_0 are the vector potentials. Because we are dealing with fermions, the wave function Ψ in the eigenvalue problem,

$$\mathcal{H}\Psi(\boldsymbol{r}_1, \ldots, \boldsymbol{r}_N) = E\Psi(\boldsymbol{r}_1, \ldots, \boldsymbol{r}_N) \ , \quad (6.2.2)$$

must be antisymmetric:

$$\begin{aligned} P\Psi(\boldsymbol{r}_1, \boldsymbol{r}_2, \ldots, \boldsymbol{r}_N) &= \Psi(\boldsymbol{r}_{P(1)}, \boldsymbol{r}_{P(2)}, \ldots, \boldsymbol{r}_{P(N)}) \\ &= (-1)^P \Psi(\boldsymbol{r}_1, \ldots, \boldsymbol{r}_N) \ . \end{aligned} \quad (6.2.3)$$

Here, P acts on $(1, 2, \ldots, N)$ as the permutation operator $(P(1), P(2), \ldots, P(N))$, and $(-1)^P$ is $+1$ in the case when P can be written with an even number of transmutations (exchanging of two elements), and -1 in the odd number case. As a special case of (6.2.3), we obtain

$$\Psi(\boldsymbol{r}_1, \ldots, \boldsymbol{r}_i, \ldots, \boldsymbol{r}_j, \ldots, \boldsymbol{r}_N) = -\Psi(\boldsymbol{r}_1, \ldots, \boldsymbol{r}_j, \ldots, \boldsymbol{r}_i, \ldots, \boldsymbol{r}_N) \ . \quad (6.2.4)$$

Conversely, when (6.2.4) holds for arbitrary i and j, then (6.2.3) can be derived.

Next, we consider the bosonic problem. However, this time the Hamiltonian is not (6.2.1), so we introduce a new vector potential $\boldsymbol{a}$ to write

$$\mathcal{H}' = \sum_{i=1}^{N} \frac{1}{2m} \left[\boldsymbol{p}_i + e\boldsymbol{A}(\boldsymbol{r}_i) + e\boldsymbol{a}(\boldsymbol{r}_i)\right]^2 + \sum_{i=1}^{N} eA_0(\boldsymbol{r}_i) + \sum_{i<j} V(\boldsymbol{r}_i - \boldsymbol{r}_j) \ . \quad (6.2.5)$$

Here, $\boldsymbol{a}(\boldsymbol{r}_i)$ is given by

$$\boldsymbol{a}(\boldsymbol{r}_i) = \frac{\phi_0}{2\pi} \frac{\theta}{\pi} \sum_{j(\neq i)} \nabla_i \alpha_{ij} \ , \quad (6.2.6)$$

with $\phi_0 = 2\pi/e$. α_{ij} is the angle between the x axis and the difference vector between $\boldsymbol{r}_j$ and $\boldsymbol{r}_i$ (that is, $\boldsymbol{r}_j - \boldsymbol{r}_i$), as shown in Fig. 6.1. Here, the value is determined only up to 2π, but when acting with ∇, the result becomes unambiguous. α_{ij} is explicitly given by

$$\alpha_{ij} = \tan^{-1} \frac{y_j - y_i}{x_j - x_i} \equiv \alpha(\boldsymbol{r}_j - \boldsymbol{r}_i) \quad (6.2.7)$$

and (6.2.6) becomes

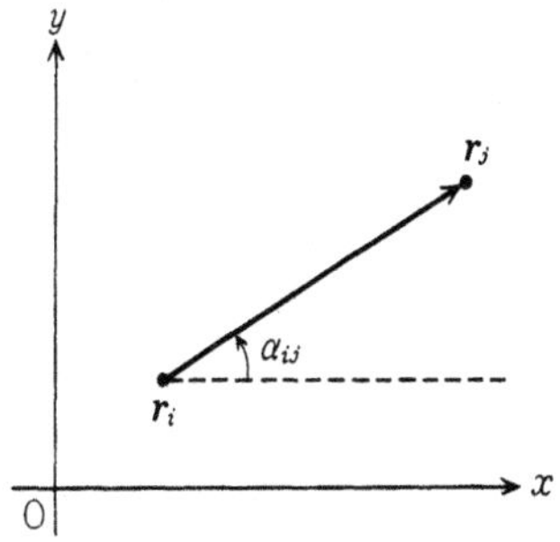

Fig. 6.1. The definition of α_{ij}

$$\boldsymbol{a}(\boldsymbol{r}_i) = \left(\frac{\phi_0}{2\pi} \frac{\theta}{\pi} \sum_{j(\neq i)} \frac{y_j - y_i}{|\boldsymbol{r}_j - \boldsymbol{r}_i|^2}, -\frac{\phi_0}{2\pi} \frac{\theta}{\pi} \sum_{j(\neq i)} \frac{x_j - x_i}{|\boldsymbol{r}_j - \boldsymbol{r}_i|^2} \right) \tag{6.2.8a}$$

or

$$a_\alpha(\boldsymbol{r}_i) = \frac{\phi_0}{2\pi} \frac{\theta}{\pi} \varepsilon_{\alpha\beta} \sum_{j(\neq i)} \frac{(x_j - x_i)_\beta}{|\boldsymbol{r}_j - \boldsymbol{r}_i|^2} \ . \tag{6.2.8b}$$

Here, we set $\varepsilon_{12} = -\varepsilon_{21} = 1$ (the other components are zero) and $(\boldsymbol{r}_j)_1 = x_j$, $(\boldsymbol{r}_j)_2 = y_j$. Notice that $\boldsymbol{p}_i$ and $\boldsymbol{r}_i$ appearing in (6.2.5) are both operators, and that $\alpha(\boldsymbol{r}_i)$ is the operator given by (6.2.8). Because we are dealing with bosons, the wave function of the eigenvalue problem,

$$\mathcal{H}' \phi(\boldsymbol{r}_1, \ldots, \boldsymbol{r}_N) = E' \phi(\boldsymbol{r}_1, \ldots, \boldsymbol{r}_n) \ , \tag{6.2.9}$$

must be totally symmetric:

$$P\phi(\boldsymbol{r}_1, \ldots, \boldsymbol{r}_N) = \phi(\boldsymbol{r}_1, \ldots, \boldsymbol{r}_N) \ . \tag{6.2.10}$$

In order to connect the fermionic system and the bosonic system described above, we introduce the following unitary operator:

$$U = \exp\left[\mathrm{i} \sum_{i<j} \frac{\theta}{\pi} \alpha_{ij} \right] \ . \tag{6.2.11}$$

Because α_{ij} is a function of $\boldsymbol{r}_j - \boldsymbol{r}_i$ as indicated in (6.2.7), this operator is independent of $\boldsymbol{p}_i$. We calculate $U\boldsymbol{p}_i U^{-1}$ as

$$\begin{aligned} U\boldsymbol{p}_i U^{-1} &= U(-\mathrm{i}\nabla_i U^{-1}) + UU^{-1}(-\mathrm{i}\nabla_i) \\ &= \exp\left[\mathrm{i} \sum_{i<j} \frac{\theta}{\pi} \alpha_{ij} \right] \left(-\sum_{j(\neq i)} \frac{\theta}{\pi} \nabla_i \alpha_{ij} \right) \exp\left[-\mathrm{i} \sum_{i<j} \frac{\theta}{\pi} \alpha_{ij} \right] + \boldsymbol{p}_i \\ &= -e\boldsymbol{a}(\boldsymbol{r}_i) + \boldsymbol{p}_i \ . \end{aligned} \tag{6.2.12}$$

Therefore, we obtain

$$\begin{aligned}U\left[\boldsymbol{p}_i + e\boldsymbol{A}(\boldsymbol{r}_i) + e\boldsymbol{a}(\boldsymbol{r}_1)\right]^2 U^{-1} \\ &= U\left[\boldsymbol{p}_i + e\boldsymbol{A}(\boldsymbol{r}_i) + e\boldsymbol{a}(\boldsymbol{r}_i)\right] U^{-1} U\left[\boldsymbol{p}_i + e\boldsymbol{A}(\boldsymbol{r}_i) + e\boldsymbol{a}(\boldsymbol{r}_i)\right] U^{-1} \\ &= \left[\boldsymbol{p}_i + e\boldsymbol{A}(\boldsymbol{r}_i)\right]^2 \ . \end{aligned} \tag{6.2.13}$$

Because U does not contain $\boldsymbol{p}_i$, $A_0(\boldsymbol{r}_i)$ and $V(\boldsymbol{r}_i - \boldsymbol{r}_j)$ remain invariant when U and U^{-1} act on the right- and left-hand sides. Therefore, the following relation between (6.2.1) and (6.2.5) holds:

$$\mathcal{H} = U\mathcal{H}'U^{-1} \quad \text{or} \quad U^{-1}\mathcal{H}U = \mathcal{H}' \ . \tag{6.2.14}$$

Now, suppose that the eigenvalue problem of the bosonic system has been solved. We call this solution ϕ. Writing

$$\Psi(\boldsymbol{r}_1, \ldots, \boldsymbol{r}_N) = U\phi(\boldsymbol{r}_1, \ldots, \boldsymbol{r}_N) \tag{6.2.15}$$

we obtain

$$\begin{aligned}\mathcal{H}\Psi &= \mathcal{H}U\phi = U(U^{-1}\mathcal{H}U)\phi \\ &= U\mathcal{H}'\phi = U(E'\phi) = E'(U\phi) = E'\Psi \ . \end{aligned} \tag{6.2.16}$$

We conclude that Ψ is an eigenfunction of $\mathcal{H}$ with the same energy eigenvalue E'. We wonder if Ψ fulfils the condition (6.2.4) for fermionic wave functions. In order to see this, we write the parts in U depending on $\boldsymbol{r}_n$ and $\boldsymbol{r}_l$ explicitly (we set $n < l$):

$$\begin{aligned}&U(\boldsymbol{r}_1, \ldots, \boldsymbol{r}_n, \ldots, \boldsymbol{r}_l, \ldots, \boldsymbol{r}_N) \\ &= \exp\left[\mathrm{i}\sum_{\substack{i<j \\ (\neq n,l)}}^{N} \frac{\theta}{\pi}\alpha_{ij} + \mathrm{i}\sum_{\substack{j=n+1 \\ (j\neq l)}}^{n-1} \frac{\theta}{\pi}\alpha_{nj} + \mathrm{i}\sum_{i=1}^{n-1} \frac{\theta}{\pi}\alpha_{in} + \mathrm{i}\sum_{j=l+1}^{N} \frac{\theta}{\pi}\alpha_{ij} \right. \\ &\quad \left. + \mathrm{i}\sum_{\substack{i=1 \\ (i\neq n)}}^{i-1} \frac{\theta}{\pi}\alpha_{il} + \mathrm{i}\frac{\theta}{\pi}\alpha_{nl}\right] \ . \end{aligned} \tag{6.2.17}$$

We write $\mathrm{i}\Phi(\boldsymbol{r}_1, \ldots, \boldsymbol{r}_n, \ldots, \boldsymbol{r}_l, \ldots, \boldsymbol{r}_n)$ for the function appearing in the exponential. We examine how the phase Φ changes when $\boldsymbol{r}_n$ and $\boldsymbol{r}_l$ are interchanged:

$$\begin{aligned}\Delta\Phi_{nl} &\equiv \Phi(\boldsymbol{r}_1, \ldots, \boldsymbol{r}_l, \ldots, \boldsymbol{r}_n, \ldots, \boldsymbol{r}_N) - \Phi(\boldsymbol{r}_1, \ldots, \boldsymbol{r}_n, \ldots, \boldsymbol{r}_l, \ldots, \boldsymbol{r}_N) \\ &= \frac{\theta}{\pi}\sum_{m=n+1}^{l-1}\left[\alpha_{lm} - \alpha_{ml} + \alpha_{mn} - \alpha_{nm}\right] + \frac{\theta}{\pi}(\alpha_{ln} - \alpha_{nl}) \ . \end{aligned} \tag{6.2.18}$$

As is clear from Fig. 6.1, $\alpha_{ij} - \alpha_{ji}$ is the change in angle that arises when reversing the roles of the particles i and j, that is

$$\alpha_{ij} - \alpha_{ji} = \pm\pi \ . \tag{6.2.19}$$

Here, + corresponds to an anti-clockwise and − corresponds to a clockwise exchange of the particles. In the case when

$$\theta = (2k+1)\pi \tag{6.2.20}$$

holds, we obtain

$$\Delta\Phi_{nl} = \sum_{m=n+1}^{l-1} (2\pi \text{ or } 0 \text{ or } -2\pi)\frac{\theta}{\pi} \pm \theta \tag{6.2.21}$$

and the first term on the right-hand side does not contribute in the exponent. We conclude that

$$\begin{aligned} U(\boldsymbol{r}_1,\ldots,\boldsymbol{r}_l,\ldots,\boldsymbol{r}_n,\ldots,\boldsymbol{r}_N) &= \mathrm{e}^{\mathrm{i}\Delta\Phi_{nl}} U(\boldsymbol{r}_1,\ldots,\boldsymbol{r}_n,\ldots,\boldsymbol{r}_l,\ldots,\boldsymbol{r}_N) \\ &= \mathrm{e}^{\pm\mathrm{i}\theta} U(\boldsymbol{r}_1,\ldots,\boldsymbol{r}_n,\ldots,\boldsymbol{r}_l,\ldots,\boldsymbol{r}_N) \\ &= -U(\boldsymbol{r}_1,\ldots,\boldsymbol{r}_n,\ldots,\boldsymbol{r}_l,\ldots,\boldsymbol{r}_N) \end{aligned} \tag{6.2.22}$$

and therefore

$$\begin{aligned} &\Psi(\boldsymbol{r}_1,\ldots,\boldsymbol{r}_l,\ldots,\boldsymbol{r}_n,\ldots,\boldsymbol{r}_N) \\ &\quad = U(\boldsymbol{r}_1,\ldots,\boldsymbol{r}_l,\ldots,\boldsymbol{r}_n,\ldots,\boldsymbol{r}_N)\phi(\boldsymbol{r}_1,\ldots,\boldsymbol{r}_l,\ldots,\boldsymbol{r}_n,\ldots,\boldsymbol{r}_N) \\ &\quad = -U(\boldsymbol{r}_1,\ldots,\boldsymbol{r}_n,\ldots,\boldsymbol{r}_l,\ldots,\boldsymbol{r}_N)\phi(\boldsymbol{r}_1,\ldots,\boldsymbol{r}_N) \\ &\quad = -\Psi(\boldsymbol{r}_1,\ldots,\boldsymbol{r}_n,\ldots,\boldsymbol{r}_l,\ldots,\boldsymbol{r}_N) \ . \end{aligned} \tag{6.2.23}$$

Here, we used the fact that the wave function of bosons is totally symmetric. In such a way, we proved that for $\theta = (2k+1)\pi$, $\Psi = U\psi$ is the solution of the fermionic eigenvalue problem (6.2.1), (6.2.2) and (6.2.3).

From the above discussion, it becomes clear that U contributes a phase $\mathrm{e}^{\pm\mathrm{i}\theta}$ when two particles are exchanged. The vector potential $\boldsymbol{a}(\boldsymbol{r}_i)$ interacts with the bosons, and this phase transformation arises as the Aharonov–Bohm effect. Equation (6.2.8) can be interpreted as the vector potential created by a particle at $\boldsymbol{r}_j$, acting on a particle at position $\boldsymbol{r}_i$.

$$\mathcal{A}(\boldsymbol{r}) = \frac{\phi_0}{2\pi}\frac{\theta}{\pi}\left(\frac{-y}{|\boldsymbol{r}|^2}, \frac{x}{|\boldsymbol{r}|^2}\right) = \frac{\phi_0}{2\pi}\frac{\theta}{\pi}\frac{1}{r}\hat{\boldsymbol{e}}_\varphi \tag{6.2.24}$$

is the vector potential with strength $\phi_0(\theta/\pi)$ at the origin. We now introduce polar coordinates (Fig. 6.2).

For an arbitrary closed path C around the origin, we obtain

$$\oint_C \mathcal{A}(\boldsymbol{r})\cdot \mathrm{d}\boldsymbol{r} = \frac{\phi_0}{2\pi}\frac{\theta}{\pi}\int_0^{2\pi} r\,\mathrm{d}\varphi\frac{1}{r} = \phi_0\frac{\theta}{\pi} \tag{6.2.25}$$

and due to Stokes' theorem

$$\oint_C \mathcal{A}(\boldsymbol{r})\cdot \mathrm{d}\boldsymbol{r} = \int \mathrm{d}^2\boldsymbol{r}\, \nabla\times\mathcal{A}(\boldsymbol{r}) \tag{6.2.26}$$

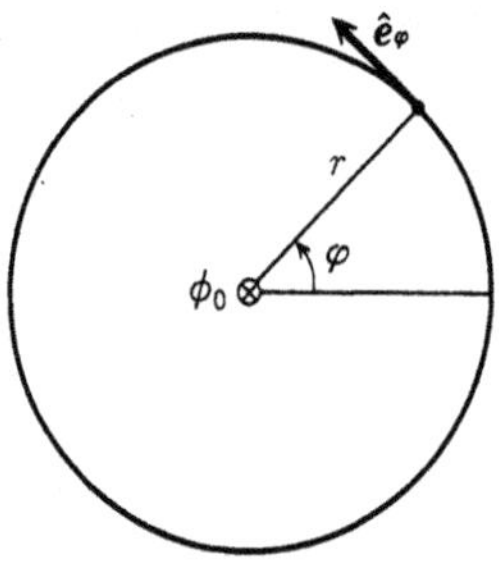

Fig. 6.2. The vector potential around the flux ϕ_0

obviously, we obtain

$$\nabla \times \mathcal{A}(\boldsymbol{r}) = \phi_0 \frac{\theta}{\pi} \delta(\boldsymbol{r}) \hat{\boldsymbol{z}} \ . \tag{6.2.27}$$

Equation (6.2.8b) can be written as

$$\boldsymbol{a}(\boldsymbol{r}_i) = \sum_{j(\neq i)} \mathcal{A}(\boldsymbol{r}_i - \boldsymbol{r}_j) \ . \tag{6.2.28}$$

The particle at position $\boldsymbol{r}_j$ can be regarded as the source of the magnetic flux $\phi_0(\theta/\pi)$, creating a vector potential at $\boldsymbol{r}_i$. Because, as indicated in Fig. 6.3, the exchange of two particles is equal to half a rotation of particle 2 around particle 1, the phase change $\Delta\Theta$ is given by

$$\Delta\Theta = \frac{e}{2} \oint_C \mathcal{A}(\boldsymbol{r}_2 - \boldsymbol{r}_1) \cdot \mathrm{d}\boldsymbol{r}_2 = \frac{e}{2} \phi_0 \frac{\theta}{\pi} = \theta \ . \tag{6.2.29}$$

This result agrees with the above considerations. This is exactly the idea of the statistical transmutation: the change in sign (in general the phase transformation) when two particles are exchanged is described by introducing a fictitious gauge field $\boldsymbol{a}$ and using the Aharonov–Bohm effect. So, what about the time component a_0 of the gauge field? For its description, it is more conveniant to formulate the problem in second quantization.

Using the standard formalism, (6.2.5) can be rewritten as

$$\begin{aligned} \mathcal{H}' = & \int \mathrm{d}^2\boldsymbol{r}\, \phi^\dagger(\boldsymbol{r}) \left(\frac{1}{2m} \left[-\mathrm{i}\nabla + e\boldsymbol{A}(\boldsymbol{r}) + e\boldsymbol{a}(\boldsymbol{r}) \right]^2 - \mu + eA_0(\boldsymbol{r}) \right) \phi(\boldsymbol{r}) \\ & + \frac{1}{2} \int \mathrm{d}^2\boldsymbol{r} \int \mathrm{d}^2\boldsymbol{r}'\, \rho(\boldsymbol{r}) V(\boldsymbol{r} - \boldsymbol{r}') \rho(\boldsymbol{r}') \ . \end{aligned} \tag{6.2.30}$$

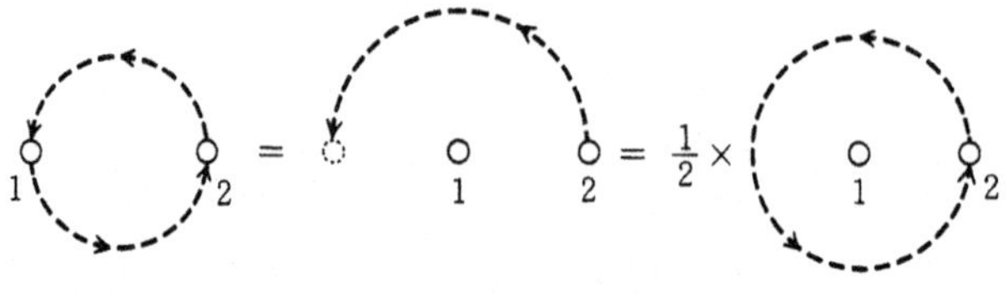

Fig. 6.3. The relation between the exchange of two particles and the rotation around one particle

Here, $\phi^\dagger(\boldsymbol{r})$ and $\phi(\boldsymbol{r})$ are the bosonic creation and annihilation operators, fulfilling the commutation relation

$$\left[\phi(\boldsymbol{r}), \phi^\dagger(\boldsymbol{r}')\right] = \delta(\boldsymbol{r} - \boldsymbol{r}') \ , \tag{6.2.31}$$

and $\rho(\boldsymbol{r}) = \phi^\dagger(\boldsymbol{r})\phi(\boldsymbol{r})$ is the particle density. (6.2.28) becomes

$$\boldsymbol{a}(\boldsymbol{r}) = \int \mathrm{d}^2\boldsymbol{r}' \, \mathcal{A}(\boldsymbol{r} - \boldsymbol{r}')\rho(\boldsymbol{r}') \ . \tag{6.2.32}$$

Using (6.2.27), we obtain

$$\begin{aligned} \hat{\boldsymbol{z}} b(\boldsymbol{r}) \equiv \nabla \times \boldsymbol{a}(\boldsymbol{r}) &= \int \mathrm{d}^2\boldsymbol{r}' \, \nabla_{\boldsymbol{r}} \times \mathcal{A}(\boldsymbol{r} - \boldsymbol{r}')\rho(\boldsymbol{r}') \\ &= \int \mathrm{d}^2\boldsymbol{r}' \, \phi_0 \frac{\theta}{\pi} \delta(\boldsymbol{r} - \boldsymbol{r}')\rho(\boldsymbol{r}')\hat{\boldsymbol{z}} \\ &= \phi_0 \frac{\theta}{\pi} \rho(\boldsymbol{r})\hat{\boldsymbol{z}} \end{aligned} \tag{6.2.33}$$

and therefore the flux density $b(\boldsymbol{r})$ is proportional to $\rho(\boldsymbol{r})$. That is, for one particle, one gauge flux θ/π is "attached", and the combination of the boson with the gauge flux string corresponds to a fermion.

Taking the divergence of (6.2.32), we obtain

$$\nabla \cdot \boldsymbol{a}(\boldsymbol{r}) = \int \mathrm{d}^2\boldsymbol{r}' \, \nabla_{\boldsymbol{r}} \cdot \mathcal{A}(\boldsymbol{r} - \boldsymbol{r}')\rho(\boldsymbol{r}') = 0 \ . \tag{6.2.34}$$

Equation (6.2.34) corresponds exactly to the Coulomb gauge condition. Conversely, using (6.2.33) and (6.2.34), it is possible to derive (6.2.32) (under the boundary condition that $\boldsymbol{a}$ is zero at infinity). Now, we interpret (6.2.34) as the gauge fixing condition, and implement only (6.2.33) explicitly as the constraint condition into the path integral. In order to do so, it is sufficient to use

$$\begin{aligned} &\prod_{\boldsymbol{r},t} \delta\left(\frac{\pi e}{\theta \phi_0} \varepsilon_{\alpha\beta}\partial_\alpha a_\beta(\boldsymbol{r},t) - e\rho(\boldsymbol{r},t)\right) \\ &\quad = \int \mathcal{D}a_0(\boldsymbol{r},t) \exp\left[\mathrm{i}\int \mathrm{d}t \int \mathrm{d}^2\boldsymbol{r} \, a_0(\boldsymbol{r},t)\left(\frac{\pi e}{\theta\phi_0}\varepsilon_{\alpha\beta}\partial_\alpha a_\beta(\boldsymbol{r},t) - e\rho(\boldsymbol{r},t)\right)\right] . \end{aligned} \tag{6.2.35}$$

Finally, under the presence of an external field A_μ, the path integral of the system reads

$$Z[A_\mu] = \int \mathcal{D}a_\mu \mathcal{D}\phi \exp[\,\mathrm{i}S_a[a_\mu] + \mathrm{i}S_\phi[A_\mu + a_\mu, \phi]\,] \ , \tag{6.2.36}$$

where S_a and S_ϕ are given by

$$S_{\mathrm{a}}[a_\mu] = \int \mathrm{d}t \int \mathrm{d}^2\boldsymbol{r} \frac{\pi e}{\theta \phi_0} \varepsilon_{\alpha\beta} a_0 \partial_\alpha a_\beta \tag{6.2.37}$$

$$\begin{aligned} S_\phi[A_\mu + a_\mu, \phi] = \int \mathrm{d}t \int \mathrm{d}^2\boldsymbol{r} \Big[& \phi^\dagger (\mathrm{i}\partial_t - e(A_0 + a_0) + \mu)\phi \\ & - \frac{1}{2m} |(-\mathrm{i}\nabla + e(\boldsymbol{A} + \boldsymbol{a}))\phi|^2 \Big] \\ & - \frac{1}{2} \int \mathrm{d}t \int \mathrm{d}^2\boldsymbol{r} \int \mathrm{d}^2\boldsymbol{r}'\, \rho(\boldsymbol{r}) V(\boldsymbol{r} - \boldsymbol{r}') \rho(\boldsymbol{r}') \ . \end{aligned} \tag{6.2.38}$$

Equations (6.2.36), (6.2.37) and (6.2.38) are valid only in the Coulomb gauge (6.2.34), and the system as it stands is not gauge invariant.

In a general gauge, the first term S_a has to be replaced by the so-called Chern–Simons term

$$S_{\mathrm{C.S.}}[a_\mu] = \int \mathrm{d}t \int \mathrm{d}^2\boldsymbol{r} \frac{\pi e}{2\theta\phi_0} \varepsilon^{\mu\nu\lambda} a_\mu \partial_\nu a_\lambda \ . \tag{6.2.39}$$

Here, μ, ν, λ represent the three components x, y, t, and $\varepsilon^{\mu\nu\lambda}$ is the totally anti-symmetric tensor in three dimensions. It is simple to prove that (6.2.39) becomes (6.2.37) under the condition (6.2.34). To summarize, the two-dimensional fermionic system can be reformulated using the action (6.2.39) and (6.2.38) by choosing a gauge and performing the functional integral with respect to the gauge field and the bosonic field ϕ. The gauge field that is introduced in this manner is called the Chern–Simons gauge field. In what follows we will discuss this functional integral using approximation techniques.

We just learned that instead of fermions, we can consider a bosonic system feeling a short-range repulsive force (that is, two particles cannot enter the same place). These bosons interact with the sum of an external field A_μ and the Chern–Simons gauge field a_μ (6.2.38). Now, we write for A_μ the sum of $\bar{A}_\mu$ corresponding to a static magnetic field $-B$ in the z direction and an infinitesimal test field A'_μ that is introduced to study the linear response of the system, $\bar{A}_\mu + A'_\mu$. On the other hand, a_μ is an internal degree of freedom that is correlated to the fluctuations of the particle density $\rho(\boldsymbol{r})$ due to (6.2.33). We represent it as the sum of the average value $\bar{a}_\mu$ and fluctuations a'_μ around it, $\bar{a}_\mu + a'_\mu$. $\bar{a}_\mu$ fulfils (6.2.33) for the average field

$$\nabla \times \bar{\boldsymbol{a}}(\boldsymbol{r}) = \phi_0 \frac{\theta}{\pi} \bar{\rho} \hat{\boldsymbol{z}} \tag{6.2.40}$$

and $\bar{a}_0 = 0$. We conclude that the average value of the gauge flux $\boldsymbol{B}_{\mathrm{eff}} = \boldsymbol{B} + \boldsymbol{b} = \nabla \times (\boldsymbol{A} + \boldsymbol{a})$ that the bosons are exposed to is given by

$$\bar{\boldsymbol{B}}_{\mathrm{eff}} = -B\hat{\boldsymbol{z}} + \phi_0 \frac{\theta}{\pi} \bar{\rho} \hat{\boldsymbol{z}} \ . \tag{6.2.41}$$

Here, θ is given by (6.2.20), and k can be an arbitrary integer number. Conversely, choosing k in such a way that the right-hand side of (6.2.41) becomes zero, the bosons feel in average no gauge flux.

Using (6.2.41), (6.1.35) and (6.1.47), we conclude that this condition is equal to the condition

$$\nu = \frac{1}{2k+1} \tag{6.2.42}$$

for the filling ratio ν. This can be easily understood using the following argument. (6.1.35) can also be expressed as

$$G = \frac{eB}{2\pi}S = \frac{B}{\phi_0}S = \frac{\Phi}{\phi_0} \quad . \tag{6.2.43}$$

Here, $\Phi = BS$ is the total magnetic flux. The above equation signifies that the degeneracy G of every Landau level equals the total number of magnetic flux measured in units of the gauge flux quantum ϕ_0. Therefore, $\nu = 1$ (that is, $N = G$) corresponds to the case where in average one magnetic flux quantum is in the vicinity of one electron. For a general ν (< 1), the number increases to $1/\nu$ quanta. When ν fulfils (6.2.42), an odd number of (external) gauge flux quanta is assigned to one electron, cancelling with an odd number of Chern–Simons gauge field flux quanta, that have been glued to the bosons leading to the phase factor of the fermionic commutation relation. In what follows, A_μ and a_μ represent the remaining parts after this cancellation of the average value, that is, A'_μ and a'_μ.

For such special cases where ν fulfils (6.2.42), in zeroth-order approximation we may consider a bosonic system without a magnetic field, where we naturally expect Bose condensation or superfluidity. Superfluidity has already been discussed in detail in Sect. 4.2. Here, we will make a simple recapitulation.

We start by representing the bosonic field $\phi(\boldsymbol{r})$ by its phase $\theta(\boldsymbol{r})$ and amplitude $\rho(\boldsymbol{r})$:

$$\phi(\boldsymbol{r}) = \rho^{1/2}(\boldsymbol{r})\,\mathrm{e}^{\mathrm{i}\theta(\boldsymbol{r})} \quad . \tag{6.2.44}$$

Inserting this into S_ϕ (6.2.38), we obtain

$$\begin{aligned} S_\phi = & \int \mathrm{d}t \int \mathrm{d}^2\boldsymbol{r} \Big\{ \rho(-\partial_t\theta - e(A_0 + a_0) + \mu) \\ & - \frac{1}{2m}\rho(\nabla\theta + e(\boldsymbol{A} + \boldsymbol{a}))^2 - \frac{1}{8m}\rho^{-1}(\nabla\rho)^2 \Big\} \\ & - \frac{1}{2}\int \mathrm{d}t \int \mathrm{d}^2\boldsymbol{r} \int \mathrm{d}^2\boldsymbol{r}'\, \rho(\boldsymbol{r})V(\boldsymbol{r}-\boldsymbol{r}')\rho(\boldsymbol{r}') \quad . \end{aligned} \tag{6.2.45}$$

Assuming that the saddle-point sulution is independent of space and time, we obtain

$$S_\phi \to T \cdot S \left[\mu\rho_0 - \tfrac{1}{2}\tilde{V}(0)\rho_0^2 \right] \ . \tag{6.2.46}$$

Here, $\tilde{V}(q)$ is the Fourier transformation of $V(\boldsymbol{r})$. The value of ρ_0 that corresponds to the extremum of (6.2.46) is

$$\rho_0 = \frac{\mu}{\tilde{V}(0)} > 0 \ . \tag{6.2.47}$$

We now express ρ as the sum of this ρ_0 and some fluctuation $\delta\rho$ and expand the action up to second order in $\delta\rho$ and $\partial_\mu\theta + e(A_\mu + a_\mu)$:

$$\begin{aligned} S_\phi = \int \mathrm{d}t \int \mathrm{d}^2\boldsymbol{r} \Big\{ & \delta\rho(-\partial_t\theta - e(A_0 + a_0)) \\ & - \frac{\rho_0}{2m}(\nabla\theta + e(\boldsymbol{A} + \boldsymbol{a}))^2 - \frac{1}{8m}\rho_0^{-1}(\nabla\delta\rho)^2 \Big\} \\ & - \frac{1}{2}\int \mathrm{d}t \int \mathrm{d}^2\boldsymbol{r} \int \mathrm{d}^2\boldsymbol{r}'\, V(\boldsymbol{r} - \boldsymbol{r}')\delta\rho(\boldsymbol{r})\delta\rho(\boldsymbol{r}') \ . \end{aligned} \tag{6.2.48}$$

Because in the long wavelength limit the term proportional to $(\nabla\delta\rho)^2$ makes only a small contribution compared with the interaction of the third term, we ignore it. After Fourier transformation and $\delta\rho$-integration (that is, completing the square) we obtain finally for the phase degree of freedom the action

$$\begin{aligned} S_\theta = \sum_\omega \sum_{\boldsymbol{k}} \Big\{ & \frac{1}{2\tilde{V}(\boldsymbol{k})} [-\mathrm{i}\omega\theta(-k) - e(A_0(-k) + a_0(-k))] \\ & \times [\mathrm{i}\omega\theta(k) - e(A_0(k) + a_0(k))] \\ & - \frac{\rho_0}{2m} [-\mathrm{i}\boldsymbol{k}\theta(-k) + e(\boldsymbol{A}(-k) + \boldsymbol{a}(-k))] \\ & \times [\mathrm{i}\boldsymbol{k}\theta(k) + e(\boldsymbol{A}(k) + \boldsymbol{a}(k))] \Big\} \ . \end{aligned} \tag{6.2.49}$$

Here, we used the notation $k = (\boldsymbol{k}, \omega)$. The sum of S_θ and $S_{\mathrm{C.S.}}$ given in (6.2.39) is the effective action of the quantum Hall liquid. The principal point is the fact that the Bose system does not show the Meissner effect with respect to A_μ, but with respect to $A_\mu + a_\mu$. Now, we will use this fact to derive $\sigma_{xy} = \nu(e^2/h)$.

In order to do so, we consider a donought ring, where a magnetic flux Φ flows through the hole in the middle. A constant magnetic field B acts on the ring itself. We consider the following two types of line integrals along the closed path C around the ring:

$$I_a(C) = \oint_C \boldsymbol{a} \cdot \mathrm{d}\boldsymbol{l} \ , \tag{6.2.50}$$

$$I_A(C) = \oint_C \boldsymbol{A} \cdot \mathrm{d}\boldsymbol{l} \ . \tag{6.2.51}$$

The Meissner effect of the bosonic system signifies that "magnetic flux quantization" with respect to $\boldsymbol{A}+\boldsymbol{a}$ occurs, and that the minima in the free energy corresponding to different magnetic flux are separated by microscopic energy barriers. Therefore, even when Φ is adiabatically changed, $I_a(C) + I_A(C)$ remains constant during this process. We conclude therefore

$$\frac{\mathrm{d}I_A(C)}{\mathrm{d}t} = \frac{\mathrm{d}\Phi}{\mathrm{d}t} = -\frac{\mathrm{d}I_a(C)}{\mathrm{d}t} \tag{6.2.52}$$

and that $\mathrm{d}\Phi/\mathrm{d}t$ of the Chern–Simons gauge flux penetrating the surface enclosed by the line C is repelled outside of it.

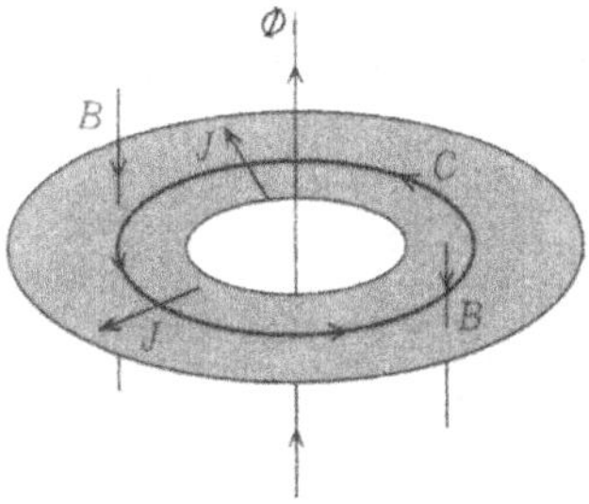

Fig. 6.4. The ring configuration

On the other hand, because the Chern–Simons gauge field is proportional to the particle density as given by (6.2.33), the current J that passes C in one unit of time is given by

$$J = (-e)\frac{\pi}{\phi_0\theta}\frac{\mathrm{d}\Phi}{\mathrm{d}t} \ . \tag{6.2.53}$$

Therefore, the potential difference at the line C can be calculated using the Maxwell equation

$$\nabla \times \boldsymbol{E} = -\frac{\partial \boldsymbol{B}}{\partial t} \tag{6.2.54}$$

and we obtain

$$\begin{aligned}
[\Delta V]_C &= \oint_C \boldsymbol{E}\cdot \mathrm{d}\boldsymbol{l} \\
&= \int_{\text{surface enclosed by } C} (\nabla \times \boldsymbol{E})_z \, \mathrm{d}S \\
&= -\int_{\text{surface enclosed by } C} \left(\frac{\partial \boldsymbol{B}}{\partial t}\right)_z \mathrm{d}S \\
&= -\frac{\mathrm{d}\Phi}{\mathrm{d}t} \ .
\end{aligned} \tag{6.2.55}$$

Because σ_{xy} is the ratio of J and ΔV, we obtain finally

$$\sigma_{xy} = \frac{J}{[\Delta V]} = e\frac{\pi}{\phi_0 \theta} = \frac{\pi}{\theta}\frac{e^2}{h} = \nu\frac{e^2}{h} \ . \tag{6.2.56}$$

We conclude that the Meissner effect or the quantization of the magnetic flux leads via the Chern–Simons gauge field to (6.2.56). Now, we need to derive this fact from (6.2.49) using functional integrals. It is sufficient to integrate out θ and a_μ and to construct an effective action $S[A_\mu]$ of A_μ. The electric current j_μ can then be determined by $j_\mu = \delta S[A_\mu]/\delta A_\mu$. Because the action $S = S_{\text{C.S.}} + S_\theta$ is quadratic in θ, a_μ and A_μ, the integration can be performed simply by completing the square. When the gauge field is integrated out, it is necessary to fix the gauge (that is, only three of the components of θ and a_x, a_y, a_0 are independent). Of course, the final result $S[A_\mu]$ is independent of the choice of the gauge; however, the physical picture appears in a different light.

First, one possible choice is

$$ea_\mu \to ea_\mu - \partial_\mu \theta \ . \tag{6.2.57}$$

Here, θ disappears, and it is possible to integrate over the three components of a_μ. Notice that (6.2.57) is nothing but a gauge fixing. The gauge condition that θ does not appear explicitly is called the unitary gauge. The longitudinal component (fulfilling $\partial_\mu a_\nu - \partial_\nu a_\mu = 0$) as it stands has no physical meaning, but because it "swallowed" θ, it transforms into a physical field. Then, in the action a term corresponding to a "mass term" $m^2 a_\mu a^\mu$ appears (that is, because θ has been swallowed, the gauge field becomes "heavy"). This phenomenon is called the Higgs mechanism.

This is the consequence of "gauge symmetry breaking" appearing with superconductivity of the bosonic system. However, notice that "symmetry breaking" does not mean that the symmetry under the gauge transformation is lost. The appearance of the mass term $m^2 a_\mu a^\mu$ is related to the choice of the unitary gauge. (Therefore in the superconducting phase also, gauge invariance and the related Ward–Takahashi identities remain valid.)

It is possible to proceed with the calculation using the unitary gauge; however, we will use the Coulomb gauge ($\nabla \cdot \boldsymbol{a} = 0$) that appeared when we derived (6.2.36)–(6.2.38). In this gauge, θ can be interpreted as the phase of the bosonic field; later, this will become useful when deriving the Laughlin wave function. [For example, in the gauge $a_0 = 0$, no clear meaning can be given to θ. This can, for example, be understood from the fact that the Green function $\langle e^{i\theta(\boldsymbol{r},t)} e^{-i\theta(0,0)} \rangle$ becomes zero for $\boldsymbol{r} \neq \boldsymbol{0}$. This is due to the fact that for time-independent gauge transformations $\theta(\boldsymbol{r},t) \to \theta(\boldsymbol{r},t) - \varphi(\boldsymbol{r})$ and $ea_\alpha(\boldsymbol{r},t) \to ea_\alpha(\boldsymbol{r},t) + \partial_\alpha \varphi(\boldsymbol{r})$, a_0 does not change, and therefore the condition $a_0 = 0$ does not fix the gauge completely. We conclude that $\langle e^{\pm\varphi(\boldsymbol{r})} \rangle = 0$ when integrated over $\varphi(\boldsymbol{r})$.]

In the Coulomb gauge, due to $\nabla \cdot \boldsymbol{a} = 0$, the Fourier components of the space components are given by

$$a_\alpha(\boldsymbol{k},\omega) = \frac{\mathrm{i}\varepsilon^{\alpha\beta}k_\beta}{|\boldsymbol{k}|}a_{\mathrm{T}}(\boldsymbol{k},\omega) \tag{6.2.58}$$

and can be expressed only by $a_{\mathrm{T}}(\boldsymbol{k},\omega)$ (T means transversal component). In what follows we will use the units $\hbar = c = e = 1$. Then, $\phi_0 = 2\pi$ holds. The action $S_{\mathrm{C.S.}}$ (6.2.39) in Fourier components reads

$$\begin{aligned} S_{\mathrm{C.S.}}[a_\mu] &= \frac{1}{2\theta}\sum_k |\boldsymbol{k}|a_0(-k)a_{\mathrm{T}}(k) \\ &= \frac{1}{4\theta}\sum_k |\boldsymbol{k}|(a_0(-k)a_{\mathrm{T}}(k) + a_{\mathrm{T}}(-k)a_0(k)) \ . \end{aligned} \tag{6.2.59}$$

The sum of this action with the action S_θ (6.2.49) of the matter field, that is $S = S_{\mathrm{C.S.}} + S_\theta$, is the starting point of the following discussion. Notice that the matter field in S_θ interacts with $a_\mu + A_\mu$. That is, the arguments appear in the combination

$$S[\theta, a_\mu, A_\mu] = S_{\mathrm{C.S.}}[a_\mu] + S_\theta[\theta, a_\mu + A_\mu] \ . \tag{6.2.60}$$

The next point to notice concerns the order of integration. One possibility is to integrate out a_μ first and to determine $S[\theta, A_\mu] = S[\partial_\mu\theta + A_\mu]$, and as the next step to integrate out θ to obtain $S[A_\mu]$. Another possibility is to integrate out θ first and to obtain $S[a_\mu, A_\mu]$ and to integrate out a_μ afterwards to obtain $S[A_\mu]$. If the only aim is to derive $S[A_\mu]$, then the second method is much more straightforward. When also information about θ, that is, the bosonic system, is required, we have to stop at the step $S[\partial_\mu\theta + A_\mu]$ and investigate this action further.

First, we choose the second way. The θ-integration in S_θ (6.2.49) can be performed by completing the square in the argument of the exponential of the following formula:

$$\begin{aligned} S_\theta = \sum_k \Bigg\{ & \left(\frac{\omega^2}{2\tilde{V}(\boldsymbol{k})} - \frac{\rho_0}{2m}|\boldsymbol{k}|^2\right)\theta(-k)\theta(k) \\ & + \theta(-k)\left\{\frac{\mathrm{i}\omega}{2\tilde{V}(\boldsymbol{k})}[a_0(k) + A_0(k)] + \frac{\rho_0}{2m}\mathrm{i}\boldsymbol{k}\cdot[\boldsymbol{a}(k) + \boldsymbol{A}(k)]\right\} \\ & + (k \to -k) \\ & + \frac{1}{2\tilde{V}(\boldsymbol{k})}(a_0(-k) + A_0(-k))(a_0(k) + A_0(k)) \\ & - \frac{\rho_0}{2m}(\boldsymbol{a}(-k) + \boldsymbol{A}(-k))\cdot(\boldsymbol{a}(k) + \boldsymbol{A}(k))\Bigg\} \ . \end{aligned} \tag{6.2.61}$$

As a result, we obtain

$$S_{\mathrm{eff}} = \sum_{k,\mu,\nu} \frac{1}{2}\pi_{\mu\nu}(k)\delta a_\mu(-k)\delta a_\nu(k) \ . \tag{6.2.62}$$

Here, $\delta a_\mu = a_\mu + A_\mu$. The coefficients $\pi_{\mu\nu}$ are given by

$$\pi_{00}(k) = \frac{1}{\tilde{V}(\boldsymbol{k})} \frac{-(\rho_0/2m)|\boldsymbol{k}|^2}{[\omega^2/2\tilde{V}(\boldsymbol{k}) - (\rho_0/2m)|\boldsymbol{k}|^2]} , \tag{6.2.63}$$

$$\pi_{\alpha\beta}(k) = -\frac{\rho_0}{m}\left[\delta_{\alpha\beta} + \frac{(\rho_0/2m)k_\alpha k_\beta}{\omega^2/2\tilde{V}(\boldsymbol{k}) - (\rho_0/2m)|\boldsymbol{k}|^2}\right] , \tag{6.2.64}$$

$$\begin{aligned}\pi_{0\alpha}(k) &= \pi_{\alpha 0}(k)\\ &= -\frac{\rho_0}{2m}\frac{\omega k_\alpha}{\tilde{V}(\boldsymbol{k})}\frac{1}{[\omega^2/2\tilde{V}(\boldsymbol{k}) - (\rho_0/2m)|\boldsymbol{k}|^2]} .\end{aligned} \tag{6.2.65}$$

Generally speaking, (6.2.62) represents the (linear) electromagnetic response of the bosonic field.

The gauge transformation of the gauge field δa_μ can be written as

$$\begin{aligned}\delta a_0(k) &\to \delta a_0(k) - \mathrm{i}\omega\varphi(k) ,\\ \delta a_\alpha(k) &\to \delta a_\alpha(k) + \mathrm{i}k_\alpha\varphi(k) .\end{aligned} \tag{6.2.66}$$

The condition that S_{eff} (6.2.62) is invariant under this transformation reads

$$-\omega\pi_{\pi 0}(k) + k_\alpha\pi_{\mu\alpha}(k) = 0 . \tag{6.2.67}$$

And indeed, this condition is fulfilled by (6.2.63), (6.2.64) and (6.2.65). In order to make this more evident, we rewrite (6.2.62) in the following manner:

$$\begin{aligned}S_{\text{eff}} = &\frac{1}{2}\sum_k \pi_0(k)\left[\delta a_0(-k) + \frac{\omega}{|\boldsymbol{k}|^2}k_\alpha\delta a_\alpha(-k)\right]\left[\delta a_0(k) + \frac{\omega}{|\boldsymbol{k}|^2}k_\beta\delta a_\beta(k)\right]\\ &+\frac{1}{2}\sum_k\sum_{\alpha,\beta}\pi_1(k)\left[\delta_{\alpha\beta} - \frac{k_\alpha k_\beta}{|\boldsymbol{k}|^2}\right]\delta a_\alpha(-k)\delta a_\beta(k) .\end{aligned} \tag{6.2.68}$$

The first term on the right-hand side in (6.2.68) expresses the longitudinal response

$$\begin{aligned}\pi_0(k) &= \pi_{00}(k)\\ &= \frac{1}{\tilde{V}(\boldsymbol{k})}\frac{-(\rho_0/2m)|\boldsymbol{k}|^2}{[\omega^2/2\tilde{V}(\boldsymbol{k}) - (\rho_0/2m)|\boldsymbol{k}|^2]} .\end{aligned} \tag{6.2.69}$$

Recalling that A_0 couples to the particle density ρ in the form $A_0\rho$, we understand that $\pi_0(k)$ is the density–density correlation, describing the density response of the scalar potential.

On the other hand, the second term on the right-hand side of (6.2.68) is the transversal response, and $\pi_1(k)$ is given by

$$\pi_1(k) = -\frac{\rho_0}{m} . \tag{6.2.70}$$

Now, what is the significance of the factor $\delta_{\alpha\beta} - k_\alpha k_\beta / |\boldsymbol{k}^2|$ (the so-called transversal factor)? We assume that a "magnetic field" $\delta b = \partial_x \delta a_y - \partial_y \delta a_x = b + B$ leads to the energy increase $\chi(\delta b)^2/2$. Written in the action, this term would be

$$
\begin{aligned}
-\frac{1}{2}\int \mathrm{d}t \int \mathrm{d}\boldsymbol{r}\, \chi(\delta b)^2
&= -\frac{\chi}{4}\sum_k \sum_{\alpha,\beta} [-\mathrm{i}k_\alpha \delta a_\beta(-k) + \mathrm{i}k_\beta \delta a_\alpha(-k)]\,[\mathrm{i}k_\alpha \delta a_\beta(k) - \mathrm{i}k_\beta \delta a_\alpha(k)] \\
&= -\frac{\chi}{4}\sum_k \sum_{\alpha,\beta} \{k_\alpha^2 \delta a_\beta(-k)\delta a_\beta(k) + k_\beta^2 \delta a_\alpha(-k)\delta a_\alpha(k) \\
&\quad - k_\alpha k_\beta\, [\delta a_\alpha(-k)\delta a_\beta(k) + \delta a_\beta(-k)\delta a_\alpha(k)]\,\} \\
&= -\frac{\chi}{2}\sum_k \sum_{\alpha,\beta} \left(|\boldsymbol{k}|^2 \delta_{\alpha\beta} - k_\alpha k_\beta\right) \delta a_\alpha(-k)\delta a_\beta(k) \ . \qquad (6.2.71)
\end{aligned}
$$

This represents the diamagnetic response of the system (because as the energy increases, the system tries to suppress δb), and χ is the so-called diamagnetic susceptibility. (6.2.71) has the same form as the second term in (6.2.68), and $\pi_1(k)$ can be interpreted as

$$
\pi_1^{(\mathrm{dia})}(k) = -\chi|\boldsymbol{k}|^2 \ . \qquad (6.2.72)
$$

Different from (6.2.70), this expression becomes zero for $|\boldsymbol{k}| \to 0$. In this sense, (6.2.70) is an even "stronger" diamagnetism than the normal diamagnetic response. As the reader may have already suspected, (6.2.70) represents the Meissner effect (perfect diamagnetism), signifying that δb cannot enter the probe. [We already used this fact after equation (6.2.51). (6.2.70) provides the mathematical description.]

We obtained the sum of $S_{\mathrm{C.S.}}$ (6.2.59) and S_{eff} (6.2.68) as the action

$$
S[a_\mu, A_\mu] = S_{\mathrm{C.S.}}[a_\mu] + S_{\mathrm{eff}}[a_\mu + A_\mu] \ . \qquad (6.2.73)
$$

Now, we integrate out a_μ. (6.2.68) is gauge independent, and when imposing the Coulomb gauge condition (6.2.58) to a_μ as well as A_μ, $S[a_\mu, A_\mu]$ becomes

$$
\begin{aligned}
S &= \sum_k \frac{1}{4\theta}|\boldsymbol{k}|(a_0(-k)a_{\mathrm{T}}(k) + a_{\mathrm{T}}(-k)a_0(k)) \\
&\quad + \sum_k \frac{1}{2}\pi_0(k)[a_0(-k) + A_0(-k)][a_0(k) + A_0(k)] \\
&\quad + \sum_k \frac{1}{2}\pi_1(k)[a_{\mathrm{T}}(-k) + A_{\mathrm{T}}(-k)][a_{\mathrm{T}}(k) + A_{\mathrm{T}}(k)] \ . \qquad (6.2.74)
\end{aligned}
$$

Completing the square with regard to a_0 and a_{T}, we obtain

$$S[A_\mu] = \frac{1}{2}\sum_k \tilde{\pi}_0(k)A_0(-k)A_0(k) + \frac{1}{2}\sum_k \tilde{\pi}_1(k)A_{\mathrm{T}}(-k)A_{\mathrm{T}}(k)$$
$$+\frac{1}{2}\sum_k \{\tilde{\pi}_{01}(k)A_0(-k)A_{\mathrm{T}}(k) + \tilde{\pi}_0 A_{\mathrm{T}}(-k)A_0(k)\} \quad . \tag{6.2.75}$$

Here, we defined

$$\tilde{\pi}_0(k) = \frac{-\pi_0(k)(|\boldsymbol{k}|/2\theta)^2}{\pi_0(k)\pi_1(k) - (|\boldsymbol{k}|/2\theta)^2} \quad , \tag{6.2.76}$$

$$\tilde{\pi}_1(k) = \frac{-\pi_1(k)(|\boldsymbol{k}|/2\theta)^2}{\pi_0(k)\pi_1(k) - (|\boldsymbol{k}|/2\theta)^2} \quad , \tag{6.2.77}$$

$$\tilde{\pi}_{01}(k) = \tilde{\pi}_{10}(k) = \frac{\pi_0(k)\pi_1(k)(|\boldsymbol{k}|/2\theta)}{\pi_0(k)\pi_1(k) - (|\boldsymbol{k}|/2\theta)^2} \quad . \tag{6.2.78}$$

The terms $\tilde{\pi}_0$, $\tilde{\pi}_1$, $\tilde{\pi}_{01}$ and $\tilde{\pi}_{10}$ express how the responses π_0 and π_1 of the bosonic system alter due to the Chern–Simons term. In the limit $\omega, |\boldsymbol{k}| \to 0$, and ignoring the term $(|\boldsymbol{k}|/2\theta)^2$ in the denominator due to $|\pi_0(k)\pi_1(k)| \gg (|\boldsymbol{k}|/2\theta)^2$, we obtain

$$\tilde{\pi}_0(k) = -\frac{(|\boldsymbol{k}|/2\theta)^2}{\pi_1(k)} \quad , \tag{6.2.79}$$

$$\tilde{\pi}_1(k) = -\frac{(|\boldsymbol{k}|/2\theta)^2}{\pi_0(k)} \quad , \tag{6.2.80}$$

$$\tilde{\pi}_{01}(k) = \tilde{\pi}_{10}(k) = \frac{|\boldsymbol{k}|}{2\theta} \quad . \tag{6.2.81}$$

The transversal (longitudinal) response $\tilde{\pi}_1(k)$ ($\tilde{\pi}_0(k)$) of the system is determined by the longitudinal (transversal) response $\pi_0(k)$ ($\pi_1(k)$) of the bosonic system.

This fact can be understood from the constraint condition (6.2.33) deduced from the Chern–Simons term, that is, the transversal component of the gauge field a_μ is related to the density ρ, which in turn is coupled to the longitudinal component of the external field A_μ. Owing to (6.2.79), the Meissner effect of the bosonic system leads to $\tilde{\pi}_0(k) \propto |k|^2$, signifying that the system is an incompressible quantum liquid. Because the Meissner effect is the repulsion of the "magnetic field" with respect to (6.2.33), we conclude that density fluctuations (in the limit $|\boldsymbol{k}| \to 0$) cannot emerge.

Next, we discuss the significance of the diagonal terms $\tilde{\pi}_{01}$ and $\tilde{\pi}_{10}$. Solving equation (6.2.58) (and replacing a_α by A_α), we obtain

$$A_{\mathrm{T}}(k) = \frac{1}{|\boldsymbol{k}|}(\mathrm{i}k_x A_y(k) - \mathrm{i}k_y A_x(k)) \quad , \tag{6.2.82}$$

and replacing $A_0(k)$ by the general gauge invariant term,

$$A_0(k) + \frac{\omega}{|\boldsymbol{k}|^2} k_\alpha A_\alpha(k) \ , \tag{6.2.83}$$

and finally inserting (6.2.81) into the third term of (6.2.75), the result is

$$\begin{aligned}
&\frac{1}{2}\sum_k \frac{|\boldsymbol{k}|}{2\theta}\left[A_0(-k) + \frac{\omega}{|\boldsymbol{k}|^2}(k_x A_x(-k) + k_y A_y(-k))\right] \\
&\quad \times \frac{1}{|\boldsymbol{k}|}\left[\mathrm{i}k_x A_y(k) - \mathrm{i}k_y A_x(k)\right] + (k \to -k) \\
&= \frac{1}{2}\sum_k \frac{1}{2\theta}\Big\{A_0(-k)\left[\mathrm{i}k_x A_y(k) - \mathrm{i}k_y A_x(k)\right] \\
&\quad + (k \to -k) + \mathrm{i}\omega\left[A_x(-k)A_y(k) - A_y(-k)A_x(k)\right]\Big\} \ .
\end{aligned} \tag{6.2.84}$$

In the limit $|\boldsymbol{k}|$, $\omega \to 0$ and simultaneously $|\boldsymbol{k}|/\omega \to 0$ (this is the limit that is used when determining the transport coefficient) all terms in (6.2.75) as well as the terms in (6.2.84) can be ignored. Then, (6.2.84) can be regarded as $S[A_\mu]$. The current j_x in the x direction is determined by

$$j_x(k) = \frac{\partial S[A_\mu]}{\partial A_x(-k)} = \frac{1}{2\theta}\left[\mathrm{i}k_y A_0(k) + \mathrm{i}\omega A_y(k)\right] \ . \tag{6.2.85}$$

In real space and real time, this reads

$$\begin{aligned}
j_x(r,t) &= \frac{1}{2\theta}\left[\partial_y A_0(r,t) - \frac{\partial}{\partial t}A_y(r,t)\right] \\
&= \frac{1}{2\theta}E_y(r,t) \ .
\end{aligned} \tag{6.2.86}$$

Here, E_y is the y component of the electric field. Comparing this expression with the definition $j_x = \sigma_{xy}E_y$, for σ_{xy} we obtain

$$\sigma_{xy} = \frac{1}{2\theta} = \frac{1}{2\pi(2k+1)} = \frac{\nu}{2\pi} = \frac{\nu e^2}{2\pi\hbar} = \nu\frac{e^2}{h} \ . \tag{6.2.87}$$

Here, we revovered the unit $e^2/\hbar$ of the conductance. This result agrees with (6.2.56).

Above, we discussed the electromagnetic response of the system, where only the Gaussian approximation for the small fluctuations of the bosonic field have been taken into account. However, as was mentioned in Sect. 5.3, vortices play an important role in two-dimensional superconductors. The same thing can be expected for the two-dimensional quantum Hall system which is discussed here. In this case, the vortices correspond to quasi-particles; however, the special feature of these particles is that they obey a fractional number statistics and have fractional charge. In what follows, this will be explained. To conform with Sect. 5.3, we will now use the imaginary time formalism for a while.

For simplicity, we consider a short-range force, and set $\tilde{V}(\boldsymbol{k})$ constant. Then, much the same as in Sect. 5.3, after rescaling, the action becomes (using the unit system $e = \hbar = c = 1$)

$$S_\theta = \int \mathrm{d}^3x \frac{J}{2}(\partial_\mu \theta + a_\mu + A_\mu)^2 \ . \tag{6.2.88}$$

Here, $x = (\boldsymbol{r}, \tau)$ is the three-dimensional coordinate, and a_μ and A_μ are measured from the average value that cancelled out. In the same manner as in (5.3.29), we split

$$\theta(x) = \theta_0(x) + \theta_\mathrm{v}(x) \tag{6.2.89}$$

and define the "vortex current density" $j_\mu(x)$ as was done in (5.3.35):

$$j_\mu(x) = \frac{1}{2\pi}\varepsilon_{\mu\nu\lambda}\partial_\nu\partial_\lambda\theta_\mathrm{v}(x) \ . \tag{6.2.90}$$

Introducing the bosonic current J_μ, using the Stratonovich–Hubbard transformation, and performing the θ_0 integration as in (5.3.44),

$$\partial_\mu J_\mu(x) = 0 \tag{6.2.91}$$

can be deduced.

Introducing the gauge field $\tilde{a}_\mu$ that expresses the bosonic current density, by writing

$$J_\mu(x) = \frac{1}{2\pi}\varepsilon_{\mu\nu\lambda}\partial_\nu\tilde{a}_\lambda \tag{6.2.92}$$

then, in analogy to (5.3.49), the action becomes

$$\begin{aligned} S_\theta = {} & A\sum_i \int \mathrm{d}s_i + \mathrm{i}\int \tilde{a}_\mu(x) j_\mu(x)\,\mathrm{d}^3x + \int \frac{1}{8\pi^2 J}(\nabla_{3\mathrm{D}} \times \tilde{\boldsymbol{a}}_{3\mathrm{D}}(x))^2\,\mathrm{d}^3x \\ & + \mathrm{i}\int \frac{1}{2\pi}\varepsilon_{\mu\nu\lambda}\partial_\nu\tilde{a}_\lambda(a_\mu + A_\mu)\,\mathrm{d}^3x \ . \end{aligned} \tag{6.2.93}$$

The last term on the right-hand side of the above equation is a new term that arises due to the interaction with the gauge field $a_\mu + A_\mu$. The sum of this term with the Chern–Simons term $S_\mathrm{C.S.}$ (6.2.39) in the imaginary time formalism

$$S_\mathrm{C.S.} = -\int \mathrm{d}^3x \frac{\mathrm{i}}{4\theta}\varepsilon_{\mu\nu\lambda}a_\mu\partial_\nu a_\lambda \tag{6.2.94}$$

is the action that will be considered. The action (6.2.93) is expressed in terms of the Chern–Simons gauge field a_μ and the gauge field $\tilde{a}_\mu$ representing the bosonic field, the vortex current density j_μ and the external field A_μ.

Now, we integrate out the Chern–Simons gauge field in the action $S = S_\theta + S_\mathrm{C.S.}$. Because this can be performed using a simple Gaussian integral, it is sufficient to perform a Fourier transformation and to complete the square. The resulting action is given by

$$S = A\sum_i \int \mathrm{d}s_i + \mathrm{i}\int \tilde{a}_\mu(x) j_\mu(x)\,\mathrm{d}^3x - \mathrm{i}\int \frac{\theta}{4\pi^2}\varepsilon_{\mu\nu\lambda}\tilde{a}_\mu\partial_\lambda\tilde{a}_\nu$$
$$+\mathrm{i}\int \frac{1}{2\pi}\varepsilon_{\mu\nu\lambda}\partial_\nu\tilde{a}_\lambda(x)A_\mu(x)\ . \qquad (6.2.95)$$

Here, the term proportional to $(\nabla_{3\mathrm{D}}\times\tilde{\boldsymbol{a}})^2$ in (6.2.93) has been ignored because in the long-wavelength limit, this term becomes small compared with the other terms. The third term on the right-hand side of (6.2.95) corresponds to the Chern–Simons term for the gauge field $\tilde{a}_\mu$ representing the bosonic field. It interacts through $\tilde{a}_\mu j_\mu$ with $\tilde{a}_\mu$, leading to a change in statistics of the vortex. That is, when two vortices are exchanged, owing to the Aharonov–Bohm effect, a factor $\mathrm{e}^{\pm\mathrm{i}\theta_\mathrm{V}}$ emerges. Following our earlier discussion, the phase θ_v is determined by the coefficient of the Chern–Simons term. Therefore, comparing equation (6.2.94) with the third term of (6.2.95), we obtain

$$\frac{1}{4\theta_\mathrm{v}} = \frac{\theta}{4\pi^2}\ , \qquad (6.2.96)$$

and using (6.2.20) we obtain

$$\theta_\mathrm{v} = \frac{\pi}{2k+1}\ . \qquad (6.2.97)$$

Particles, where the phase change is different from 0 or π when two are exchanged, are called anyons.

The action of the vortex can be determined by integrating out $\tilde{a}_\mu$ in (6.2.95):

$$S_\mathrm{v} = \int \mathrm{d}^3x \left\{ -\frac{\mathrm{i}}{4\theta}\varepsilon_{\mu\nu\lambda}A_\mu\partial_\nu A_\lambda - \frac{\mathrm{i}\pi}{\theta}A_\mu j_\mu + \frac{\mathrm{i}\pi^2}{\theta}\varepsilon_{\mu\nu\lambda}j_\mu\frac{\partial_\nu}{\partial^2}j_\lambda \right\}\ . \qquad (6.2.98)$$

For the derivation, it is useful to express $j_\mu = \frac{1}{2\pi}\varepsilon_{\mu\nu\lambda}\partial_\nu b_\lambda$ with b_λ using $\partial_\mu j_\mu = 0$, and notice that $\varepsilon_{\mu\nu\lambda}\partial_\nu j_\lambda = -\frac{1}{2\pi}\partial^2 b_\mu$ holds. The second term on the right-hand side represents the interaction with the electromagnetic field, and from the coefficient we read off the charge of the vortex:

$$e_\mathrm{v} = -\frac{\pi}{\theta} = -\frac{1}{2k+1}\left(= -\frac{e}{2k+1}\right)\ . \qquad (6.2.99)$$

In such a manner, the charge becomes a fractional value. The third term on the right-hand side of (6.2.98) is the so-called Hopf interaction representing the statistics transmutation through the non-local interaction, without explicit use of the Chern–Simons gauge field. The above considerations go through in a totally parallel manner for the antivortex, and in this case the charge becomes $e_\mathrm{v} = e/(2k+1)$.

We conclude that the quasi-particles of the fractional quantum Hall system are anyons with fractional charge $e_{\rm v} = \mp e/(2k+1)$ that change their phase only by $\theta_{\rm v} = \pi/(2k+1)$ when two particles are exchanged. In a direct manner, this can be interpreted as follows. Setting $A_\mu = 0$, then due to the Meissner effect of the bosonic system, the gauge flux b is quantized as a multiple of ϕ_0. However, due to (6.2.33), the charge that is associated with $\pm\phi_0$ is $e_{\rm v} = \mp e(2k+1)$, and therefore the Aharonov–Bohm phase also becomes $1/(2k+1)$, which is the characteristic of anyons. Because one electron and one hole correspond to $\pm(2k+1)\phi_0$, respectively, the quasi-particle can be regarded as being constructed from one electron or one hole divided in $(2k+1)$ parts.

In this section, the effective theory of the quantum Hall system has been discussed. In the next section, we demonstrate that it is possible to construct the Laughlin wave function that was introduced at the end of Sect. 6.1 using this effective theory.

6.3 The Derivation of the Laughlin Wave Function

As has been mentioned after equation (6.2.60), when the order of integration is changed, and a_μ is integrated out first, $S[\theta, A_\mu]$ can be determined. In what follows we set the external field to zero, $A_\mu = 0$. Again, we start with the action

$$S = S_{\rm C.S.}[a_\mu] = S_\theta[\theta, a_\mu, A_\mu = 0] \ . \tag{6.3.1}$$

With the knowledge gained so far, the integration of a_μ should be simple. The derivation is left as an exercise to the reader; the result $S[\theta, A_\mu = 0]$ is given by

$$S[\theta, A_\mu = 0] = \sum_{\boldsymbol{k}} \frac{1}{2} \left[\frac{\omega^2}{\tilde{V}(\boldsymbol{k}) + (\rho_0/4m)(4\theta/|\boldsymbol{k}|)^2} - \frac{\rho_0}{m}|\boldsymbol{k}|^2 \right] \theta(-\boldsymbol{k})\theta(\boldsymbol{k}) \ . \tag{6.3.2}$$

We now return, with respect to ω, to the real time formalism and obtain

$$\begin{aligned} S[\theta, A_\mu = 0] = \int \mathrm{d}t \sum_{\boldsymbol{k}} \frac{1}{2} \Bigg\{ & \frac{\partial_t\theta(-\boldsymbol{k},t)\partial_t\theta(\boldsymbol{k},t)}{\tilde{V}(\boldsymbol{k}) + (\rho_0/4m)(4\theta/|\boldsymbol{k}|)^2} \\ & -\frac{\rho_0}{m}|\boldsymbol{k}|^2\theta(-\boldsymbol{k},t)\theta(\boldsymbol{k},t) \Bigg\} \ . \end{aligned} \tag{6.3.3}$$

Because, originally, $\theta(\boldsymbol{r},t)$ was a real field, the relation $\theta(-\boldsymbol{k},t) = [\theta(\boldsymbol{k},t)]^*$ is valid. Now, we restrict the summation range of $\boldsymbol{k}$ to the half and introduce the real part θ' and the imaginary part θ'':

$$\begin{aligned} \theta(\boldsymbol{k},t) &= \theta'(\boldsymbol{k},t) + \mathrm{i}\theta''(\boldsymbol{k},t) \ , \\ \theta(-\boldsymbol{k},t) &= \theta'(\boldsymbol{k},t) - \mathrm{i}\theta''(\boldsymbol{k},t) \ . \end{aligned} \tag{6.3.4}$$

Then, (6.3.3) becomes

$$S[\theta, A_\mu = 0] = \int \mathrm{d}t \sum_{\boldsymbol{k}:\,\text{half}} \left\{A(\boldsymbol{k})[\partial_t\theta'(\boldsymbol{k},t)]^2 - B(\boldsymbol{k})[\theta'(\boldsymbol{k},t)]^2\right\}$$
$$+ \int \mathrm{d}t \sum_{\boldsymbol{k}:\,\text{half}} \left\{A(\boldsymbol{k})[\partial_t\theta''(\boldsymbol{k},t)]^2 - B(\boldsymbol{k})[\theta''(\boldsymbol{k},t)]^2\right\} \quad (6.3.5)$$

the sum of harmonic oscillators. Here, we defined

$$A(\boldsymbol{k}) = \frac{1}{\tilde{V}(\boldsymbol{k}) + (\rho_0/4m)(4\theta/|\boldsymbol{k}|)^2} \ , \quad (6.3.6)$$

$$B(\boldsymbol{k}) = \frac{\rho_0}{m}|\boldsymbol{k}|^2 \ . \quad (6.3.7)$$

We now derive the Hamiltonian from (6.3.5). The canonical conjugate momenta of θ' and θ'' are given by, respectively

$$\begin{aligned} P'(\boldsymbol{k}) &= \frac{\partial L}{\partial(\dot{\theta}'(\boldsymbol{k}))} = 2A(\boldsymbol{k})\dot{\theta}'(\boldsymbol{k}) \ , \\ P''(\boldsymbol{k}) &= \frac{\partial L}{\partial(\dot{\theta}''(\boldsymbol{k}))} = 2A(\boldsymbol{k})\dot{\theta}''(\boldsymbol{k}) \ . \end{aligned} \quad (6.3.8)$$

The Hamiltonian becomes

$$\begin{aligned} H &= \sum_{\boldsymbol{k}:\,\text{half}} \left\{P'(\boldsymbol{k})\dot{\theta}'(\boldsymbol{k}) + P''(\boldsymbol{k})\dot{\theta}''(\boldsymbol{k})\right\} - L \\ &= \sum_{\boldsymbol{k}:\,\text{half}} \left\{\frac{1}{4A(\boldsymbol{k})}\left[P'(\boldsymbol{k})\right]^2 + B(\boldsymbol{k})\left[\theta'(\boldsymbol{k})\right]^2\right\} \\ &\quad + \sum_{\boldsymbol{k}:\,\text{half}} \left\{\frac{1}{4A(\boldsymbol{k})}\left[P''(\boldsymbol{k})\right]^2 + B(\boldsymbol{k})\left[\theta''(\boldsymbol{k})\right]^2\right\} \ . \end{aligned} \quad (6.3.9)$$

We interpret this Hamiltonian as an operator and want to derive its ground state. θ' and P', θ'' and P'' also become operators, with the commutation relations

$$\begin{aligned} [\theta'(\boldsymbol{k}), P'(\boldsymbol{k}')] &= \mathrm{i}\delta_{\boldsymbol{k},\boldsymbol{k}'} \ , \\ [\theta''(\boldsymbol{k}), P''(\boldsymbol{k}')] &= \mathrm{i}\delta_{\boldsymbol{k},\boldsymbol{k}'} \ . \end{aligned} \quad (6.3.10)$$

Therefore, it is possible to express θ' and θ'' as

$$\begin{aligned} \theta'(\boldsymbol{k}) &= \mathrm{i}\frac{\partial}{\partial P'(\boldsymbol{k})} \ , \\ \theta''(\boldsymbol{k}) &= \mathrm{i}\frac{\partial}{\partial P''(\boldsymbol{k})} \ . \end{aligned} \quad (6.3.11)$$

The Hamiltonian can then be written as

$$H = \sum_{\boldsymbol{k}:\,\text{half}} \left\{ -B(\boldsymbol{k}) \left(\frac{\partial}{\partial P'(\boldsymbol{k})} \right)^2 + \frac{1}{4A(\boldsymbol{k})} [P'(\boldsymbol{k})]^2 \right\} + \sum_{\boldsymbol{k}:\,\text{half}} \left\{ -B(\boldsymbol{k}) \left(\frac{\partial}{\partial P''(\boldsymbol{k})} \right)^2 + \frac{1}{4A(\boldsymbol{k})} [P''(\boldsymbol{k})]^2 \right\} . \quad (6.3.12)$$

Expressed in $P'(\boldsymbol{k})$ and $P''(\boldsymbol{k})$, the wave function of the ground state is given by

$$\Psi_{\text{g}} [\{P'(\boldsymbol{k})\}, \{P''(\boldsymbol{k})\}] \propto \exp \left[-\frac{1}{2} \sum_{\boldsymbol{k}:\,\text{half}} \frac{1}{2\sqrt{A(\boldsymbol{k})B(\boldsymbol{k})}} \left\{ [P'(\boldsymbol{k})]^2 + [P''(\boldsymbol{k})]^2 \right\} \right] . \quad (6.3.13)$$

Especially for the long wavelength approximation, due to (6.3.6) and (6.3.7)

$$A(\boldsymbol{k})B(\boldsymbol{k}) \cong \frac{|\boldsymbol{k}|^4}{(2\theta)^2} \quad (6.3.14)$$

holds, and we obtain

$$\Psi_{\text{g}} [\{P'(\boldsymbol{k})\}, \{P''(\boldsymbol{k})\}] \propto \exp \left[-\frac{\theta}{2} \sum_{\boldsymbol{k}:\,\text{half}} \frac{1}{|\boldsymbol{k}|^2} \left\{ [P'(\boldsymbol{k})]^2 + [P''(\boldsymbol{k})]^2 \right\} \right] = \exp \left[-\frac{\theta}{4} \sum_{\boldsymbol{k}} \frac{1}{|\boldsymbol{k}|^2} P(-\boldsymbol{k}) P(\boldsymbol{k}) \right] . \quad (6.3.15)$$

Here, we introduced

$$P(\boldsymbol{k}) = P'(\boldsymbol{k}) + \mathrm{i} P''(\boldsymbol{k}) . \quad (6.3.16)$$

The real field $P(\boldsymbol{r})$ is obtained by inverse Fourier transformation.

So, what might be the form of $P(\boldsymbol{r})$ itself? In order to find the answer, we calculate the following commutator. Recalling $P(-\boldsymbol{k}) = P'(\boldsymbol{k}) - \mathrm{i}P''(\boldsymbol{k})$, we obtain

$$[\theta(\boldsymbol{k}), P(\boldsymbol{k}')] = [\theta'(\boldsymbol{k}) + \mathrm{i}\theta''(\boldsymbol{k}), P'(\boldsymbol{k}') + \mathrm{i}P''(\boldsymbol{k}')] = 2\mathrm{i}\delta_{\boldsymbol{k},-\boldsymbol{k}'} \quad (6.3.17)$$

and therefore

$$[\theta(\boldsymbol{r}), P(\boldsymbol{r}')] = 2\mathrm{i}\delta(\boldsymbol{r} - \boldsymbol{r}') . \quad (6.3.18)$$

On the other hand, as has been emphasized already, between the phase $\theta(\boldsymbol{r})$ and the density $\rho(\boldsymbol{r})$ the canonical commutation relation

$$[\rho(\boldsymbol{r}'), \theta(\boldsymbol{r})] = \mathrm{i}\delta(\boldsymbol{r} - \boldsymbol{r}') \tag{6.3.19}$$

holds. Comparing (6.3.18) with (6.3.19), we obtain

$$P(\boldsymbol{r}) = -2\rho(\boldsymbol{r}) \ . \tag{6.3.20}$$

In the framework of first quantization, we have

$$\rho(\boldsymbol{r}) = \sum_{i=1}^{N} \delta(\boldsymbol{r} - \boldsymbol{r}_i) \tag{6.3.21}$$

and the Fourier transformation is given by

$$\rho(\boldsymbol{k}) = \frac{1}{\sqrt{S}} \sum_{i=1}^{N} \mathrm{e}^{-\mathrm{i}\boldsymbol{k}\cdot\boldsymbol{r}_i} \ . \tag{6.3.22}$$

Here, S is the surface of the system. Inserting (6.3.20) into (6.3.15) we obtain

$$\Psi_{\mathrm{g}}(\boldsymbol{r}_1, \ldots, \boldsymbol{r}_N) \propto \exp\left[-\theta \sum_{\boldsymbol{k}} \frac{1}{|\boldsymbol{k}|^2} \rho(-\boldsymbol{k})\rho(\boldsymbol{k})\right] \ . \tag{6.3.23}$$

We rewrite the integral as an integral in real space. For this purpose, we consider the following Fourier integral:

$$\int \mathrm{d}^2\boldsymbol{k} \frac{\mathrm{e}^{\mathrm{i}\boldsymbol{k}\cdot\boldsymbol{r}}}{|\boldsymbol{k}|^2} = \int_0^{2\pi} \mathrm{d}\theta \int_0^{\infty} k\, \mathrm{d}k \frac{\mathrm{e}^{\mathrm{i}kr\cos\theta}}{k^2} \equiv \int_0^{2\pi} \mathrm{d}\theta\, f(r\cos\theta) \ . \tag{6.3.24}$$

We set $r = |\boldsymbol{r}|$. Taking the derivative of $f(\xi)$ with respect to ξ, we obtain

$$\frac{\partial}{\partial\xi} f(\xi) = \frac{\partial}{\mathrm{d}\xi} \int_0^{\infty} \frac{\mathrm{e}^{\mathrm{i}k\xi}}{k} \,\mathrm{d}k = \lim_{\varepsilon\to 0+} \int_0^{\infty} \mathrm{i}\, \mathrm{e}^{\mathrm{i}k\xi - \varepsilon k}\, \mathrm{d}k = -\frac{1}{\xi} \ . \tag{6.3.25}$$

However, because the k-integral in $f(\xi)$ is infrared divergent, it is not well defined. When the divergence is fixed, for example in $f(\xi_0)$ at $\xi = \xi_0$, we obtain

$$f(\xi) = f(\xi_0) + \ln \frac{\xi_0}{\xi} \ . \tag{6.3.26}$$

Therefore, with r_0 being some constant we obtain

$$\int \mathrm{d}^2\boldsymbol{k} \frac{\mathrm{e}^{\mathrm{i}\boldsymbol{k}\cdot\boldsymbol{r}}}{|\boldsymbol{k}|^2} = 2\pi \ln\left(\frac{r_0}{r}\right) \ . \tag{6.3.27}$$

Inverting the Fourier transformation in (6.3.27), we obtain

$$\frac{1}{|\boldsymbol{k}|^2} = \int \frac{\mathrm{d}^2\boldsymbol{r}}{2\pi} \ln \frac{r_0}{r}\, \mathrm{e}^{-\mathrm{i}\boldsymbol{k}\cdot\boldsymbol{r}} \ . \tag{6.3.28}$$

We insert this expression and

$$\rho(\boldsymbol{k}) = \frac{1}{\sqrt{S}} \int \mathrm{d}^2\boldsymbol{r}\, \mathrm{e}^{-\mathrm{i}\boldsymbol{k}\cdot\boldsymbol{r}} [\rho(\boldsymbol{r}) - \bar{\rho}] \tag{6.3.29}$$

into the argument of the exponential in (6.3.23). (The average density value $\bar{\rho}$ has been subtracted in (6.3.29) because then, in the $\boldsymbol{k}$-summation in (6.2.23), the term $\boldsymbol{k} = 0$ is omitted.)

$$\begin{aligned}
-\theta \sum_{\boldsymbol{k}} \frac{1}{|\boldsymbol{k}|^2} \rho(-\boldsymbol{k})\rho(\boldsymbol{k}) &= -\theta \int \frac{\mathrm{d}^2\boldsymbol{k}}{(2\pi)^2} \int \mathrm{d}^2\boldsymbol{r}'' \int \mathrm{d}^2\boldsymbol{r} \int \mathrm{d}^2\boldsymbol{r}' \frac{1}{2\pi} \ln \frac{r_0}{|\boldsymbol{r}''|} \\
&\quad \times [\rho(\boldsymbol{r}) - \bar{\rho}][\rho(\boldsymbol{r}') - \bar{\rho}]\, \mathrm{e}^{-\mathrm{i}\boldsymbol{k}\cdot\boldsymbol{r}'' + \mathrm{i}\boldsymbol{k}\cdot\boldsymbol{r} - \mathrm{i}\boldsymbol{k}\cdot\boldsymbol{r}'} \\
&= -\frac{\theta}{2\pi} \int \mathrm{d}^2\boldsymbol{r}'' \int \mathrm{d}^2\boldsymbol{r} \int \mathrm{d}^2\boldsymbol{r}'\, \delta^2(\boldsymbol{r}'' - \boldsymbol{r} + \boldsymbol{r}') \\
&\quad \times \ln \frac{r_0}{|\boldsymbol{r}''|} [\rho(\boldsymbol{r}) - \bar{\rho}][\rho(\boldsymbol{r}') - \bar{\rho}] \\
&= \frac{\theta}{2\pi} \int \mathrm{d}^2\boldsymbol{r} \int \mathrm{d}^2\boldsymbol{r}' [\rho(\boldsymbol{r}) - \bar{\rho}] \ln |\boldsymbol{r} - \boldsymbol{r}'| [\rho(\boldsymbol{r}') - \bar{\rho}] \ .
\end{aligned} \tag{6.3.30}$$

Inserting now (6.3.21), we obtain

$$\text{Eq. (6.3.30)} = \frac{\theta}{2\pi} \sum_{\substack{i,j \\ (i \neq j)}} \ln |\boldsymbol{r}_i - \boldsymbol{r}_j| - \frac{\theta\bar{\rho}}{\pi} \sum_i \int \mathrm{d}^2\boldsymbol{r} \ln |\boldsymbol{r} - \boldsymbol{r}_i| + \text{const.} \tag{6.3.31}$$

We now perform the integration of the second term at the right-hand side. Because in the case when the integral is calculated as it stands the contribution for $|\boldsymbol{r}| \to \infty$ is divergent, we consider instead

$$I = \int \mathrm{d}^2\boldsymbol{r} \ln \frac{|\boldsymbol{r} - \boldsymbol{r}_i|}{|\boldsymbol{r}|} \ . \tag{6.3.32}$$

Using polar coordinates, we obtain

$$I = \int_0^\infty r\, \mathrm{d}r \int_0^{2\pi} \mathrm{d}\theta \frac{1}{2} \ln \left(\frac{r^2 + r_i^2 - 2rr_i \cos\theta}{r^2} \right) \ . \tag{6.3.33}$$

The angular variable can be integrated as follows. The derivative of

$$f(a) = \int_0^{2\pi} \mathrm{d}\theta \ln(1 + a^2 - 2a\cos\theta) \tag{6.3.34}$$

with respect to a is given by

$$\frac{\mathrm{d}f(a)}{\mathrm{d}a} = \int_0^{2\pi} \mathrm{d}\theta \frac{2(a - \cos\theta)}{1 + a^2 - 2a\cos\theta} \ . \tag{6.3.35}$$

This integral can be performed in the complex plane by setting $z = \mathrm{e}^{\mathrm{i}\theta}$. The result is

$$\frac{\mathrm{d}f(a)}{\mathrm{d}a}=\frac{2\pi}{a}\left[1+\frac{a^2-1}{|a^2-1|}\right]=\begin{cases}0 & (|a|<1)\\ \dfrac{4\pi}{a} & (|a|>1)\end{cases}\ . \tag{6.3.36}$$

With $f(0)=0$ and (6.3.36), we obtain

$$f(a)=\begin{cases}0 & (0<a<1)\\ 4\pi\ln a & (a>1)\end{cases}\ . \tag{6.3.37}$$

Applying this formula to (6.3.33) by choosing $a=r/r_i$, the result is

$$I=\int_0^{r_i} r\,\mathrm{d}r\cdot 2\pi\ln\left(\frac{r_i}{r}\right)=2\pi r_i^2\int_0^1 x\,\mathrm{d}x\cdot\ln\frac{1}{x}=\frac{\pi}{2}r_i^2\ , \tag{6.3.38}$$

and finally (6.3.31) becomes

$$\frac{\theta}{\pi}\sum_{i<j}\ln|\boldsymbol{r}_i-\boldsymbol{r}_j|-\frac{\theta\bar{\rho}}{2}\sum_i|\boldsymbol{r}_i|^2+\text{const.} \tag{6.3.39}$$

Because, due to (6.2.40) the relations $\theta\bar{\rho}=B/2$ and $\theta=\pi(2k+1)$ hold, with C being a constant, Ψ_{g} becomes

$$\Psi_{\mathrm{g}}(\boldsymbol{r}_1,\dots,\boldsymbol{r}_N)=C\prod_{i<j}|\boldsymbol{r}_i-\boldsymbol{r}_j|^{1/\nu}\exp\left[-\frac{eB}{4}\sum_i|\boldsymbol{r}_i|^2\right]\ . \tag{6.3.40}$$

Here, we reintroduced e. In terms of the complex coordinates $z_i=x_i+iy_i$, the expression becomes

$$\Psi_{\mathrm{g}}(z_1,\dots,z_N)=C\prod_{i<j}|z_i-z_j|^{1/\nu}\exp\left[-\frac{eB}{4}\sum_i|z_i|^2\right]\ . \tag{6.3.41}$$

This is the wave function of the bosons.

The wave function of the fermions can be obtained from this expression by applying the unitary transformation U defined in (6.2.11). In terms of complex coordinates, U can be expressed as

$$U=\prod_{i>j}\frac{(z_i-z_j)^{2k+1}}{|z_i-z_j|^{2k+1}}\ . \tag{6.3.42}$$

Therefore, the wave function of the fermions is given by

$$\begin{aligned}\Psi_{(\text{fermion})}(z_1,\dots,z_N)&=U\Psi_{\mathrm{g}}(z_1,\dots,z_N)\\ &=C\prod_{i>j}(z_i-z_j)^{2k+1}\exp\left[-\frac{eB}{4}\sum_i|z_i|^2\right]\ .\end{aligned} \tag{6.3.43}$$

This is nothing but the wave function (6.1.49) given by Laughlin.

Therefore, we understood that the quantum Hall liquid is an new kind of incompressible quantum liquid that is connected through a statistical transformation to phase coherence (superfluid, superconductor), and that the Laughlin wave function is the correct description for it.

Appendix

A. Fourier Transformation

We consider a function $f(x)$ with periodicity L, $f(x+L) = f(x)$. Therefore, the function is determined for all x when it is fixed in the region $0 \leq x \leq L$.

Now, we discretize x

$$x_i = \mathrm{i}\Delta x \qquad (i = 0, 1, 2, \ldots, N) \ . \tag{A.1}$$

We set $x_N = N\Delta x = L$. Then, for the N different values $x_1, \ldots, x_N$ of x, the value of the function $f(x)$ is determined. We arrange these values in one vector $\boldsymbol{f} = (f(x_1), \ldots, f(x_N))^t$. We transform this vector with the following unitary matrix U:

$$U_{lj} = \frac{1}{\sqrt{N}} \exp\left[-\mathrm{i}\frac{2\pi l}{N} j\right] \ . \tag{A.2}$$

The proof of the unitarity of U follows from

$$\begin{aligned}\sum_{j=1}^{N} U_{lj} U^{\dagger}_{jl'} &= \sum_{j=1}^{N} U_{lj} U^{*}_{l'j} \\ &= \sum_{j=1}^{N} \frac{1}{N} \exp\left[-\mathrm{i}\frac{2\pi(l-l')}{N} j\right] = \delta_{l,l'} \ .\end{aligned} \tag{A.3}$$

Using this transformation, a new vector is obtained:

$$\boldsymbol{F} = U\boldsymbol{f} \ . \tag{A.4}$$

In components, we obtain

$$F_l = \sum_{j=1}^{N} \frac{1}{\sqrt{N}} \exp\left[-\mathrm{i}\frac{2\pi l}{N} j\right] f(x_j) \ . \tag{A.5}$$

Introducing the wave number $k_l = 2\pi l/L = 2\pi l/N\Delta x$, we obtain, instead of (A.5)

$$F(k_l) = \sum_{j=1}^{N} \frac{1}{\sqrt{N}} \exp[-\mathrm{i}k_l \cdot x_j] f(x_j) \ . \tag{A.6}$$

Writing the inverse transformation to (A.4)

$$\boldsymbol{f} = U^{-1}\boldsymbol{F} = U^{\dagger}\boldsymbol{F} \tag{A.7}$$

in components, we obtain

$$f(x_j) = \sum_{l=1}^{N} \frac{1}{\sqrt{N}} \exp[\mathrm{i}k_l \cdot x_j] F(k_l) \ . \tag{A.8}$$

Equations (A.6) and (A.8) are the (discrete) Fourier transformation of the function $f(x)$. Now, we take the limits $\Delta x \to \infty$ and $N \to \infty$. The wave number k is discretized in steps of $2\pi/L$, and the integer l reaches all values from $-\infty$ to $+\infty$. (When performing naively the limit $N \to \infty$ for $l = 1, \ldots, N$, the reader might think that it is sufficient to take only positive integers for l. However, a more precise mathematical proof shows that all integers for l have to be included. Here, we accept this as a matter of fact.)

Then, (A.8) becomes

$$f(x) = \sum_{l=-\infty}^{\infty} \frac{1}{\sqrt{L}} \mathrm{e}^{\mathrm{i}k_l x} \tilde{F}(k_l) \ . \tag{A.9}$$

Here, we have

$$\tilde{F}(k_l) = \sqrt{\Delta x} F(k_l) = \Delta x \sum_{j=1}^{N} \frac{1}{\sqrt{L}} \mathrm{e}^{-\mathrm{i}k_l x_j} f(x_j) \ . \tag{A.10}$$

Taking the limit $\Delta x \to 0$ in this equation, we obtain

$$\tilde{F}(k_l) = \int_0^L \mathrm{d}x \frac{1}{\sqrt{L}} \mathrm{e}^{-\mathrm{i}k_l x} f(x) \ . \tag{A.11}$$

Equation (A.9) as well as (A.11) is the Fourier transformation in a finite space interval.

Next, we derive the Fourier transformation in infinite space. Symmetrizing (A.11) around the origin of x

$$\hat{F}(k_l) = \sqrt{\frac{L}{2\pi}} \tilde{F}(k_l) = \int_{-L/2}^{L/2} \frac{\mathrm{d}x}{\sqrt{2\pi}} \mathrm{e}^{-\mathrm{i}k_l x} f(x) \tag{A.12}$$

and taking the limit $L \to \infty$, we define

$$\hat{F}(k) = \int_{-\infty}^{\infty} \frac{\mathrm{d}x}{\sqrt{2\pi}} \mathrm{e}^{-\mathrm{i}kx} f(x) \ . \tag{A.13}$$

Equation (A.9) becomes

$$f(x) = \frac{2\pi}{L}\sum_{l=-\infty}^{\infty} e^{ik_l x}\frac{1}{\sqrt{2\pi}}\sqrt{\frac{L}{2\pi}}\tilde{F}(k_l)$$
$$= \frac{2\pi}{L}\sum_{l=-\infty}^{\infty} \frac{e^{ik_l x}}{\sqrt{2\pi}}\hat{F}(k_l) \tag{A.14}$$

and again taking the limit $L \to \infty$, we obtain

$$f(x) = \int_{-\infty}^{\infty} \frac{dk}{\sqrt{2\pi}} e^{ikx}\hat{F}(k) \ . \tag{A.15}$$

Equations (A.13) and (A.15) are the Fourier transformation in infinite space.

B. Functionals and the Variation Principle

Usually, a function f provides a correspondence between a real number x and a real number $f(x)$:

$$f : x \in R \to f(x) \in R \ . \tag{B.1}$$

On the other hand, a real number $F(\{f(x)\})$ that corresponds to a function $f(x)$ defined in the interval $\alpha \leq x \leq \beta$ is called a functional:

$$F : f(x) \in \mathcal{H} \to F(\{f(x)\}) \in R \ . \tag{B.2}$$

Here, $\mathcal{H}$ is a space the elements of which are functions, i.e. a functional space. This space can be considered as the Hilbert space introduced in Sect. 1.1. When x takes discrete values $x_1 = \alpha < x_2 < \cdots < x_N = \beta$, then the functional can be considered as a function of the N numbers $f_i = f(x_i)$

$$F(\{f(x)\}) \to F(f_1, \ldots, f_N) \ . \tag{B.3}$$

Conversely, in the limit where the distance between x_i and x_{i+1} becomes small, that is, for $N \to \infty$, $F(f_1, \ldots, f_N)$ approaches $F(\{f(x)\})$. In this sense, $F(\{f(x)\})$ can be interpreted as a function of infinitely many variables.

Next, for physical applications, the following integral functional given in the form of an integral is of importance:

$$F(\{f(x)\}) = \int_{\alpha}^{\beta} K\left(f(x), \frac{df(x)}{dx}, x\right) dx \ . \tag{B.4}$$

Here, K is a function $K(a, b, c)$ of three variables, and we obtain $F(\{f(x)\})$ when we insert $a = f(x)$, $b = df(x)/dx$, $c = x$ and integrate with respect to x. Considering again the discretized version, we obtain

$$F(f_1,\ldots,f_N)=\sum_{i=1}^{N-1}K\left(f_i,\frac{f_{i+1}-f_i}{x_{i+1}-x_i},x_i\right)(x_{i+1}-x_i)\ . \tag{B.5}$$

It should be clear that the f_i dependence is quite restricted.

Now, we alter slightly the function $f(x)$ in (B.4) to $f(x)+\delta f(x)$. In terms of (B.5), this corresponds to $f_i \to f_i+\delta f_i$. We examine how F is altered by this infinitesimal transformation. First, in the discrete case, directly from (B.5) we obtain

$$\begin{aligned}\delta F &= F(f_1+\delta f_1,\ldots,f_N+\delta f_N)-F(f_1,\ldots,f_N)\\ &\simeq \sum_{i=1}^{N-1}\left\{\left.\frac{\partial K(a,b,c)}{\partial a}\right|_i \delta f_i+\left.\frac{\partial K(a,b,c)}{\partial b}\right|_i \frac{\delta f_{i+1}-\delta f_i}{x_{i+1}-x_i}\right\}(x_{i+1}-x_i)\ .\end{aligned} \tag{B.6}$$

Here, we ignored terms of second and higher order in δf_i and introduced the notation

$$\left.\frac{\partial K(a,b,c)}{\partial a}\right|_i=\left.\frac{\partial K(a,b,c)}{\partial a}\right|_{a=f_i,b=\frac{f_{i+1}-f_i}{x_{i+1}-x_i},c=x_i}\ . \tag{B.7}$$

In what follows, for simplicity we assume that $x_{i+1}-x_i=\Delta x$ is independent of i.

Rewriting (B.6) slightly, we obtain

$$\begin{aligned}\delta F &= \Delta x\sum_{i=1}^{N-1}\left.\frac{\partial K(a,b,c)}{\partial a}\right|_i\delta f_i+\sum_{i=2}^{N-1}\left(\left.\frac{\partial K(a,b,c)}{\partial b}\right|_{i-1}-\left.\frac{\partial K(a,b,c)}{\partial b}\right|_i\right)\delta f_i\\ &\quad+\left.\frac{\partial K(a,b,c)}{\partial b}\right|_{N-1}\delta f_N-\left.\frac{\partial K(a,b,c)}{\partial b}\right|_1\delta f_1\ .\end{aligned} \tag{B.8}$$

The extreme value of F is obtained for $(f_1,\ldots,f_N)$ when, for arbitrary $\delta f_1,\ldots,\delta f_N$, the condition $\delta F=0$

$$\left.\frac{\partial K(a,b,c)}{\partial a}\right|_i-\left(\left.\frac{\partial K(a,b,c)}{\partial b}\right|_i-\left.\frac{\partial K(a,b,c)}{\partial b}\right|_{i-1}\right)=0\quad(2\leqq i\leqq N-1)\ , \tag{B.9a}$$

$$\Delta x\left.\frac{\partial K(a,b,c)}{\partial a}\right|_1-\left.\frac{\partial K(a,b,c)}{\partial b}\right|_1=0\ , \tag{B.9b}$$

$$\left.\frac{\partial K(a,b,c)}{\partial b}\right|_{N-1}=0 \tag{B.9c}$$

is fulfilled. In the case when the functional is extremized under the condition that the value of f at the start point f_1 and end point f_N is fixed, (B.9b) and (B.9c) drop out, and it is sufficient to solve the $N-2$ equations for the undetermined $f_2,\ldots,f_{N-1}$.

The above consideration in the continuum limit $N \to \infty$, $\Delta x \to 0$ is called the variation principle. Analogous to (B.6)–(B.9), the variation δF of F is given by

$$\begin{aligned}\delta F &= F(\{f(x)+\delta f(x)\}) - F(\{f(x)\}) \\ &= \int_\alpha^\beta \mathrm{d}x \left\{ K\left(f(x)+\delta f(x), \frac{\mathrm{d}f(x)}{\mathrm{d}x} + \frac{\mathrm{d}\delta f(x)}{\mathrm{d}x}, x\right) \right. \\ &\quad \left. - K\left(f(x), \frac{\mathrm{d}f(x)}{\mathrm{d}x}, x\right)\right\} \\ &= \int_\alpha^\beta \mathrm{d}x \left(\left.\frac{\partial K(a,b,c)}{\partial a}\right|_x \delta f(x) + \left.\frac{\partial K(a,b,c)}{\partial b}\right|_x \frac{\mathrm{d}}{\mathrm{d}x}(\delta f(x)) \right\} . \end{aligned} \tag{B.6$'$}$$

Here, we introduced

$$\left.\frac{\partial K(a,b,c)}{\partial a}\right|_x = \left.\frac{\partial K(a,b,c)}{\partial a}\right|_{a=f(x),\, b=\frac{df(x)}{dx},\, c=x} . \tag{B.7$'$}$$

Performing partial integration, we obtain

$$\begin{aligned}\delta F = &\int_\alpha^\beta \mathrm{d}x \left\{ \left.\frac{\partial K(a,b,c)}{\partial a}\right|_x - \frac{\mathrm{d}}{\mathrm{d}x}\left(\left.\frac{\partial K(a,b,c)}{\partial b}\right|_x \right)\right\} \delta f(x) \\ &+ \left.\frac{\partial K(a,b,c)}{\partial b}\right|_\beta \delta f(\beta) - \left.\frac{\partial K(a,b,c)}{\partial b}\right|_\alpha \delta f(\alpha) . \end{aligned} \tag{B.8$'$}$$

The condition $\delta F = 0$ leads to

$$\left.\frac{\partial K(a,b,c)}{\partial a}\right|_x - \frac{\mathrm{d}}{\mathrm{d}x}\left(\left.\frac{\partial K(a,b,c)}{\partial b}\right|_x \right) = 0 , \tag{B.9$'$a}$$

$$-\left.\frac{\partial K(a,b,c)}{\partial b}\right|_\alpha = 0 , \tag{B.9$'$b}$$

$$\left.\frac{\partial K(a,b,c)}{\partial b}\right|_\beta = 0 . \tag{B.9$'$c}$$

Equation (B.9′) is called the Euler–Lagrange equation, obtained by extremizing the variation of F.

Next, we discuss the derivative of a functional. First, in the discretized case, we can write

$$\delta F = F(f_1+\delta f_1, \ldots, f_N + \delta f_N) - F(f_1, \ldots, f_N) \approx \sum_{i=1}^N \frac{\partial F(f_1, \ldots, f_N)}{\partial f_i} \delta f_i . \tag{B.10}$$

Rewriting this slightly,

$$\delta F \simeq \Delta x \sum_{i=1}^{N} \left(\frac{1}{\Delta x} \frac{\partial F(f_1, \ldots, f_N)}{\partial f_i} \right) \delta f_i \tag{B.11}$$

in the continuum limit $\Delta x \to 0$, $N \to \infty$, we obtain

$$\delta F = \int_{\alpha}^{\beta} \mathrm{d}x \frac{\delta F(\{f(x)\})}{\delta f(x)} \delta f(x) \ . \tag{B.12}$$

Here,

$$\frac{\partial F(\{f(x)\})}{\delta f(x)} \equiv \lim_{\Delta x \to 0} \frac{1}{\Delta x} \frac{\partial F(f_1, \ldots, f_N)}{\partial f_i} \tag{B.13}$$

is the so-called functional derivative. Setting, for example, $F(\{f(x)\}) = f(x_0)$ with $x_{i_0} = x_0$:

$$\frac{\partial F(\{f(x)\})}{\partial f(x)} = \lim_{\Delta x \to 0} \frac{1}{\Delta x} \frac{\partial f_{i_0}}{\partial f_i} = \lim_{\Delta x \to 0} \frac{\delta_{i_0,i}}{\Delta x} \ . \tag{B.14}$$

Owing to

$$\lim_{\substack{\Delta x \to 0 \\ N \to \infty}} \sum_{i=1}^{N-1} \delta_{i_0,i} = \lim_{\substack{\Delta x \to 0 \\ N \to \infty}} \sum_{i=1}^{N-1} \frac{\delta_{i_0,i}}{\Delta x} = \int \mathrm{d}x \lim_{\Delta x \to 0} \frac{\delta_{i_0,i}}{\Delta x} = \int \mathrm{d}x\, \delta(x - x_0) \ , \tag{B.15}$$

for (B.14) we obtain

$$\frac{\delta f(x_0)}{\delta f(x)} = \delta(x - x_0) \ . \tag{B.16}$$

Even simpler, we obtain this equation by writing

$$f(x_0) = \int \mathrm{d}x\, \delta(x - x_0) f(x) \tag{B.17}$$

and taking the variation on both sides

$$\delta f(x_0) = \int \mathrm{d}x\, \delta(x - x_0) \delta f(x) = \int \mathrm{d}x \frac{\delta f(x_0)}{\delta f(x)} \delta f(x) \tag{B.18}$$

we obtain (B.16). Applying this to (B.6′), we obtain

$$\frac{\partial F(\{f(x)\})}{\delta f(x)} = \left. \frac{\partial K(a,b,c)}{\partial a} \right|_x - \frac{\mathrm{d}}{\mathrm{d}x} \left. \frac{\partial K(a,b,c)}{\partial b} \right|_x \ . \tag{B.19}$$

Above, we discussed the functional integral $F = \int \mathrm{d}x\, K(f, \mathrm{d}f/\mathrm{d}x, x)$. The generalization to the case of many variables $(x_1, \ldots, x_M)$ or the case where K contains higher order derivatives should be simple.

C. Quantum Statistical Mechanics

We assume that the Hamiltonian H, the eigenvalues E_n and the eigenstates $|n\rangle$ of the system are known. That is, for

$$H|n\rangle = E_n|n\rangle \tag{C.1}$$

the probability P_n for the state $|n\rangle$ in the canonical distribution is given by

$$P_n = \frac{1}{Z}\,\mathrm{e}^{-\beta E_n} \quad . \tag{C.2}$$

Here, $1/\beta = k_{\mathrm{B}}T$ is the thermal energy at the temperature T, and Z is due to

$$\sum_n P_n = 1 \tag{C.3}$$

given by

$$Z = \sum_n \mathrm{e}^{-\beta E_n} \quad . \tag{C.4}$$

Z is the so-called partition function. All thermodynamical properties of the system can be deduced from the sum of states. Explicitly, the Helmholtz free energy F is given by

$$F = -k_{\mathrm{B}}T \ln Z \tag{C.5}$$

and all the other thermodynamical quantities can be deduced from F. The sum of states Z can be rewritten using (C.4) as

$$Z = \sum_n \langle n|\,\mathrm{e}^{-\beta H}|n\rangle = \mathrm{Tr}\,\mathrm{e}^{-\beta H} \quad . \tag{C.6}$$

Here, we used the fact that due to $H|n\rangle = E_n|n\rangle$, the equation $\mathrm{e}^{-\beta H}|n\rangle = \mathrm{e}^{-\beta E_n}|n\rangle$ holds. The notation Tr stands for the so-called trace, running over all diagonal matrix elements in the basis $\{|n\rangle\}$:

$$\mathrm{Tr}\,A \equiv \sum_n \langle n|A|n\rangle \quad . \tag{C.7}$$

The two most important properties of the Tr are

(1) $\mathrm{Tr}(AB) = \mathrm{Tr}(BA)$ and
(2) the trace is independent of the basis.

(1) can be proven easily by

$$\begin{aligned}\mathrm{Tr}(AB) &= \sum_n \langle n|AB|n\rangle = \sum_{n,m} \langle n|A|m\rangle\langle m|B|n\rangle \\ &= \sum_{n,m} \langle m|B|n\rangle\langle n|A|m\rangle = \sum_m \langle m|BA|m\rangle = \mathrm{Tr}(BA) \quad .\end{aligned} \tag{C.8}$$

Here, we used $\sum_n |n\rangle\langle n| = 1$.

On the other hand, because the change of basis can be expressed using a unitary matrix U, then

$$\begin{aligned} |n\rangle &= \sum_{\alpha} U_{n\alpha}|\alpha\rangle \ , \\ \langle n| &= \sum_{\beta} \langle\beta|U^*_{n\beta} = \sum_{\beta} \langle\beta|(U^\dagger)_{\beta n} \ . \end{aligned} \tag{C.9}$$

(2) can be proven with

$$\begin{aligned} \sum_n \langle n|A|n\rangle &= \sum_n \sum_{\alpha,\beta} \langle\beta|(U^\dagger)_{\beta n} A U_{n\alpha}|\alpha\rangle \\ &= \sum_{\alpha,\beta} \langle\beta|A|\alpha\rangle (U^\dagger U)_{\beta\alpha} = \sum_{\alpha} \langle\alpha|A|\alpha\rangle \ . \end{aligned} \tag{C.10}$$

Therefore, we conclude that to calculate the trace in (C.6) it is not necessary to use the basis of eigenfunctions of the Hamiltonian H. Especially in the one-particle problem, (2.1.30) expresses the partition function in the basis of eigenfunctions $|x\rangle$ of the coordinate $\hat{x}$.

References

General Recommendations

There exist many well-known textbooks on quantum field theory. Here, we present only books that are valuable and understandable for students majoring in condensed matter physics.

[G.1] P. Ramond: *Field Theory: A Modern Primer* (Addison-Wesley, 1990) [An introductory and easy to understand book]

[G.2] T.D. Lee: *Particle Physics and Introduction to Field Theory* (Harwood Academic Pub., 1981) [Contains discussions on symmetry and quark confinement]

[G.3] R. Rajaraman: *Solitons and Instantons* (North-Holland, 1987) [A self-contained textbook that is also worth reading for people studying condensed matter physics. Topological aspects of field theories are explained extensively]

[G.4] S.-K. Ma: *Modern Theory of Critical Phenomena* (Benjamin, 1976) [An excellent textbook dealing with the renormalization group]

[G.5] D.J. Amit: *Field Theory, the Renormalization Group, and Critical Phenomena* (World Scientific) [Book for specialists discussing the renormalization group and critical phenomena]

[G.6] S. Coleman: *Aspects of Symmetry* (Cambridge Univ. Press, 1985) [Contains the well-known Coleman Lectures, which are presented in a very clear and understandable manner]

[G.7] M. Creutz: *Quarks, Gluons and Lattices* (Cambridge Univ. Press, 1983) [A compact textbook dealing with lattice gauge theory]

[G.8] A.M. Polyakov: *Gauge Fields and Strings* (Harwood Academic Pub., 1987) A book for experts, but contains very interesting discussions]

[G.9] J.B Kogut: "An introduction to lattice gauge theory and spin systems". Rev. Mod. Phys. 51 (1979) 659 [An excellent review introducing lattice gauge theory. Kogut wrote this review especially for graduate students]

[G.10] J.B Kogut: "The lattice gauge theory approach to quantum chromodynamics". Rev. Mod. Phys. 55 (1983) 775

[G.11] A.A. Abrikosov, L.P. Gorkov and I.E. Dzyaloshinski: *Methods of Quantum Field Theory in Statistical Physics* (Dover, 1975)

[G.12] A.L. Fetter and J.D. Walecka: *Quantum Theory of Many-Particle Systems* (McGraw-Hill, 1971)

[G.13] G.D. Mahan: *Many-Particle Physics* (Plenum Press, 1981)

[G.14] E. Fradkin: *Field Theory of Condensed Matter Systems* (Addison-Wesley, 1991) [This book deals with high-temperature superconductors, quantum spin systems and the quantum Hall effects, being at present the main research topics in condensed matter physics]

[G.15] V.N. Popov: *Functional Integrals and Collective Excitations* (Cambridge Univ. Press, 1987) [An interesting textbook that also contains research results of the author himself]
[G.16] J.W. Negele and H. Orland: *Quantum Many Particle Systems* (Addison-Wesley, 1988)
[G.17] A.M. Tsvelik: *Quantum Field Theory in Condensed Matter Physics* (Cambridge Univ. Press, 1995)

Chapter 1

As a representative textbook on quantum mechanics, we cite

[1] L.I. Schiff: *Quantum Mechanics* (McGraw-Hill, 1968)

A new and excellent textbook on quantum mechanics is

[2] J.J. Sakurai: *Modern Quantum Mechanics* (Addison-Wesley, 1985)

The second quantization is very well explained in

[3] R.P. Feynman: *Statistical Mechanics* (Benjamin, 1972)
[4] P.W. Anderson: *Concepts in Solids* (Benjamin, 1963)

Both of which are indispensable references.

Chapter 2

An unforgettable book about path integrals is

[5] R.P. Feynman and A.R. Hibbs: *Quantum Mechanics and Path Integrals* (McGraw-Hill, 1965)

The path integral formalism for quantum field theory is explained in many of the textbooks that are cited under general recommandations.
The original paper about the Berry phase is

[6] M.V. Berry: Proc. R. Soc. Lond. A392 (1984) 45

Chapter 3

The classical theory of phase transitions is discussed in

[7] Landau-Lifschitz: *Statistical Physics* (Pergamon, 1949)
[8] K. Huang: *Statistical Mechanics* (John Wiley & Sons, 1987)

Profound discussions about symmetry breaking in condensed matter Physics can be found in

[9] P.W. Anderson: *Basic Notions of Condensed Matter Physics* (Benjamin, 1984) Chap. 2

The original paper about the Mermin–Wagner theorem is

[10] P.C. Hohenberg: Phys. Rev. 158 (1967) 383
N.T. Mermin and H. Wagner: Phys. Rev. Lett. 17 (1966) 1133

The original paper about the Kosterlitz–Thouless transition is

[11] J.M. Kosterlitz and D.J. Thouless: J. Phys. C: Solid State Phys. 6 (1973) 1181

The duality transformation can be found in

[12] J.V. Jose et al.: Phys. Rev. B16 (1977) 1217

The original paper about lattice gauge theory and the problem of colour-confinement is

[13] K.G. Wilson: Phys. Rev. D14 (1974) 2455

and the Elitzur theorem first appeared in

[14] S. Elitzur: Phys. Rev. D12 (1975) 3978

The discussion of colour-confinement in this volume is based on

[15] T. Banks, R. Myerson and J. Kogut: Nucl. Phys. B129 (1977) 493
[16] P.R. Thomas and M. Stone: Nucl. Phys. B144 (1978) 513

The interested reader might further study this problem by reading

[17] J. Kogut and L. Susskind: Phys. Rev. D11 (1975) 395
[18] A.M. Poyakov: Phys. Lett. 72B (1978) 477
[19] L. Susskind: Phys. Rev. D20 (1979) 2610

Chapter 4

The RPA theory of the electron gas is discussed in

[20] D. Pines and P. Nozières: *The Theory of Quantum Liquids I* (Addison-Wesley, 1989)
[21] D. Pines: *Elementary Excitations in Solids* (Benjamin, 1964)

The calculation of the correlation energy can be found in G12.
Bose condensation and superfluids are discussed in G11, G12, G13, G15, [10] and

[22] P. Nozières and D. Pines: *The Theory of Quantum Liquids II* (Addison-Wesley, 1989)

Chapter 5

There exist many good textbooks about superconductivity, and the following list reflects some personal preferences.

[23] J.R. Schrieffler: *Theory of Superconductivity* (Addison-Wesley, 1964)
[24] P.G. de Gennes: *Superconductivity of Metals and Alloys* (Addison-Wesley, 1966)

A paper about macroscopic quantum effects is

[25] A.O. Caldeira and A.J. Leggett: Phys. Rev. Lett. 46 (1981) 211; Ann. Phys. (N.Y.) 149 (1983) 374

A review about the tunnel contact is

[26] G. Schön and A.D. Zaikin: Phys. Rev. 198 (1990) 237

The discussion of dissipation using the renormalization group is given in

[27] M.P.A. Fisher and W. Zwerger: Phys. Rev. B32 (1985) 6190

and references listed therein. A paper about the duality transformation is

[28] A. Schmid: Phys. Rev. Lett. 51 (1983) 1506

A review about superfluid two-dimensional helium is given in

[29] B.I. Halperin: "Superfluidity, melting, and liquid-crystal phases in two dimensions" in *Physics of Low-Dimensional Systems* 1979

A paper about the universal jump is

[30] D.R. Nelson and J.M. Kosterlitz: Phys. Rev. Lett. 39 (1977) 1201

In this paper, a superliquid is considered, and ρ_s is defined as the ratio between the matter density current $j = \hbar|\psi|^2\nabla\theta$ and v_s, leading finally to the limit $\lim m^2 k_{\rm B}T/\hbar^2\rho_s = \pi/2$ as in (5.3.17).
Impurities in the superconductor are discussed in

[31] M. Tinkham: *Introduction to Superconductivity* (McGraw-Hill, 1975)

The experimental results as shown in Fig. 5.7 are taken from

[32] M.R. Beasley et al.: Phys. Rev. Lett. 42 (1979) 1165

Papers about the duality transformation of the vortex system are

[33] M. Stone and P.R. Thomas: Phys. Rev. Lett. 41 (1978) 351
[34] X.Q. Wen and A. Zee: Int. J. Mod. Phys. B4 (1990) 437

Chapter 6

A general book about the quantum Hall effect is

[35] R.F. Prange and S.M. Girvin (eds.): *The Quantum Hall Effect* (Springer-Verlag, 1987)

and references therein. The Laughlin wave-function is found in

[36] R.B. Laughlin: Phys. Rev. Lett. 50 (1983) 1395

The analogy with the Bose condensation is described in

[37] S.C. Zhang: Int. J. Mod. Phys. B6 (1992) 25

Index

Printing (Computer to Film): Saladruck, Berlin
Binding: Lüderitz & Bauer, Berlin

Printed by Printforce, the Netherlands